INTERNATIONAL ORGANIZATIONS
AND ENVIRONMENTAL PROTECTION

The Environment in History: International Perspectives

Series Editors: Dolly Jørgensen, *University of Stavanger;* Christof Mauch, *LMU Munich;* Kieko Matteson, *University of Hawai'i at Mānoa;* Helmuth Trischler, *Deutsches Museum, Munich*

eseh european
society for
environmental
history

Rachel
Carson
Center

ENVIRONMENT AND SOCIETY

International Organizations and Environmental Protection

Conservation and Globalization in the Twentieth Century

Edited by

Wolfram Kaiser and **Jan-Henrik Meyer**

berghahn
NEW YORK · OXFORD
www.berghahnbooks.com

First published in 2017 by
Berghahn Books
www.berghahnbooks.com

© 2017, 2019 Wolfram Kaiser and Jan-Henrik Meyer
First paperback edition published in 2019

Library of Congress Cataloging-in-Publication Data

A C.I.P. cataloging record is available from the Library of Congress

British Library Cataloguing in Publication Data

A catalogue record for this book is available from the British Library

ISBN 978-1-78533-362-0 hardback
ISBN 978-1-78920-090-4 paperback
ISBN 978-1-78533-363-7 ebook

Contents

Acknowledgements

From the beginning, we have conceived of this book as a collaborative endeavour to bring together and integrate the emerging research on the role of international organizations (IOs) in environmental policy making and protection in the twentieth century. We organized a workshop at an intermediate stage in the project to discuss draft papers. Our exchanges at the workshop and beyond and the subsequent editing process have also allowed us to include cross-references throughout the book to highlight the interconnectedness of different roles of IOs from facilitating the formation of transnational networks (for example, of scientists) to setting agendas and institutionalizing environmental policy making at the regional and global levels.

We would like to thank Tanja Börzel and Thomas Risse from the Research College 'The Transformative Power of Europe' at the Free University of Berlin for their intellectual and financial support for the workshop. We would also like to thank Andrew Waterman for his diligent final language check.

Portsmouth/Trondheim, January 2016

Abbreviations

ACAST	Advisory Committee on the Application of Science and Technology to Development
AFM	Archivio Fondazione Luigi Micheletti
AGGG	Advisory Group on Greenhouse Gases
AID	Agency for International Development
ANZ	Archives New Zealand
ASOC	Antarctic and Southern Ocean Coalition
ASPCA	American Society for the Prevention of Cruelty to Animals
BAS	British Antarctic Survey Archives
BIAC	Business and Industry Advisory Committee
BIOMASS	Biological Investigations of Marine Antarctic Systems and Stocks
CCAMLR	Convention on the Conservation of Antarctic Marine Living Resources
CCINP	Consultative Commission for International Nature Protection
CCOL	Coordinating Committee on the Ozone Layer
CCTA	Commission for Technical Cooperation in Africa South of the Sahara
CDC	Conservation for Development Centre
CFCs	Chlorofluorocarbons
CMEA	Council for Mutual Economic Assistance
COMITEXTIL	Coordination Committee for the Textile Industries of the EC
DAC	Development Assistance Committee
DoE	Department of Energy
EC	European Communities
ECOSOC	Economic and Social Commission (of the UN)
ECS	Equilibrium Climate Sensitivity
ECSC	European Coal and Steel Community
EEC	European Economic Community

EPA	Environmental Protection Agency, European Productivity Agency
ESPP	Environmental Science and Public Policy Archives
EU	European Union
EURATOM	European Atomic Community
FAO	Food and Agriculture Organization
FKK	Danish Research Council for Culture and Communication
FOR	Fellowship of Reconciliation
GARP	Global Atmospheric Research Program
GATT	General Agreement on Tariffs and Trade
GCMs	General Circulation Models
GHGs	Greenhouse Gases
GNP	Gross National Product
HAEC	Historical Archives of the European Commission
IAEA	International Atomic Energy Agency
ICBP	International Council for Bird Protection
ICES	International Council for the Exploration of the Sea
IGY	International Geophysical Year
IIEA	International Institute of Environmental Affairs
IIED	International Institute for Environment and Development
IMF	International Monetary Fund
INGO	International Non-Governmental Organization
IO	International Organization
IOC	Intergovernmental Oceanographic Commission
IOPN	International Office for the Protection of Nature
IPCC	Intergovernmental Panel on Climate Change
ISSC	International Social Sciences Council
IUCN	International Union for Conservation of Nature
IUPN	International Union for the Protection of Nature
IWC	International Whaling Commission
MAB	Man and the Biosphere programme
MEP	Member of the European Parliament
MIT	Massachusetts Institute of Technology
NAA	National Archives of Australia
NAS	National Academy of Sciences
NASA	National Aeronautics and Space Administration
NATO	North Atlantic Treaty Organization
NCPO	National Climate Program Office

NGO	Non-Governmental Organization
NIEO	New International Economic Order
NRDC	National Resources Defense Council
NSF	National Science Foundation
OCF	*Our Common Future*
OECD	Organisation for Economic Cooperation and Development
OECDA	Organisation for Economic Cooperation and Development Archives
OEEC	Organisation for European Economic Co-operation
PPP	Polluter Pays Principle
RSPB	Royal Society for the Protection of Birds
RSPCA	Society for the Prevention of Cruelty to Animals
SCAR	Scientific Committee on Antarctic Research
SCEP	Study of Critical Environmental Problems
SCIPR	Sierra Club International Program Records
SCOPE	Scientific Committee on Problems of the Environment
SEPC	Service of the Environment and Consumer Protection
SIPI	Scientists Institute for Public Information
SMIC	Study of Man's Impact on Climate
TUAC	Trade Union Advisory Committee
UN	United Nations
UNA	United Nations Archives
UNCLOS	United Nations Law of the Sea Conference
UNCTAD	United Nations Conference on Trade and Development
UNECE	United Nations Economic Commission for Europe
UNEP	United Nations Environment Programme
UNESCO	United Nations Educational, Scientific and Cultural Organization
UNGA	United Nations General Assembly
UNOG	United Nations Office Geneva
VDEh	Verein Deutscher Eisenhüttenleute
WCC	World Climate Conference
WCED	World Commission on Environment and Development
WCIP	World Climate Impacts Programme
WCP	World Climate Programme
WCS	World Conservation Strategy
WHO	World Health Organization
WMO	World Meteorological Organization
WWF	World Wildlife Fund
YMCA	Young Men's Christian Association

Introduction

International Organizations and Environmental Protection in the Global Twentieth Century

Wolfram Kaiser and Jan-Henrik Meyer

Carbon dioxide emissions contribute to rising average surface temperatures and the melting of arctic sea ice as well as ocean acidification, threatening precious natural habitats like coral reefs. In 2014, the Intergovernmental Panel on Climate Change (IPCC) predicted with 'high confidence' in its fifth assessment report that, even in the case of moderate global warming, many regions in the world would experience more extreme weather events in the future. Moreover, a rise in the average global temperature of around three per cent would lead to 'extensive biodiversity loss'.[1] The report outlined two extreme scenarios: one, a 'low-emission mitigation scenario', in which states worldwide reduce their emissions substantially and take coordinated systematic action to control their impact, limiting the rise in temperature to 2.6 degrees centigrade during the period from 2081 to 2100 (compared to the mean temperature between 1986 and 2005); and, two, a catastrophic 'high-emission scenario' with a temperature rise of 8.5 degrees centigrade in the same period.[2]

The fifth IPCC assessment report resulted from a global effort since 2008 to collate and interpret scientific data on climate change. A total of 259 authors from thirty-nine countries debated the physical science base of climate change, receiving 54,677 comments in the process. A total of 309 authors from seventy countries analysed issues regarding the impact of climate change, adaptation to its consequences and the vulnerability of different human societies, considering 50,444 comments in the course of their work. Finally, 235 authors from 57 countries devoted themselves to identifying ways to mitigate climate change, incorporating 38,315 comments.[3]

The scale and character of the IPCC's recent globally cooperative work aptly illustrates many of the key issues covered in this book about the role of international organizations (IOs) in addressing environmental problems in the global twentieth century. First, it highlights the growing role of scientific experts and their networks in drawing attention to and assessing environmental hazards and advocating policy solutions at both the international and national levels. Secondly, it points to the high level of insecurity not only about the quality and validity of this data, but also their interpretation and resulting predictions for the future. Ideological preferences and economic and political interests, as well as different scientific concepts, methods and data interpretations, influence the results eventually presented to policy makers and the public. Thirdly, the IPCC work demonstrates the role of IOs in structuring global cooperation on environmental protection, shaping debates and advocating policy solutions. Fourthly, it shows the ambiguous, dual role of IOs such as the IPCC as both political institutions and expert bodies.[4] Finally, it becomes clear that multiple actors are involved in these debates. IOs create platforms for cooperation and contestation, not just among diplomats from national governments – as 'realist' and 'neorealist' models of international relations suggest[5] – but also by scientific experts, international non-governmental organizations (INGOs) and journalists who can influence these debates through their media reporting.[6] It is this multifarious character of IOs and of their work on environmental protection that this book seeks to explore and better understand.

The modern notion of the 'environment' alerts us to humanity's increasingly problematic relationship with nature. It emphasizes nature's widespread degradation due to human interference and suggests that nature must be protected from humankind. The environment is a political concept that encompasses many different dimensions, from air and water pollution to the loss of habitats and biodiversity, for example, and conceives of the problem as global in scope.[7] However, such a broad notion – inspired by the postwar rise of ecology and the popularization of ecological thinking – has only existed since the late 1960s, as Jan-Henrik Meyer shows in his first chapter in this book. Until then, the media and public across the Global North discussed different phenomena of environmental degradation separately, without yet fully grasping their interrelated character that seems so obvious today. The preservation of natural habitats and the creation of national parks had been a concern since the late nineteenth century. Air and water pollution was initially seen as a hygiene and health issue before its wider impact (such as through so-called acid rain) on plants, natural habitats and biodiversity became more obvious.

This fragmentation of environmental issues was reflected in the bureaucratic organization of governments, ministries and their agencies. In a first wave between 1970 and 1973, Australia, Austria, Canada, Denmark, East Germany, France, Italy, Norway and the United Kingdom established environmental ministries. Other countries, such as the West German Federal Republic of Germany, concentrated responsibilities for (almost) all environmental issues within one ministry. Yet another form of institutionalizing environmental policy was to create separate environmental protection agencies, with often wide-ranging competences, such as the pioneering Swedish *Naturvårdsverket*, which dates back to 1967, the Environmental Protection Agency (EPA) in the United States, founded in 1970, and the Environmental Agency in Japan, established in 1971. Almost a decade later, in a second wave between 1982 and 1988, Brazil, Finland, the Netherlands, Sweden, Switzerland and West Germany, among others, followed and set up environmental ministries.[8] The progressive transformation of the environment into a distinct policy field reflected its rise on the political agenda. At the same time, this process may well have made its much-discussed 'mainstreaming' – namely, the systematic incorporation of environmental concerns in the work of all ministries – more difficult.[9]

Environmental degradation is frequently transnational in character. This is what citizens and policy makers became increasingly aware of during the postwar period. Protecting elephants, lions and cheetahs in one colonial territory in Africa was of limited use if they were shot in another territory after crossing the border, as Bernhard Grzimek forcefully demonstrated to Western audiences in his 1959 documentary *Serengeti Shall Not Die*, for instance.[10] Controlling the release of chemicals used in industrial production into rivers could improve the quality of drinking water and protect species. However, it was of limited use if another country upstream increased its own water pollution through the development of new industrial sites without imposing stricter laws. The policy of building tall chimneys as an attempt to solve air pollution problems in countries like the United Kingdom and West Germany perhaps limited its impact there. However, the generally prevailing westerly winds in the Northern Hemisphere made sure that acid rain would come down on trees and lakes in Scandinavia, for example, and destroy forests and fish there – something that encouraged Sweden to place the environment on the international agenda in the late 1960s.[11]

As a result, environmental problems have induced bilateral, transnational and international cooperation for a long time, as Jan-Henrik Meyer explains in greater detail in the first chapter of this book. From the outset of the twentieth century, colonial officials worried about African wildlife across colonies ruled by different European states – although mostly out of concern for game hunting. In 1900, European governments with a stake in Africa agreed upon the Convention for the Preservation of Wild Animals, Birds and Fish in Africa (London Convention), which, however, never entered into force due to a lack of ratification. In 1933, it was replaced by the London Convention Relative to the Preservation of Fauna and Flora in Their Natural State. The meeting in London subsequently led to the establishment of the International Office for the Protection of Nature, the predecessor of the International Union for the Protection of Nature (IUPN), in 1934.[12] At the same time, unintended consequences of global trade for human health were addressed by establishing the International Office of Epizootics (now the World Organisation for Animal Health) in Paris in 1924. Its role was to control and prevent the spread of animal diseases.[13] As humans developed technologies of industrial-style exploitation during the twentieth century, the management of global commons such as oceanic fish resources and marine wildlife, notably whales, equally called for international regulation by conventions. These conventions often established IOs, such as the International Whaling Commission created in 1946. Such regulation was routinely resented and actively undermined or disregarded outright by hunters, whalers and fishermen alike as it limited access to what they and others had long considered boundless and inexhaustible resources.[14]

Early warnings about the capacity of humankind to destroy on a global scale the very resources on which it depended, notably by excessive population growth, date back to the late 1940s. Under the impression of Hiroshima and the destruction wrought by the Second World War, Fairfield Osborn's *Our Plundered Planet* or William Vogt's *Road to Survival* painted the future in black in 1948.[15] However, it was only in the 1960s that an entire wave of much-translated bestselling books highlighted the transnational and global character of what they saw as an unprecedented environmental crisis.[16] The most influential of these books was arguably *Silent Spring* by the American biologist Rachel Carson in 1962. In it she condemned the apparently reckless use of pesticides, notably DDT, and the resulting lasting chemical contamination of the environment on a global scale.[17] Her analysis and those of other authors such as Paul Ehrlich's *Population Bomb*, Garrett Hardin's *Tragedy of the Commons* or ecologist Barry Commoner's *The Closing Circle*[18] often led them to make gloomy if not apocalyptic prophecies about the state of the global environment. These culminated in the Club of Rome's 1972 report *Limits to Growth*.

Limits to Growth presented different future scenarios of global environmental development similar to what the IPCC produces today. At the time, however, such computer-calculated global models were unprecedented. This clearly contributed to the impact they had on the public. These scenarios aimed at assessing and making predictions about the development of food and other resources and the accumulation of waste on a global scale, under varying conditions such as different levels of population growth and economic expansion. Highlighting the potentially disastrous consequences of continuing the prevalent resource-intensive way of life, the authors advocated a policy change to overcome the self-defeating logic of 'exponential growth'. Their goal was for humanity to arrive at a 'global equilibrium' to avoid cataclysm. They actually made very concrete proposals, such as the development of more efficient technology, including recycling and waste avoidance through durable and easy-to-repair consumer goods, the use of solar energy, natural pest control, better medical provision and contraception.[19] The study – and in particular the gloomy prospects it implied – made a huge impression on Western publics and a number of contemporary policy makers, such as European Commission President Sicco Mansholt.[20] At the same time, the book was quickly caricatured and dismissed by its critics as exaggerating the nature and consequences of environmental destruction.[21]

Against this backdrop, the United Nation's (UN) 1972 Conference on the Human Environment (also referred to as the Stockholm Conference in this book) created a focal point for these debates about planetary limits and the carrying capacity of the globe. Initiated in 1968, it activated natural and social scientists as formal advisors or activists working with INGOs – notably Barbara Ward and René Dubois, who published an 'unofficial report' for the conference called *Only One Earth*[22] – and brought together member state governments with the aim of coordinated global action for environmental protection. Some of the newly established environmentalist INGOs like fledgling Greenpeace and Friends of the Earth also used the Stockholm Conference to stage so-called counterconferences. They called for more radical action, attacking the prevailing economic systems and connected cultures of consumption for their responsibility for environmental degradation. They also criticized governments on very concrete contemporary global issues, such as whaling or nuclear weapons testing in the Pacific.[23]

The conference highlighted the fragmentation of the international system at the time. Soviet Bloc countries refused to participate, ostensibly in protest against the continued non-admittance of East Germany to the UN. Governments from the Global South argued that the developed countries wanted to impose on the poor countries the costs of dealing with the environmental destruction they had wrought since industrialization in order to retain an economic edge over them, as Stephen Macekura discusses in his chapter. This fragmentation continues to characterize global environmental politics under different auspices until the present day, as the pronounced reluctance of countries like China to commit to binding limits on CO_2 emissions illustrates.[24] It remains to be seen whether the 2015 21st Conference of the Parties (COP21) Paris Agreement on limiting climate change will fundamentally transform this longstanding conflict in terms of its actual implementation on the ground.[25]

In spite of these difficulties, the Stockholm Conference brought about the formation of the United Nations Environmental Programme (UNEP) as a UN agency, with its seat in Nairobi and directed for the first three years by the Canadian Maurice Strong, the conference's secretary-general. Its creation marked a considerable shift towards the globalization of coordinated (albeit limited and only partly successful) action to protect the environment. UNEP became a focal point for global environmental politics and policies at a time when many other IOs also developed a stronger interest in this emerging policy field. In 1970, for example, the Organisation for Economic Co-operation and Development (OECD), created in this form in 1961, was the first IO to institutionalize a directorate and committee for the new policy field.[26] From then onwards, IOs, including regional integration organizations like the present-day European Union (EU), became active players in environmental politics. The origins of the EU's supranational environmental policy, which came under the qualified majority voting procedure with the Maastricht Treaty in 1993, date back to the 1970s. Without a treaty competence, the then European Communities instigated two environmental action programmes in 1973 and 1977 and passed the Birds Directive for protecting endangered avian species in 1978/79.[27]

These IOs drew on the evident functional need for transnational and global coordination, which was one of their traditional tasks. They linked a variety of actors with an interest in environmental problems; collated, analysed and disseminated data and knowledge about environmental hazards and degradation; developed agendas and recommended policy solutions; fostered greater global institutionalization of environmental politics; and advocated and drafted international protocols and conventions, inducing member states to support and sign them. This in turn required domestic legislative and administrative changes to comply with new rules and regulations. In short, IOs mattered for environmental protection in the global twentieth century.

In this book we explore how they mattered. We enquire, to begin with, about who became involved in global environmental politics and how. We consider a wide range of actors from expert scientists to INGOs in a policy field that cannot be properly understood with sole reference to the bargaining of member states about their 'interests', even though national governments played a crucial role in shaping new institutions and taking binding decisions. Thus, IOs and INGOs as well as (usually a number of) governments frequently worked closely together, for instance, to propagate strict limits on whaling, which became severely curtailed, although not outlawed completely.

We also wish to explore how IOs helped to shape ways of thinking and talking about environmental issues and necessary global protective measures, or what political scientists call agenda setting.[28] IOs established links with scientists and fostered the formation of their international networks. The politics–science nexus provided IOs with scientific capital in the form of expertise and knowledge, but also with policy ideas and legitimacy for demanding internationally coordinated action for environmental protection. Although not free of friction, cooperation with INGOs, too, helped IOs shape transnational and global debates. These INGOs often saw IOs as natural partners in their attempt to overcome national resistance to substantive legislative and financial commitments. After all, IOs could suggest action, but such action largely had to be implemented and paid for at the national level.

We are further interested in how existing IOs responded to the new policy challenges and how new organizations were formed to meet them. To begin with, IOs transformed their own internal organizations for discussing environmental matters. However, bureaucratic patterns sometimes persisted across organizational changes. Thus, the new OECD directorate was initially staffed with economists from elsewhere in the organization so that economic perspectives continued to dominate its main institutional mission. They were applied to this new policy field, too. But new organizations like UNEP and specialized agencies like the IPCC were also set up and created new path dependencies.[29] Initial decisions covering, for example, the organization's mission, the location of its headquarters or the appointment of its secretary general often had a long-term impact on its environmental work.

Finally, we enquire into temporal change in the way in which IOs have addressed environmental concerns in the global twentieth century, with particular emphasis on the period since the 1960s. The time around the Stockholm Conference from the late 1960s well into the 1970s appears as a kind of *Sattelzeit*, or 'saddle period', a term originally coined by the German historian of concepts Reinhart Koselleck to denote the transition from the early modern to the modern period, which involved the invention of new political concepts and the greater politicization of societies.[30]

The new postwar saddle period brought about crucial transformations, especially new political concepts of the environment and its protection, and a more decisive globalization of debates about the environmental crisis. Arguably, it also substantially enhanced the role of IOs in the search for solutions and internationally coordinated action.[31] At the same time, only one year after the Stockholm Conference, the first oil crisis ended nearly thirty years of growth in the Western world after the Second World War.[32] In the light of rising unemployment and budget and state deficits, governments became much more preoccupied with economic policy issues again and sometimes less able and willing politically to bear the short- and medium-term costs of more far-reaching environmental protection policies.[33] To what extent discursive shifts at the international level towards new environmentalist rhetoric were actually followed up by concrete policy changes at the national and subnational levels remains an important issue for empirical research. However, such comparative research on the implementation of agreed environmental norms and regulations is only in its infancy, even in political science,[34] and is beyond the scope of this book.

Thus, our book mainly relates and contributes to four sets of literature. One of these is the transnational, international and global history of the twentieth century. This has become much more open, compared to the older diplomatic history, towards considering the role of actors other than national governments and motivations other than the rational calculation of mostly material interests.[35] Transnational history focuses on crossborder issues and action by people, networks and institutions. It has recently improved our understanding of regional integration as a special case of IO involvement.[36] It emphasizes, for instance, that the present-day EU led to the formation of a transnational society and polity of sorts, that is, far more than an intergovernmental setting for the bargaining of interests by member state governments.[37] Our book shows how such transnational networks and cooperation stretched not just across countries, but also continents, and how they helped shape IO approaches to global environmental politics.

The IOs themselves have increasingly become the focus of international history, forming a second set of more specialized literature. Mark Mazower and Akira Iriye have argued their importance in literature-based overviews of the global twentieth century.[38] Several studies based on archival research have focused on single IOs, such as the League of Nations, the OECD, the North Atlantic Treaty Organization (NATO) and others.[39] Due to its quasi-federal character with many state-like features, much has been written about the history of the present-day EU.[40] With the exception of the EU, much of this emerging literature on IOs focuses on their internal institutional dynamics. However, a few more recent studies have zoomed in on the role of IOs and experts in transnational policy making and regulation, and on transfers of ideas between IOs and INGOs.[41] Our book seeks to make a major contribution to understanding how IOs mattered for particular policy fields and political and legislative decision making globally, at different levels of government and governance. With this approach we hope to break the mould of the older literature with its heavily formal-institutional research design and focus.

With this perspective we hope, thirdly, to provide empirical insights for the ongoing debate in International Relations about the nature of the international system and global politics and policy making, and the links and multiple connections between IOs and other actors.[42] Our findings demonstrate the limited usefulness of notions of international politics as a 'two-level game'[43] of interaction among domestic politics and intergovernmental negotiations within IOs or regional integration organizations like the EU. Experts and INGOs in particular have mattered a great deal in global environmental politics. Moreover, in contrast with the notion of 'epistemic communities' of experts driven by a shared understanding of the scientific issue at hand,[44] the chapters in this book illustrate the strong normative commitments of many experts in this particular field, the heavy political contestation of their expertise and advice, and its use by actors like national governments and INGOs, for their own political agendas.

Finally, the book seeks to contribute to the study of environmental history. Environmental history first developed in the United States in the early 1970s in the context of the growing political concern about environmental degradation.[45] At the time, researchers aimed at analysing human relations with the environment in historical perspective. Two research strands emerged: one focused on changing perceptions of nature from a history of ideas perspective;[46] and another on the actual impact of humans on the environment, such as through colonization.[47] Such studies sought to trace 'the historical origins of our ecological crisis', as historian of science Lynn White phrased it in the journal *Science* as early as 1967.[48] In Europe, despite pioneering publications such as an *Annales* special issue published in 1974 under the immediate influence of the new environmental discourse, environmental history as a field emerged much more slowly.[49] Current environmental problems called for historical perspectives, so that, notably in Germany, many researchers focused on the history of pollution, offering an alternative perspective on the conventional story of industrial progress.[50] National histories prevailed,[51] with occasional comparative perspectives.[52] Although environmental movements and states addressed transnational issues and environmental historians became increasingly more connected across borders, researchers continued to focus mainly on their countries of origin or residence.

More recently, a trend has emerged towards international and global perspectives – notably in textbooks and historical overviews.[53] Some historians have begun to explore transnational links and interaction.[54] In contrast, international organizations have only recently received greater attention from environmental historians.[55] However, to understand their role in global environmental politics becomes more crucial than ever as a contribution to making sense historically of ongoing processes of globalization. Moreover, postcolonial theories and calls for decentring Europe (and 'provincializing' the EU)[56] equally impel historians towards pursuing global perspectives. A prominent example is the study of the institutionalization and spread of national parks, in which international bodies such as the International Union for the Conservation of Nature (IUCN, formerly the IUPN) played an important role.[57]

Against this background, our book analyses for the first time how IOs have influenced environmental politics in the global twentieth century. In his introductory chapter, Jan-Henrik Meyer provides a historical overview of the role of IOs in this policy field from the origins of bilateral and transnational action to combat environmental degradation before the First World War until the 1960s by sketching the origins of the Stockholm Conference. The chapter traces how existing international bodies and newly created IOs addressed various aspects of what we now consider environmental concerns. It highlights how conceptions of nature changed at the international level, and examines change and continuity in how IOs framed issues and set agendas.

The remaining nine chapters, which are all based on fresh archival research and, in some cases, interviews, fall into two categories. Three chapters focus on a variety of actors and their role in the preparation of, and negotiations during, the Stockholm Conference: from scientific experts and development economists to the Vatican and the Global South. The six subsequent chapters explore the Stockholm Conference's impact and limits and the role of a variety of IOs in environmental politics until roughly 1992. In this year, the UN Conference on the Environment and Development in Rio de Janeiro resulted in the Framework Convention on Climate Change.

In her chapter Enora Javaudin discusses the crucial role of scientific experts and expertise in propagating and preparing the Stockholm Conference. Transnational voluntary organizations and IOs had involved technology and science experts for a long time, such as in global meteorological cooperation. In the 1960s, however, many scientific experts intervened more forcefully in public debates and became activists of environmental protection. Yet, while scientists sought to develop a sound scientific basis for collating and analysing data and recommending policy solutions, they held diverging views on the nature of the crisis and necessary policy priorities. Some worked for IOs like the UN, while others cooperated with INGOs and participated in their counterconferences at Stockholm.

Development economists constituted another influential group in the preparation of the Stockholm Conference. In his chapter Michael W. Manulak identifies three different approaches to the environment–development nexus that clashed prior to the Stockholm Conference. Eventually, however, a small group of interventionist economists came to dominate the Founex seminar and its report in 1971, which influenced the Stockholm Conference process, its agenda and outcomes. These economists favoured robust government intervention and saw environmental policies primarily as an instrument for limiting environmental disruption and improving human living conditions. Drawing on an influential social science concept, Manulak characterizes this small group as an 'epistemic community'[58] whose members shared a similar professional background, academic viewpoint and objectives for global environmental politics.

The preparation for the Stockholm Conference also activated participants who had not previously taken any interest in the environment as a distinctive issue. The Vatican, with its hybrid identity, is a case in point. A microstate and formal participant in the negotiations, the Vatican was at the same time a global religious organization. As Luigi Piccioni shows in his chapter, the Holy See only developed an interest in environmental issues as a result of the UN initiative. It pleaded for environmental protection in line with the 'progressive' Atlantic position of states like Sweden and Canada. At the same time, it sided with the countries from the Global South – some of them predominantly Catholic, such as Brazil – in defending their interest in development, which these countries perceived as the only route to combat poverty. To complicate matters further, the Vatican was largely isolated in the debate over population control, which many scientists and governments advocated as a strategy for limiting environmental degradation,[59] but which the Holy See rejected outright on doctrinal grounds.

Several countries from the Global South actually considered a boycott of the Stockholm Conference, though they eventually participated. Some of these countries defended their development agenda especially vigorously in the face of the challenge of the environmental issue. Thus, Brazil, whose dictatorship gained much of its legitimacy from its policy of accelerated economic development, became the leader of a coalition that sought to water down what they considered the excessive environmentalist fervour of the Northern countries. They sought to ensure continued support at the international level for traditional growth-oriented policies – as a means to overcome poverty and 'underdevelopment'. From this perspective, pollution almost seemed desirable rather than a problem.

Starting the second section on IO activities in environmental politics, Wolfram Kaiser discusses the case of the steel industry from the 1950s to the late 1980s, one of the leading air polluting sectors. IOs dealing with or regulating (in the case of the present-day EU) the steel sector were primarily concerned with reducing the industry's energy consumption to save costs. New process technologies, introduced from the 1950s onwards, reduced consumption and, as a result, emissions. Demand for steel stagnated or fell during the steel crisis in Europe and North America after 1974, which further limited emissions there, but new production facilities and air pollution grew rapidly in developing countries, especially in Asia. However, IOs like the OECD remained focused on the economic and social costs associated with the sector's crisis, transformation and globalization. They concentrated on studies of new technologies and energy reduction, but as an economic rather than an environmental concern – something that highlights the limited 'mainstreaming' of the environmental protection agenda.

In his chapter, Jan-Henrik Meyer explores the role that competition among IOs as well as transfers of concepts played when the European Communities first set up an environmental policy in the context and wake of the Stockholm Conference in the 1970s. Drawing on the example of the polluter pays principle, he traces how the EC transferred and assimilated this concept and how it became established (in the non-binding legal form of a recommendation) and adapted to the needs of its common market.

Iris Borowy discusses the role of the OECD in global environmental politics. She traces the origins of the organization's commitment to embedded liberalism and free trade policies in the Western world and the shift towards greater attention to the apparent conflict between its agenda for economic growth and the increasing environmental degradation, which resulted from such growth. In 1971, the OECD created a separate environmental division and committee. As Borowy shows, the organization had (and still has) few means of tangible policy influence. Its main role is that of a think tank drawing on internal and external expertise in formalized and informal relationships. In this way, it has been able to significantly influence political agendas, not least by contributing to the development of the concept of 'sustainable development' as an attempt to harmonize its economic growth and development priority with global environmental protection.

The promise of 'sustainable development' is also at the heart of Stephen Macekura's analysis of the origins of the World Conservation Strategy (WCS). As he shows in his chapter, the WCS resulted from close cooperation between two INGOs, the International Union for the Conservation of Nature and the World Wildlife Fund, and UNEP. Much of this cooperation took place in informal networks where environmentalists and development experts sought to integrate environmental protection and economic development needs. While these networks helped shape the meaning of the term 'sustainable development', they did not succeed in inducing national governments to comply with the associated objectives more fully. Limited funds meant that the WCS was never properly implemented on a global scale.

As Alessandro Antonello demonstrates in his chapter on the Southern Ocean ecosystem, not every environmental protection measure necessarily reflects, in the first instance, concerns about the environment. The 1959 Antarctic Treaty created important path dependencies for negotiations about the preservation of this ecosystem with its immense importance for individual species like whales and oceanic life more generally. Antonello shows how the original signatories were initially interested in protecting their own privileges through preserving the existing institutional setup. However, they, and a variety of IOs, disagreed over the actual approach to Antarctica. States like the Soviet Union and Japan were keener on its exploitation – especially for krill as a food resource – while other signatories led by the US prevailed in the end with their agenda of prioritising its conservation.

In the last chapter, David G. Hirst delves into the most prominent issue of international environmental politics of the past three decades. He explores the global politics of climate change, tracing the origins of global action from the 1985 Villach Conference on the 'Assessment of the Role of Carbon Dioxide and of Other Greenhouse Gases in Climate Variations and Associated Impacts' to the creation of the IPCC by the World Metrological Association and UNEP in 1988. He argues that a group of scientists initially favoured what he calls a 'scientized' approach to climate change. However, instead of this rather technocratic approach, the United States in particular favoured and secured an intergovernmental assessment mechanism, which effectively politicized the global politics of climate change and subjected it to the traditional diplomatic logic of intergovernmental bargaining.

Not just climate change but also many other transnational environmental issues remain high on the agenda of IOs well into the twenty-first century. IOs continue to cooperate with scientists, INGOs and other actors, influence debates and agendas, and provide a platform for the contestation over environmental problems and solutions. This transnational and global contestation remains driven by fear over the future of humankind, prevailing societal norms as well as short-term financial cost-benefit calculations in times of global economic competition – and the resulting individual and collective preferences that are deeply linked to the functioning of consumer societies and our way of life. To understand the history of this contestation and deep ambiguity, the role of IOs in global environmental protection can provide us with insights that may well be useful for addressing environmental problems in the future.

Wolfram Kaiser is Professor of European Studies, University of Portsmouth, United Kingdom and Visiting Professor at the College of Europe, Bruges, Belgium, and at NTNU Trondheim, Norway.

Jan-Henrik Meyer is a senior researcher at the Max-Planck-Institute for the History of European Law, Frankfurt, Germany and an associate researcher at the Centre for Contemporary History, Potsdam.

Notes

1. For a critical view on the emergence of biodiversity as a central environmental concept, see: Libby Robin, 'The Rise of the Idea of Biodiversity: Crises, Responses and Expertise', *Quaderni. Communication, technologies, pouvoir* 76: 3 (2011): 25–37; Timothy J. Farnham, *Saving Nature's Legacy: The Origins of the Idea of Biodiversity* (New Haven: Yale University Press, 2007).

2. IPCC, 'Summary for Policymakers', in *Climate Change 2014: Impacts, Adaptation, and Vulnerability. Part A: Global and Sectoral Aspects. Contribution of Working Group II to the Fifth Assessment Report of the Intergovernmental Panel on Climate Change*, C.B. Field et al. (eds) (Cambridge: Cambridge University Press, 2014), 1–32.

3. IPCC, Fifth Assessment Report (AR5), 2014, http://www.ipcc.ch/report/ar5/index.shtml, accessed 28 May 2016.

4. It is particularly this dual – and seemingly contradictory – role that has been attacked by the so-called climate sceptics who argue that the IPCC compromises its scientific credibility by engaging in diplomatic consensus seeking. Achim Brunnengräber, 'Klimaskeptiker in Deutschland und ihr Kampf gegen die Energiewende', *FFU-Report*, 03 (2013), http://edocs.fu-berlin.de/docs/receive/FUDOCS_document_000000017134, accessed 28 May 2016, 23; Riley E. Dunlap and Aaron M. McCright, 'Organized Climate Change Denial', in *Oxford Handbook of Climate Change and Society*, John S. Dryzek, Richard B. Norgaard and David Schlosberg (eds) (Oxford: Oxford University Press, 2013), 144–60.

5. See e.g. Hans J. Morgenthau, *Politics among Nations: The Struggle for Power and Peace* (New York: Knopf, 1948); Kenneth N. Waltz, *Theory of International Politics* (Reading, MA: Addison-Wesley, 1979).

6. For an earlier example in the field of health policy, see Norman Howard-Jones, 'The Scientific Background of the International Sanitary Conferences 1851–1938', 1975, World Health Organization, http://apps.who.int/iris/bitstream/10665/62873/1/14549_eng.pdf?ua=1, accessed 28 May 2016.

7. Jens Ivo Engels, 'Modern Environmentalism', in *The Turning Points of Environmental History*, Frank Uekötter (ed.) (Pittsburgh: University of Pittsburgh Press, 2010), 119–31.

8. Luigi Piccioni, 'Un punto d'arrivo, un punto di partenza. Discutendo di Paesaggio Costituzione cemento (di Salvatore Setti)', *Storia* 18(52) (2012): 87–

114, 111. See also Michael Bess, *The Light Green Society: Ecology and Technological Modernity in France, 1960-2000* (Chicago: University of Chicago Press, 2003), 83; John McCormick, *British Politics and the Environment* (London: Earthscan, 1991), 16 f.; Frank Uekötter, *The Greenest Nation? A New History of German Environmentalism* (Cambridge, MA: MIT Press, 2014), 88.

9. On the effects of mainstreaming, or the lack thereof, see, e.g. Peter M. Haas, *Saving the Mediterranean: The Politics of International Environmental Cooperation* (New York: Columbia University Press, 1990).

10. Thomas Lekan, 'Serengeti *Shall Not Die*: Bernhard Grzimek, Wildlife Film, and the Making of a Tourist Landscape in East Africa', *German History* 29(2) (2011): 224-64.

11. John McCormick, *The Global Environmental Movement* (Chichester: John Wiley, 1995), 110.

12. Mark Cioc, *The Game of Conservation: International Treaties to Protect the World's Migratory Species* (Athens, OH: Ohio University Press, 2009), 34-40, 47-57.

13. Cornelia Knab, 'Infectious Rats and Dangerous Cows: Transnational Perspectives on Animal Diseases in the First Half of the Twentieth Century', *Contemporary European History* 20(3) (2011): 281-306.

14. Kurkpatrick Dorsey, *Whales and Nations: Environmental Diplomacy on the High Seas* (Seattle: University of Washington Press, 2013); Dean L.Y. Bavington, *Managed Annihilation: An Unnatural History of the Newfoundland Cod Collapse* (Vancouver: UBC Press, 2010); Cioc, *The Game of Conservation*; Anna Katharina Wöbse, 'Der Schutz der Natur im Völkerbund–Anfänge einer Weltumweltpolitik', *Archiv für Sozialgeschichte* 43(1) (2003): 177-91; Kurk Dorsey, *The Dawn of Conservation Diplomacy: U.S.-Canadian Wildlife Protection Treaties in the Progressive Era* (Seattle: University of Washington Press, 1998).

15. Thomas Robertson, 'Total War and the Total Environment: Fairfield Osborn, William Vogt, and the Birth of Global Ecology', *Environmental History* 17(2) (2012): 336-64; Fairfield Osborn, *Our Plundered Planet* (Boston: Little Brown, 1948); William Vogt, *Road to Survival* (New York: Sloane Associates, 1948).

16. Yannick Mahrane et al., 'De la nature à la biosphere. L'invention politique de l'environnement global, 1945-1972', *Vingtième Siècle* 113(1) (2012): 127-41.

17. Christof Mauch, 'Blick durchs Ökoskop. Rachel Carsons Klassiker und die Anfänge des modernen Umweltbewusstseins', *Studies in Contemporary History* 9(1) (2012): 1-4; Sarah L. Thomas, 'A Call to Action: Silent Spring, Public Discourse and the Rise of Modern Environmentalism', in *Natural Protest: Essays on the History of American Environmentalism*, Michael Egan and Jeff Crane (eds) (New York: Routledge, 2009), 185-204; Rachel Carson, *Silent Spring* (Greenwich, CT: Fawcett, 1962).

18. Barry Commoner, *The Closing Circle: Nature, Man and Technology* (New York: Knopf, 1971); Paul Ehrlich, *The Population Bomb* (New York: Ballentine Books, 1968); Garrett Hardin, 'The Tragedy of the Commons', *Science* 162(3859) (1968): 1243-48.

19. Dennis Meadows et al., *The Limits to Growth* (New York: Universe Books, 1972), 177.

20. Laura Scichilone, 'The Origins of the Common Environmental Policy: The Contributions of Spinelli and Mansholt in the *Ad Hoc* Group of the European Commission', in *The Road to a United Europe: Interpretations of the Process of European Integration*, Morten Rasmussen and Ann-Christina Lauring Knudsen (eds) (Brussels: PIE-Peter Lang, 2009), 335–348.

21. Samuel P. Hays, 'The Limits-to-Growth Issue: A Historical Perspective', in *Explorations in Environmental History*, Samuel P. Hays (ed.) (Pittsburgh: University of Pittsburgh Press, 1998), 3–23; Meadows et al., *The Limits to Growth*; Mauricio Schoijet, 'Limits to Growth and the Rise of Catastrophism', *Environmental History* 4(4) (1999): 515–30; Ugo Bardi, *The Limits to Growth Revisited* (New York: Springer, 2011).

22. Barbara Ward and René Dubos, *Only One Earth: The Care and Maintenance of a Small Planet* (Harmondsworth: Penguin, 1972).

23. Frank Zelko, *Make it a Green Peace!: The Rise of Countercultural Environmentalism* (Oxford: Oxford University Press, 2013), 135 f. For an overview of the INGOs' activities at Stockholm, see Ecologist and Friends of the Earth, *Stockholm Conference Eco*, 6–16 June 1972.

24. Stephen J. Macekura, *Of Limits and Growth: The Rise of Global Sustainable Development in the Twentieth Century* (Cambridge: Cambridge University Press, 2015)

25. http://www.theguardian.com/environment/climate-consensus-97-per-cent/ 2015/dec/30/why-we-need-the-next-to-impossible-15c-temperature-target, accessed 28 May 2016.

26. Iris Borowy, *Defining Sustainable Development for Our Common Future: A History of the World Commission on Environment and Development (Brundtland Commission)* (Abingdon: Routledge, 2014); Matthias Schmelzer, 'The Crisis before the Crisis: The "Problems of Modern Society" and the OECD, 1968–74', *European Review of History: Revue europeenne d'histoire* 19(6) (2012): 999–1020.

27. Jan-Henrik Meyer, 'Saving Migrants: A Transnational Network Supporting Supranational Bird Protection Policy in the 1970s', in *Transnational Networks in Regional Integration. Governing Europe 1945–83*, Wolfram Kaiser, Brigitte Leucht and Michael Gehler (eds) (Basingstoke: Palgrave Macmillan, 2010), 176–98.

28. Jan-Henrik Meyer, 'Getting Started: Agenda-Setting in European Environmental Policy in the 1970s', in *The Institutions and Dynamics of the European Community, 1973–83*, Johnny Laursen (ed.) (Baden-Baden: Nomos, 2014), 221–42; Sebastiaan Princen, *Agenda Setting in the European Union* (Basingstoke: Palgrave Macmillan, 2009); Sarah B. Pralle, 'Agenda-Setting and Climate Change', *Environmental Politics* 18(5) (2009): 781–99.

29. Paul Pierson, *Politics in Time: History, Institutions, and Social Analysis* (Princeton: Princeton University Press, 2004).

30. Reinhart Koselleck, ‚Einleitung', in *Geschichtliche Grundbegriffe*, vol. 1, Otto Brunner, Werner Conze, Reinhart Koselleck (eds) (Stuttgart: Klett Cotta, 1979), XV; see also, relating to the Stockholm Conference, Thorsten Schulz-Walden, *Anfänge globaler Umweltpolitik. Umweltsicherheit in der internationalen Politik (1969–1975)* (Munich: Oldenbourg, 2013), 260.

31. Kai F. Hünemörder, ‘1972 – Epochenschwelle der Umweltgeschichte?', in *Natur- und Umweltschutz nach 1945. Konzepte, Konflikte, Kompetenzen*, Franz-Josef Brüggemeier and Jens Ivo Engels (eds) (Frankfurt: Campus, 2005), 125–44; Patrick Kupper, ‘Die "1970er-Diagnose". Grundsätzliche Überlegungen zu einem Wendepunkt der Umweltgeschichte', *Archiv für Sozialgeschichte* 43(1) (2003): 325–48.

32. Hendrik Ehrhardt and Thomas Kroll, eds., *Energie in der modernen Gesellschaft. Zeithistorische Perspektiven* (Göttingen: Vandenhoeck & Ruprecht, 2012); Niall Ferguson, et al., eds., *The Shock of the Global. The 1970s in Perspective* (Cambridge, MA: Belknap Press, 2010); Anselm Doering-Manteuffel and Lutz Raphael (eds), *Nach dem Boom. Perspektiven auf die Zeitgeschichte seit 1970* (Göttingen: Vandenhoeck & Ruprecht, 2008).

33. Schulz-Walden, *Anfänge globaler Umweltpolitik*, 293–330.

34. For Europe, see e.g. Tanja Börzel, *Environmental Leaders and Laggards in Europe: Why There is (Not) a Southern Problem* (Aldershot: Ashgate, 2003).

35. Jost Dülffer and Wilfried Loth (eds), *Dimensionen internationaler Geschichte* (Munich: Oldenbourg, 2012); Madeleine Herren, *Internationale Organisationen seit 1865. Eine Globalgeschichte der internationalen Ordnung* (Darmstadt: Wissenschaftliche Buchgesellschaft, 2009); Pierre-Yves Saunier, ‘International Non-governmental Organizations (INGOs)', in *The Palgrave Dictionary of Transnational History*, Akira Iriye and Pierre-Yves Saunier (eds) (Basingstoke: Palgrave Macmillan, 2009), 573–80.

36. See e.g. Davide Rodogno, Bernhard Struck and Jakob Vogel (eds), *Shaping the Transnational Sphere: Experts, Networks and Issues from the 1840s to the 1930s* (New York: Berghahn Books, 2015).

37. Wolfram Kaiser and Jan-Henrik Meyer (eds), *Societal Actors in European Integration. Polity-Building and Policy-Making 1958–1992* (Basingstoke: Palgrave Macmillan, 2013); Wolfram Kaiser, Morten Rasmussen and Brigitte Leucht (eds), *The History of the European Union: Origins of a Trans- and Supranational Polity 1950–72* (Abingdon: Routledge, 2009).

38. Akira Iriye, Jürgen Osterhammel and Wilfried Loth (eds), *Global Interdependence: The World after 1945 (History of the World)* (Cambridge, MA: Harvard University Press, 2014); Mark Mazower, *Governing the World: The History of an Idea* (London: Allen Lane, 2013); Akira Iriye, *Global Community: The Role of International Organizations in the Making of the Contemporary World* (Berkeley: University of California Press, 2002).

39. Patricia Clavin, *Securing the World Economy: The Reinvention of the League of Nations, 1920–1946* (Oxford: Oxford University Press, 2013); Schmelzer, ‘The Crisis before the Crisis'; Thorsten Schulz, ‘Transatlantic Environmental Security in the 1970s? NATO's "Third Dimension" as an Early Environmental

and Human Security Approach', *Historical Social Research* 35(4) (2010): 309–28; Jacob Darwin Hamblin, 'Environmentalism for the Atlantic Alliance: NATO's Experiment with the "Challenges of Modern Society"', *Environmental History* 15(1) (2010): 54–75; Iris Borowy, 'The Brundtland Commission: Sustainable Development as Health Issue', *Michael Quarterly* 10(2) (2013): 196–206; Iris Borowy, *Coming to Terms with World Health: The League of Nations Health Organisation 1921–1946* (Frankfurt: Peter Lang, 2009).

40. For an overview of the historiography, see Wolfram Kaiser and Antonio Varsori (eds), *European Union History: Themes and Debates* (Basingstoke: Palgrave Macmillan, 2010). The relations between the EEC/EC/EU and the UN system are explored in Lorenzo Mechi, Guia Migani and Francesco Petrini (eds), *Networks of Global Governance: International Organisations and European Integration in Historical Perspective* (Cambridge: Cambridge Scholars, 2014).

41. Wolfram Kaiser and Johan Schot, *Writing the Rules for Europe: Experts, Cartels, International Organizations* (Basingstoke: Palgrave Macmillan, 2014); Martin Kohlrausch and Helmuth Trischler, *Building Europe on Expertise. Innovators, Organizers, Networkers* (Basingstoke: Palgrave Macmillan, 2014); Astrid Mignon Kirchhof and Jan-Henrik Meyer, 'Global Protest against Nuclear Power: Transfer and Transnational Exchange in the 1970s and 1980s. Focus Issue', *Historical Social Research* 39(1) (2014): 163–273; Jan-Henrik Meyer, 'Appropriating the Environment: How the European Institutions Received the Novel Idea of the Environment and Made it Their Own', *KFG Working Paper* 31 (2011): 1–33, http://edocs.fu-berlin.de/docs/receive/FUDOCS_document_000000012522, accessed 18 May 2016.

42. Gerda Falkner, *EU Policies in a Global Perspective: Shaping or Taking International Regimes?* (Abingdon: Routledge, 2014); Frank Biermann, Bernd Siebenhüner and Anna Schreyögg, *International Organizations in Global Environmental Governance: Routledge Research in Environmental Politics* (London: Routledge, 2009); Thomas Risse-Kappen (ed.), *Bringing Transnational Relations Back in: Non-state Actors, Domestic Structures and International Institutions* (Cambridge: Cambridge University Press, 1995); Oran R. Young, *International Cooperation: Building Regimes for Natural Resources and the Environment* (Ithaca: Cornell University Press, 1989).

43. Robert Putnam, 'Diplomacy and Domestic Politics: The Logic of Two-Level Games', *International Organization* 42(3) (1988): 427–60.

44. Peter M. Haas, 'Introduction: Epistemic Communities and International Policy Coordination', *International Organization* 46(1) (1992): 1–35.

45. Paul S. Sutter, 'The World with Us: The State of American Environmental History', *Journal of American History* 100(1) (2013): 94–119; Peter Coates, 'Emerging from the Wilderness (or, from Redwoods to Bananas): Recent Environmental History in the United States and the Rest of the Americas', *Environment and History* 10(4) (2004): 407–38.

46. E.g. Donald Worster, *Nature's Economy. A History of Ecological Ideas* (Cambridge: Cambridge University Press, 1994 [1977]); Roderick Nash,

Wilderness and the American Mind (New Haven: Yale University Press, 1982 [1967]).

47. Alfred W. Crosby, *The Columbian Exchange: Biological and Cultural Consequences of 1492* (Westport, CT: Praeger, 2003 [1972]); William Cronon, *Changes in the Land: Indians, Colonists, and the Ecology of New England* (New York: Hill and Wang, 1983); Donald Worster, *Dust Bowl: The Southern Plains in the 1930s* (New York: Oxford University Press, 1979).

48. Lynn White, 'The Historical Roots of Our Ecologic Crisis', *Science* 155(3767) (1967): 1203–7.

49. Emmanuel Le Roy Ladurie, 'Présentation "Histoire et Environnement"' (Special Issue)', *Annales. Économies, Sociétés, Civilisations* 29(3) (1974): 537; Geneviève Massard-Guilbaud, 'De la "part du milieu" à l'histoire de l'environnement', *Le Mouvement social* 200 (2002): 64–72; Verena Winiwarter et al., 'Environmental History in Europe from 1994 to 2004: Enthusiasm and Consolidation', *Environment and History* 10(4) (2004): 501–30; Finn Arne Jørgensen et al., 'Entangled Environments: Historians and Nature in the Nordic Countries', *Historisk Tidsskrift* 92(1) (2013): 9–34.

50. Franz-Josef Brüggemeier, *Das unendliche Meer der Lüfte: Luftverschmutzung, Industrialisierung und Risikodebatten im 19. Jahrhundert* (Essen: Klartext, 1996); Franz-Josef Brüggemeier, *Besiegte Natur: Geschichte der Umwelt im 19. und 20. Jahrhundert* (Munich: Beck, 1989); Geneviève Massard-Guilbaud, *Histoire de la pollution industrielle: France, 1789–1914* (Paris: Ecole des hautes études en sciences sociales, 2010).

51. Marco Armiero and Marcus Hall (eds), *Nature and History in Modern Italy* (Athens, OH: Ohio University Press, 2010); Jean-François Mouhot and Charles-François Mathis (eds), *Une protection de l'environnement à la française, XIXe–XXe siècles* (Seyssel: Editions Champ Vallon, 2013); Uekötter, *The Greenest Nation?*; Franz-Josef Brüggemeier, *Die Schranken der Natur. Umwelt, Gesellschaft, Experimente* (Essen: Klartext, 2014).

52. Karl Ditt, 'Nature Conservation in England and Germany 1900–70: Forerunner of Environmental Protection?', *Contemporary European History* 5(1) (1996): 1–28; David Stradling and Peter Thorsheim, 'The Smoke of Great Cities: British and American Efforts to Control Air Pollution, 1860–1914', *Environmental History* 4(1) (1999): 6–31; Frank Uekötter, *The Age of Smoke: Environmental Policy in Germany and the United States, 1880–1970* (Pittsburgh: University of Pittsburgh Press, 2009 [originally in German 2003]); François Walter, *Les figures paysagères de la nation: territoire et paysage en Europe (16e–20e siècle)* (Paris: Ecole des hautes études en sciences sociales, 2004); Heike Weber, 'Les ordures ménagères et l'apparition de la consommation de masse: une comparaison franco-allemande (1945–1975)', in *Une protection de l'environnement à la française, XIXe–XXe siècles*, Jean-François Mouhot and Charles-François Mathis (eds) (Seyssel: Editions Champ Vallon, 2013), 141–56.

53. Joachim Radkau, *The Age of Ecology: A Global History* (London: Polity Press, 2014); Erika Marie Bsumek, David Kinkela and Mark Atwood Lawrence (eds),

Nation-States and the Global Environment: New Approaches to International Environmental History (Oxford: Oxford University Press, 2013); Steven Mosley, *The Environment in World History* (Abingdon: Routledge, 2010); Johnson Donald Hughes, *An Environmental History of the World: Humankind's Changing Role in Community of Life.* 2nd edn (London: Routledge, 2009); I.G. Simmons, *Global Environmental History. 10,000 BC to AD 2000* (Edinburgh: Edinburgh University Press, 2008); Joachim Radkau, *Nature and Power: A Global History of the Environment* (Cambridge: Cambridge University Press, 2008); John R. McNeill, *Something New under the Sun: An Environmental History of the Twentieth-Century World* (New York: Norton, 2000).

54. John R. McNeill and Peter Engelke, 'Mensch und Umwelt im Zeitalter des Anthropozän', in *Geschichte der Welt 1945 bis heute: Die globalisierte Welt*, Akira Iriye, et al. (eds) (Munich: Beck, 2013), 357–534; Patrick Kupper, 'Translating Yellowstone: Early European National Parks, Weltnaturschutz and the Swiss Model', in *Civilizing Nature. National Parks in Global Historical Perspective*, Bernhard Gissibl, Sabine Höhler and Patrick Kupper (eds) (New York: Berghahn Books, 2012), 123–39; Lekan, 'Serengeti Shall Not Die'; Luigi Piccioni, 'L'influence de la France dans la protection de la nature en Italie au début du XXe siècle', in *Une protection de l'environnement à la française*, 97–107; Anna Katharina Wöbse, 'Les liaisons sinueuses: les relations franco-allemandes en matière de protection de la nature dans la première moitié due XXe siècle', in *Une protection de l'environnement à la française*, 108–19.

55. E.g. Schulz-Walden, *Anfänge globaler Umweltpolitik*; Anna-Katharina Wöbse, 'Oil on Troubled Waters? Environmental Diplomacy in the League of Nations', *Diplomatic History* 32(4) (2008): 519–37; Kurk Dorsey and Mark Lytle, 'Introduction. [Special Issue on Environmental and Diplomatic History]', *Diplomatic History* 32(4) (2008): 517–18; Jan-Henrik Meyer, 'Un faux départ? Les acteurs français dans la politique environnementale européenne des années 1970', in *Une protection de l'environnement à la française*, 120–30; Wöbse, 'Les liaisons sinueuses'.

56. Kiran Klaus Patel, 'Provincialising European Union: Co-operation and Integration in Europe in a Historical Perspective', *Contemporary European History* 22(4) (2013): 649–73; Dipesh Chakrabarty, *Provincializing Europe: Postcolonial Thought and Historical Difference* (Princeton: Princeton University Press, 2000).

57. Bernhard Gissibl, Sabine Höhler and Patrick Kupper (eds), *Civilizing Nature: National Parks in Global Historical Perspective* (New York: Berghahn Books, 2012), Macekura, *Of Limits and Growth*, 54–90.

58. Peter M. Haas, 'Do Regimes Matter? Epistemic Communities and Mediterranean Pollution Control', *International Organization* 43(3) (1989): 377–403.

59. Thomas Robertson, *The Malthusian Moment: Global Population Growth and the Birth of American Environmentalism* (New Brunswick: Rutgers University Press, 2012); Matthew James Connelly, *Fatal Misconception: The Struggle to Control World Population* (Cambridge, MA: Belknap Press, 2008); Marc Frey,

'Experten, Stiftungen und Politik: Zur Genese des globalen Diskurses über Bevölkerung seit 1945', *Studies in Contemporary History*, 4(1–2) (2007), http://www.zeithistorische-forschungen.de/16126041-Frey-2-2007, accessed 28 May 2016; Björn-Ola Linnér, *The Return of Malthus. Environmentalism and Postwar Population-Resource Crises* (Isle of Harris: White Horse Press, 2003).

Bibliography

Armiero, Marco and Marcus Hall (eds), *Nature and History in Modern Italy* (Athens, OH: Ohio University Press, 2010).

Bardi, Ugo, *The Limits to Growth Revisited* (New York: Springer, 2011).

Bavington, Dean L.Y., *Managed Annihilation: An Unnatural History of the Newfoundland Cod Collapse* (Vancouver: UBC Press, 2010).

Bess, Michael, *The Light Green Society: Ecology and Technological Modernity in France, 1960–2000* (Chicago: University of Chicago Press, 2003).

Biermann, Frank, Bernd Siebenhüner and Anna Schreyögg, *International Organizations in Global Environmental Governance: Routledge Research in Environmental Politics* (London: Routledge, 2009).

Börzel, Tanja, *Environmental Leaders and Laggards in Europe: Why There is (Not) a Southern Problem* (Aldershot: Ashgate, 2003).

Borowy, Iris, *Coming to Terms with World Health: The League of Nations Health Organization 1921–1946* (Frankfurt: Peter Lang, 2009).

———. 'The Brundtland Commission: Sustainable Development as Health Issue', *Michael Quarterly* 10(2) (2013): 196–206.

———. *Defining Sustainable Development for Our Common Future: A History of the World Commission on Environment and Development (Brundtland Commission)* (Abingdon: Routledge, 2014).

Brüggemeier, Franz-Josef, *Besiegte Natur: Geschichte der Umwelt im 19. und 20. Jahrhundert* (Munich: Beck, 1989).

———. *Das unendliche Meer der Lüfte: Luftverschmutzung, Industrialisierung und Risikodebatten im 19. Jahrhundert* (Essen: Klartext, 1996).

———. *Die Schranken der Natur. Umwelt, Gesellschaft, Experimente* (Essen: Klartext, 2014).

Brunnengräber, Achim, 'Klimaskeptiker in Deutschland und ihr Kampf gegen die Energiewende', *FFU-Report*, 03 (2013), http://edocs.fu-berlin.de/docs/receive/FUDOCS_document_000000017134, accessed 28 May 2016.

Bsumek, Erika Marie, David Kinkela and Mark Atwood Lawrence (eds), *Nation-States and the Global Environment: New Approaches to International Environmental History* (Oxford: Oxford University Press, 2013).

Carson, Rachel, *Silent Spring* (Greenwich, CT: Fawcett, 1962).

Chakrabarty, Dipesh, *Provincializing Europe: Postcolonial Thought and Historical Difference* (Princeton: Princeton University Press, 2000).

Cioc, Mark, *The Game of Conservation. International Treaties to Protect the World's Migratory Species* (Athens, OH: Ohio University Press, 2009).

Clavin, Patricia, *Securing the World Economy. The Reinvention of the League of Nations, 1920–1946* (Oxford: Oxford University Press, 2013).

Coates, Peter, 'Emerging from the Wilderness (or, from Redwoods to Bananas): Recent Environmental History in the United States and the Rest of the Americas', *Environment and History* 10(4) (2004): 407–38.

Commoner, Barry, *The Closing Circle: Nature, Man and Technology* (New York: Knopf, 1971).

Connelly, Matthew James, *Fatal Misconception: The Struggle to Control World Population* (Cambridge, MA: Belknap Press, 2008).

Cronon, William, *Changes in the Land: Indians, Colonists, and the Ecology of New England* (New York: Hill and Wang, 1983).

Crosby, Alfred W., *The Columbian Exchange: Biological and Cultural Consequences of 1492* (Westport, CT: Praeger, 2003 [1972]).

Ditt, Karl, 'Nature Conservation in England and Germany 1900–70: Forerunner of Environmental Protection?', *Contemporary European History* 5(1) (1996): 1–28.

Doering-Manteuffel, Anselm and Lutz Raphael (eds), *Nach dem Boom. Perspektiven auf die Zeitgeschichte seit 1970* (Göttingen: Vandenhoeck & Ruprecht, 2008).

Dorsey, Kurk, *The Dawn of Conservation Diplomacy: U.S.-Canadian Wildlife Protection Treaties in the Progressive Era* (Seattle: University of Washington Press, 1998).

———. *Whales and Nations: Environmental Diplomacy on the High Seas* (Seattle: University of Washington Press, 2014).

Dorsey, Kurk and Mark Lytle, 'Introduction. [Special Issue on Environmental and Diplomatic History]', *Diplomatic History* 32(4) (2008): 517–18.

Dülffer, Jost and Wilfried Loth (eds), *Dimensionen internationaler Geschichte* (Munich: Oldenbourg, 2012).

Dunlap, Riley E. and Aaron M. McCright, 'Organized Climate Change Denial', in *Oxford Handbook of Climate Change and Society*, John S. Dryzek, Richard B. Norgaard and David Schlosberg (eds) (Oxford: Oxford University Press, 2013), 144–60.

Ecologist and Friends of the Earth, *Stockholm Conference Eco*, 6–16 June 1972.

Ehrhardt, Hendrik and Thomas Kroll (eds), *Energie in der modernen Gesellschaft. Zeithistorische Perspektiven* (Göttingen: Vandenhoeck & Ruprecht, 2012).

Ehrlich, Paul, *The Population Bomb* (New York: Ballentine Books, 1968).

Engels, Jens Ivo, 'Modern Environmentalism', in *The Turning Points of Environmental History*, Frank Uekötter (ed.) (Pittsburgh: University of Pittsburgh Press, 2010), 119–31.

Falkner, Gerda, *EU Policies in a Global Perspective: Shaping or Taking International Regimes?* (Abingdon: Routledge, 2014).

Farnham, Timothy J., *Saving Nature's Legacy: The Origins of the Idea of Biodiversity* (New Haven: Yale University Press, 2007).

Ferguson, Niall et al. (eds), *The Shock of the Global: The 1970s in Perspective* (Cambridge, MA: Belknap Press, 2010).

Frey, Marc, 'Experten, Stiftungen und Politik: Zur Genese des globalen Diskurses über Bevölkerung seit 1945', *Studies in Contemporary History* 4(1–2) (2007), http://www.zeithistorische-forschungen.de/16126041-Frey-2-2007, accessed 28 May 2016.

Gissibl, Bernhard, Sabine Höhler and Patrick Kupper (eds), *Civilizing Nature: National Parks in Global Historical Perspective* (New York: Berghahn Books, 2012).

Haas, Peter M., 'Do Regimes Matter? Epistemic Communities and Mediterranean Pollution Control', *International Organization* 43(3) (1989): 377–403.

———. *Saving the Mediterranean: The Politics of International Environmental Cooperation* (New York: Columbia University Press, 1990).

———. 'Introduction: Epistemic Communities and International Policy Coordination', *International Organization* 46(1) (1992): 1–35.

Hamblin, Jacob Darwin, 'Environmentalism for the Atlantic Alliance: NATO's Experiment with the "Challenges of Modern Society"', *Environmental History* 15(1) (2010): 54–75.

Hardin, Garrett, 'The Tragedy of the Commons', *Science* 162(3859) (1968): 1243–48.

Hays, Samuel P., 'The Limits-to-Growth Issue: A Historical Perspective', in *Explorations in Environmental History*, Samuel P. Hays (ed.) (Pittsburgh: University of Pittsburgh Press, 1998), 3–23.

Herren, Madeleine, *Internationale Organisationen seit 1865. Eine Globalgeschichte der internationalen Ordnung* (Darmstadt: Wissenschaftliche Buchgesellschaft, 2009).

Howard-Jones, Norman, 'The Scientific Background of the International Sanitary Conferences 1851–1938', 1975, World Health Organization, http://apps.who.int/iris/bitstream/10665/62873/1/14549_eng.pdf?ua=1, accessed 28 May 2016.

Hughes, Johnson Donald, *An Environmental History of the World: Humankind's Changing Role in Community of Life*. 2nd edn (London: Routledge, 2009).

Hünemörder, Kai F., '1972 – Epochenschwelle der Umweltgeschichte?', in *Natur- und Umweltschutz nach 1945. Konzepte, Konflikte, Kompetenzen*, Franz-Josef Brüggemeier and Jens Ivo Engels (eds) (Frankfurt: Campus, 2005), 125–44.

IPCC, Fifth Assessment Report (AR5), 2014, http://www.ipcc.ch/report/ar5/index.shtml, accessed 28 May 2016.

———. 'Summary for Policymakers', in *Climate Change 2014: Impacts, Adaptation, and Vulnerability. Part A: Global and Sectoral Aspects. Contribution of Working Group II to the Fifth Assessment Report of the Intergovernmental Panel on Climate Change*, C.B. Field et al. (eds) (Cambridge: Cambridge University Press, 2014), 1–32.

Iriye, Akira, *Global Community: The Role of International Organizations in the Making of the Contemporary World* (Berkeley: University of California Press, 2002).

Iriye, Akira, Jürgen Osterhammel and Wilfried Loth (eds), *Global Interdependence: The World after 1945 (History of the World)* (Cambridge, MA: Harvard University Press, 2014).

Jørgensen, Finn Arne et al., 'Entangled Environments: Historians and Nature in the Nordic Countries', *Historisk tidsskrift* 92(1) (2013): 9–34.

Kaiser, Wolfram and Jan-Henrik Meyer (eds), *Societal Actors in European Integration: Polity-Building and Policy-Making 1958-1992* (Basingstoke: Palgrave Macmillan, 2013).

Kaiser, Wolfram, Morten Rasmussen and Brigitte Leucht (eds), *The History of the European Union: Origins of a Trans- and Supranational Polity 1950-72* (Abingdon: Routledge, 2009).

Kaiser, Wolfram and Johan Schot, *Writing the Rules for Europe: Experts, Cartels, International Organizations* (Basingstoke: Palgrave Macmillan, 2014).

Kaiser, Wolfram and Antonio Varsori (eds), *European Union History: Themes and Debates* (Basingstoke: Palgrave Macmillan, 2010).

Kirchhof, Astrid Mignon and Jan-Henrik Meyer, 'Global Protest against Nuclear Power: Transfer and Transnational Exchange in the 1970s and 1980s. Focus Issue', *Historical Social Research* 39(1) (2014): 163–273.

Knab, Cornelia, 'Infectious Rats and Dangerous Cows: Transnational Perspectives on Animal Diseases in the First Half of the Twentieth Century', *Contemporary European History* 20(3) (2011): 281–306.

Kohlrausch, Martin and Helmuth Trischler, *Building Europe on Expertise: Innovators, Organizers, Networkers* (Basingstoke: Palgrave Macmillan, 2014).

Koselleck, Reinhart, *Einleitung*, in *Geschichtliche Grundbegriffe*, vol. 1, Otto Brunner, Werner Conze, Reinhart Koselleck (eds) (Stuttgart: Klett Cotta, 1979), XV.

Kupper, Patrick, 'Die "1970er-Diagnose". Grundsätzliche Überlegungen zu einem Wendepunkt der Umweltgeschichte', *Archiv für Sozialgeschichte* 43(1) (2003): 325–48.

———. 'Translating Yellowstone: Early European National Parks, Weltnaturschutz and the Swiss Model', in *Civilizing Nature: National Parks in Global Historical Perspective*, Bernhard Gissibl, Sabine Höhler and Patrick Kupper (eds) (New York: Berghahn Books, 2012), 123–39.

Le Roy Ladurie, Emmanuel, 'Présentation "Histoire et Environnement" (Special Issue)', *Annales. Économies, Sociétés, Civilisations* 29(3) (1974): 537.

Lekan, Thomas, '*Serengeti Shall Not Die*: Bernhard Grzimek, Wildlife Film, and the Making of a Tourist Landscape in East Africa', *German History* 29(2) (2011): 224–64.

Linnér, Björn-Ola, *The Return of Malthus: Environmentalism and Post-war Population-Resource Crises* (Isle of Harris: White Horse Press, 2003).

Macekura, Stephen J., *Of Limits and Growth: The Rise of Global Sustainable Development in the Twentieth Century* (Cambridge: Cambridge University Press, 2015).

Mahrane, Yannick et al., 'De la nature à la biosphere. L'invention politique de l'environnement global, 1945–1972', *Vingtième Siècle* 113(1) (2012): 127–41.

Massard-Guilbaud, Geneviève, 'De la "part du milieu" à l'histoire de l'environnement', *Le Mouvement social* 200 (2002): 64–72.

————. *Histoire de la pollution industrielle: France, 1789–1914* (Paris: Ecole des hautes études en sciences sociales, 2010).

Mauch, Christof, 'Blick durchs Ökoskop. Rachel Carsons Klassiker und die Anfänge des modernen Umweltbewusstseins', *Studies in Contemporary History* 9(1) (2012): 1–4.

Mazower, Mark, *Governing the World: The History of an Idea* (London: Allen Lane, 2013).

McCormick, John, *British Politics and the Environment* (London: Earthscan, 1991).

————. *The Global Environmental Movement* (Chichester: John Wiley, 1995).

McCormick, John R. and Peter Engelke, 'Mensch und Umwelt im Zeitalter des Anthropozän', in *Geschichte der Welt 1945 bis heute: Die globalisierte Welt*, Akira Iriye et al. (eds) (Munich: Beck, 2013), 357–534.

McNeill, John R., *Something New under the Sun: An Environmental History of the Twentieth-Century World* (New York: Norton, 2000).

Meadows, Dennis et al., *The Limits to Growth* (New York: Universe Books, 1972).

Mechi, Lorenzo, Guia Migani and Francesco Petrini (eds), *Networks of Global Governance: International Organisations and European Integration in Historical Perspective* (Cambridge: Cambridge Scholars, 2014).

Meyer, Jan-Henrik, 'Saving Migrants: A Transnational Network Supporting Supranational Bird Protection Policy in the 1970s', in *Transnational Networks in Regional Integration. Governing Europe 1945–83*, Wolfram Kaiser, Brigitte Leucht and Michael Gehler (eds) (Basingstoke: Palgrave Macmillan, 2010), 176–98.

————. 'Appropriating the Environment: How the European Institutions Received the Novel Idea of the Environment and Made it Their Own', *KFG Working Paper* 31 (2011): 1–33, http://edocs.fu-berlin.de/docs/receive/ FUDOCS_document_000000012522, accessed 18 May 2016.

————. 'Un faux départ? Les acteurs français dans la politique environnementale européenne des années 1970', in *Une protection de l'environnement à la française, XIXe–XXe siècles*, Jean-François Mouhot and Charles-François Mathis (eds) (Seyssel: Editions Champ Vallon, 2013), 120–30.

————. 'Getting Started: Agenda-Setting in European Environmental Policy in the 1970s', in *The Institutions and Dynamics of the European Community, 1973–83*, Johnny Laursen (ed.) (Baden-Baden: Nomos, 2014), 221–242.

Morgenthau, Hans J., *Politics among Nations: The Struggle for Power and Peace* (New York: Knopf, 1948).

Mosley, Steven, *The Environment in World History* (Abingdon: Routledge, 2010).

Mouhot, Jean-François and Charles-François Mathis (eds), *Une protection de l'environnement à la française, XIXe–XXe siècles* (Seyssel: Editions Champ Vallon, 2013).

Nash, Roderick, *Wilderness and the American Mind* (New Haven: Yale University Press, 1982 [1967]).

Osborn, Fairfield, *Our Plundered Planet* (Boston: Little Brown, 1948).

Patel, Kiran Klaus, 'Provincialising European Union: Co-operation and Integration in Europe in a Historical Perspective', *Contemporary European History* 22(4) (2013): 649–73.

Piccioni, Luigi, 'Un punto d'arrivo, un punto di partenza. Discutendo di Paesaggio Costituzione cemento (di Salvatore Setti)', *Storia* 18(52) (2012): 87–114.

———. 'L'influence de la France dans la protection de la nature en Italie au début du XXe siècle', in *Une protection de l'environnement à la française, XIXe–XXe siècles*, Jean-François Mouhot and Charles-François Mathis (eds) (Seyssel: Editions Champ Vallon, 2013), 97–107.

Pierson, Paul, *Politics in Time: History, Institutions, and Social Analysis* (Princeton: Princeton University Press, 2004).

Pralle, Sarah B., 'Agenda-Setting and Climate Change', *Environmental Politics* 18(5) (2009): 781–99.

Princen, Sebastiaan, *Agenda Setting in the European Union* (Basingstoke: Palgrave Macmillan, 2009).

Putnam, Robert, 'Diplomacy and Domestic Politics: The Logic of Two-Level Games', *International Organization* 42(3) (1988): 427–60.

Radkau, Joachim, *Nature and Power: A Global History of the Environment* (Cambridge: Cambridge University Press, 2008).

———. *The Age of Ecology: A Global History* (London: Polity Press, 2014).

Risse-Kappen, Thomas (ed.), *Bringing Transnational Relations Back in: Non-state Actors, Domestic Structures and International Institutions* (Cambridge: Cambridge University Press, 1995).

Robertson, Thomas, *The Malthusian Moment: Global Population Growth and the Birth of American Environmentalism* (New Brunswick: Rutgers University Press, 2012).

———. 'Total War and the Total Environment: Fairfield Osborn, William Vogt, and the Birth of Global Ecology', *Environmental History* 17(2) (2012): 336–64.

Robin, Libby, 'The Rise of the Idea of Biodiversity: Crises, Responses and Expertise', *Quaderni. Communication, technologies, pouvoir* 76(3) (2011): 25–37.

Rodogno, Davide, Bernhard Struck and Jakob Vogel (eds), *Shaping the Transnational Sphere: Experts, Networks and Issues from the 1840s to the 1930s* (New York: Berghahn Books, 2015).

Saunier, Pierre-Yves, 'International Non-governmental Organizations (INGOs)', in *The Palgrave Dictionary of Transnational History*, Akira Iriye and Pierre-Yves Saunier (eds) (Basingstoke: Palgrave Macmillan, 2009), 573–80.

Schmelzer, Matthias, 'The Crisis before the Crisis: The "Problems of Modern Society" and the OECD, 1968–74', *European Review of History: Revue europeenne d'histoire* 19(6) (2012): 999–1020.

Schoijet, Mauricio, 'Limits to Growth and the Rise of Catastrophism', *Environmental History* 4(4) (1999): 515–30.

Schulz, Thorsten, 'Transatlantic Environmental Security in the 1970s? NATO's "Third Dimension" as an Early Environmental and Human Security Approach', *Historical Social Research* 35(4) (2010): 309–28.

Schulz-Walden, Thorsten, *Anfänge globaler Umweltpolitik. Umweltsicherheit in der internationalen Politik (1969–1975)* (Munich: Oldenbourg, 2013).

Scichilone, Laura, 'The Origins of the Common Environmental Policy. The Contributions of Spinelli and Mansholt in the *Ad Hoc* Group of the European

Commission', in *The Road to a United Europe: Interpretations of the Process of European Integration*, Morten Rasmussen and Ann-Christina Lauring Knudsen (eds) (Brussels: PIE-Peter Lang, 2009), 335–48.

Simmons, I.G., *Global Environmental History. 10,000 BC to AD 2000* (Edinburgh: Edinburgh University Press, 2008).

Stradling, David and Peter Thorsheim, 'The Smoke of Great Cities: British and American Efforts to Control Air Pollution, 1860–1914', *Environmental History* 4(1) (1999): 6–31.

Sutter, Paul S., 'The World with Us: The State of American Environmental History', *Journal of American History* 100(1) (2013): 94–119.

Thomas, Sarah L., 'A Call to Action: Silent Spring, Public Discourse and the Rise of Modern Environmentalism', in *Natural Protest: Essays on the History of American Environmentalism*, Michael Egan and Jeff Crane (eds) (New York: Routledge, 2009), 185–204.

Uekötter, Frank, *The Age of Smoke: Environmental Policy in Germany and the United States, 1880–1970* (Pittsburgh: University of Pittsburgh Press, 2009 [originally in German 2003]).

Uekötter, Frank, *The Greenest Nation? A New History of German Environmentalism* (Cambridge, MA: MIT Press, 2014).

Vogt, William, *Road to Survival* (New York: Sloane Associates, 1948).

Walter, François, *Les figures paysagères de la nation: territoire et paysage en Europe (16e–20e siècle)* (Paris: Ecole des hautes études en sciences sociales, 2004).

Waltz, Kenneth N., *Theory of International Politics* (Reading, MA: Addison-Wesley, 1979).

Ward, Barbara and René Dubos, *Only One Earth: The Care and Maintenance of a Small Planet* (Harmondsworth: Penguin, 1972).

Weber, Heike, 'Les ordures ménagères et l'apparition de la consommation de masse: une comparaison franco-allemande (1945–1975)', in *Une protection de l'environnement à la française, XIXe–XXe siècles*, Jean-François Mouhot and Charles-François Mathis (eds) (Seyssel: Editions Champ Vallon, 2013), 141–56.

White, Lynn, 'The Historical Roots of Our Ecologic Crisis', *Science* 155(3767) (1967): 1203–7.

Winiwarter, Verena et al., 'Environmental History in Europe from 1994 to 2004: Enthusiasm and Consolidation', *Environment and History* 10(4) (2004): 501–30.

Wöbse, Anna-Katharina, 'Der Schutz der Natur im Völkerbund – Anfänge einer Weltumweltpolitik', *Archiv für Sozialgeschichte* 43(1) (2003): 177–91.

———. 'Oil on Troubled Waters? Environmental Diplomacy in the League of Nations', *Diplomatic History* 32(4) (2008): 519–37.

———. 'Les liaisons sinueuses: les relations franco-allemandes en matière de protection de la nature dans la première moitié due XXe siècle', in *Une protection de l'environnement à la française, XIXe–XXe siècles*, Jean-François Mouhot and Charles-François Mathis (eds) (Seyssel: Editions Champ Vallon, 2013), 108–19.

Worster, Donald, *Dust Bowl: The Southern Plains in the 1930s* (New York: Oxford University Press, 1979).

———. *Nature's Economy: A History of Ecological Ideas* (Cambridge: Cambridge University Press, 1994 [1977]).

Young, Oran R., *International Cooperation: Building Regimes for Natural Resources and the Environment* (Ithaca: Cornell University Press, 1989).

Zelko, Frank, *Make it a Green Peace!: The Rise of Countercultural Environmentalism* (Oxford: Oxford University Press, 2013).

CHAPTER 1

From Nature to Environment

International Organizations and Environmental Protection before Stockholm

Jan-Henrik Meyer

International political efforts to deal with what we consider environmental issues today began in the early years of the twentieth century – a period commonly considered to be one of extensive transnational exchange among scientists and activists, of internationalism and early forms of globalization.[1] It was also a period of growing awareness, in Europe and North America, of the impact that modern humanity, equipped with the new technology of the age of steam, had on nature, notably on its animal species. Two examples loomed large in the contemporary debate: the mass killing of the vast herds of the American bison and the extinction of the once seemingly indestructible American passenger pigeon, with the last specimen dying in a zoo in Cincinnati in 1914.[2] The period before the First World War was also characterized by European colonial rule over large swathes of the world, including regions featuring what many contemporaries (often erroneously) considered to be pristine, Eden-like places of nature so far untouched by man.[3] These perceptions influenced which kinds of natural environments first made it onto the international agenda.

To discuss the preservation of animal species, the British Foreign Office invited representatives of the colonial powers of Germany, France, Italy, Portugal, Spain and the Congo Free State (then in the private possession of the Belgian King Leopold) to the International Conference on the Preservation of Wild Animals, Birds and Fish in Africa held in London from 24 April to 19 May 1900. Its purpose was to address a problem that is still making news today: the protection of African 'charismatic megafauna' including lions or elephants.[4] African elephants were hunted for ivory that was traded across colonial borders throughout the continent. Harvesting and selling ivory was a sizeable industry that left its mark on elephant populations across different African colonies already in the nineteenth century.[5] At a time when big game hunting was a prestigious sport among European and American elites, it was not only colonial officers who worried about the survival of their most favoured game species, but also upper-class hunters, naturalists and zoologists in the metropolis.[6]

Initially British and German colonial authorities in Africa had developed the idea of holding such an international conference. They realized that their own efforts at preservation and introducing European-style 'ethics' into African hunting were doomed to fail as long as they could not get the other colonial powers on board. Hence, conference participants discussed proposals for common rules aiming both at preservation and conservation. They included trade restrictions, such as a minimum weight of elephant tusks for export to protect immature animals, establishing reserves and closed seasons, and licences both for European and African hunters. In the spirit of colonialism, access by indigenous people to ammunition would be restricted. The negotiating parties severely watered down the proposals. The convention was never ratified by all its signatories and never entered into force. Nonetheless, the event itself and the convention are still widely considered an important precedent for international environmental rule making,[7] and they served as a model for regulation in other parts of the world, such as in British Malaya.[8]

A second major international event actually led to the foundation of the first international organization (IO) dealing with environmental protection. The *Weltnaturschutzkonferenz*, or 'Conference for the International Protection of Nature', took place thirteen years after the London conference, in November 1913 in Bern, the capital of landlocked Switzerland. The Swiss government supported the event as part of its internationally oriented foreign policy that had contributed to facilitating international cooperation across different sectors, such as in railway transportation.[9]

Nearly sixty years elapsed between the Conference for the International Protection of Nature in Bern and the Conference on the Human Environment in Stockholm in 1972, when the United Nations (UN) decided to establish its own environmental programme, commonly known by its acronym UNEP, as a separate UN organization located in Nairobi, Kenya. This chapter traces international debates about and the institutionalization of environmental protection beyond the national level during this period in order to provide background and context for the remaining chapters in this book that focus on the period around and beyond the Stockholm Conference.[10] It does so chronologically, focusing especially on the issues that IOs decided to take up and place on the political agenda; the solutions they promoted, notably with a view to establishing new institutions; and the actors who pushed for environmental action.

Between the conference in Bern in 1913 and the 1972 Stockholm Conference, the way in which a variety of actors, from governments to IOs, scientists as well as nationally based non-governmental organizations (NGOs) and newly created international non-governmental organizations (INGOs) understood, talked about and framed environmental issues changed quite fundamentally. This changing language of environmental protection is crucially important for understanding the transformation of IOs and environmental protection during the roughly sixty years until 1972.

The Language of the Environment

The term 'environment' as a political concept, relating to both a bounded set of problems and the policies and measures developed to resolve these problems, is a rather recent arrival to modern politics. For almost the entire period until 1972, no one understood or called these problems 'environmental'. A variety of actors treated them as problems of protecting animals and animal species or, more generally, nature and natural beauty, or they conceived of them as issues of pollution, waste or public health. While some of these issues seemed related, they were not commonly considered to be part of one global, comprehensive problem.

Ecological thinking, which spread among biologists during the postwar period, acted as a catalyst in this respect. Ecological ideas encouraged thinking in large systems, and the conception of the environment as global in scope.[11] Ecology's basic assumption was – and is – that all life on this planet is interrelated in a complex manner.[12] All substances brought into the system of life – including the toxic ones – do not simply decompose, dissolve and disappear, as many scientists and policy makers routinely assumed, for instance, when they authorized the dumping of nuclear waste at sea.[13] Rachel Carson's scathing critique of the reckless use of pesticides such as DDT in her 1962 book *Silent Spring* did not simply scandalize these practices, but by highlighting their – often unintended – environmental consequences, she popularized some of the central insights of ecology.[14] These tenets included, as Barry Commoner famously put them in 1971, 'everything is connected to everything else' and 'everything needs to go somewhere'.[15] Indeed, DDT – as well as heavy metals or nuclear isotopes – accumulated in the fatty tissues of animals, and came back with unforeseen and harmful side-effects, such as disturbing the buildup of birds' eggshells. Increasingly, the impact of new technology was understood as leading to new risks to human health and ultimately to human survival. In the course of the 1960s, this ecological thinking progressively turned political.[16]

The 'environment' became a hallmark term and was increasingly used to signify the entire array of issues relating to humanity's problematic relations with nature, including the older issues of nature conservation and animal protection, as well as the newer concerns about air, water and noise pollution and land use issues. Moreover, as scientists' prestige and presence grew in nature protection NGOs, government administrations as well as the public in general, rational science supplanted aesthetics as the single legitimate criterion by which problems were to be judged and addressed. Solving these issues seemed urgent, given the apparent size and scope of the human impact on nature in an age of technology, prosperity and mass consumption. The environment was a political term right from the start. In an era of state activism, it seemed self-evident to scientists, governments and the nascent environmental movement that environmental protection was to be achieved by collective, political means.[17]

From the late 1960s onwards, policy makers introduced measures that were explicitly called environmental policies, first in Sweden[18] and the United States,[19] and then more widely, even east of the Iron Curtain.[20] The environment became a category of local, national and international practice. The term was routinely translated into different languages, as states introduced environmental policies. For instance, in the case of Germany, the responsible minister literally translated the American terminology of environmental protection into *Umweltschutz*.[21]

Views and priorities concerning their goals and preferred instruments differed between the various groups and NGOs that emerged since the late nineteenth century in Europe and North America to promote environmental issues. This is reflected in the terminological distinction between preservation, conservation and protection, and the movements they represented. Preservation aimed at keeping scenic landscapes, places and wild animals in as much of a natural state as possible, protecting them against destructive human intervention. Preservationists – who played an important role in setting up national parks – admired nature primarily for its beauty. Next to aesthetics, moral and recreational values played an important role.[22]

Conversely, conservationists had a more utilitarian approach to nature. Their goal was to prevent wasteful exploitation of nature and to support the rational use and management of nature and natural resources, for example through scientific forest management or the multifunctional use of watercourses.[23] Conservation's central standards of judgement were utility and science. Conservationist ideas were politically very influential, with many conservationists working for the U.S. government until the 1960s, for example. The main bone of contention between both camps was dam building, which seemed a rational practice to conservationists, to harness the forces of water for electricity generation, flood prevention and providing cities with clean fresh water. However, preservationists opposed the flooding of beautiful river gorges as the home and breeding grounds of fish and pristine wildlife.[24]

In Europe, this ideal-typical juxtaposition seemed less relevant than in the U.S. Moreover, the usage of these concepts varied across different languages. What Europeans mostly called nature protection or even nature conservation (e.g. *Naturschutz, protection de la nature*) was closer to the preservation camp. During the first six decades of the twentieth century, aesthetic motivations remained very important along with patriotic, touristic and scientific ones.[25]

The struggle over meaning was also reflected in the naming of international bodies and programmes. Originally, upon its creation in 1948, the postwar IO dealing with nature was called the International Union for the Protection of Nature (IUPN). During the course of the 1950s, conservationists and ecologists alike criticized protection as erroneously suggesting 'a more limited and perhaps more defensive or sentimental image' of their organization's work.[26] Hence, in 1956, the IUPN was renamed the International Union for Conservation of Nature and Natural Resources (IUCN).[27] Nevertheless, protection was still the term of choice to describe the new Environmental Protection Agency (EPA) founded by the Nixon administration in the United States in December 1970 'to protect and safeguard public health and the environment'.[28]

The most common current usage in both environmental politics and environmental history is to speak of nature conservation or nature protection when referring more narrowly to nature, such as species and habitats. Environmental protection or environmental policy usually designates the broader range of problems included in the new political concept of the environment, of which nature protection is but a part.

Weltnaturschutz: Environmental Protection before 1914

While Switzerland was not a colonial power, the main promoter of holding a conference on *Weltnaturschutz*,[29] Paul Sarasin, the son of a wealthy industrialist from Basel, was very much a product of the colonial age and of the internationalism of the decades before the First World War.[30] During his voyages as a young zoologist in the 1880s and 1890s, he had travelled the British and Dutch colonies in Southern and Southeastern Asia, including present-day Sri Lanka and Indonesia, engaging in the study of geography, geology and animals as well as indigenous human populations. The changes to and the exploitation of nature he encountered left a deep impression on him.[31] Back in Switzerland, he supported the protection of nature in his own country and became the first president of the committee for nature conservation within the Swiss Society for the Study of Nature in 1906.[32] As such, he was actively involved in setting up the first Swiss national park in the Engadin Alps in 1909. This park drew on and modified the American model of Yellowstone. Encouraged by the apparent success of nature conservation in his country and in the spirit of internationalism, Sarasin – by the time already a well-connected scholar – set out to push for a practicable solution to what seemed to him an increasingly urgent problem: namely, protecting nature on a global scale.[33]

International meetings of the newly established nature protection groups and of scientists were the breeding ground for such ideas during the early years of the twentieth century. Already in 1909, nature protection groups from all over Europe had met in Paris at an International Congress for the Protection of Nature. It was at this meeting that demands for a multinational approach and for setting up an international body first came up, even if no practical steps were taken.[34]

Sarasin, who had attended the Paris meeting, promoted new forms of institutionalization and used another international scientific forum to launch his ideas. When he spoke at the International Zoological Association's congress in the Austrian city of Graz in August 1910, he went beyond a purely scientific debate and presented his political vision of *Weltnaturschutz*, the protection of nature on a global scale. He argued that such an approach was urgent and necessary and he provided an overview of the most pressing problems.[35]

Sarasin scandalized the increasing destruction of nature across the globe for the sake of profit making, including the large-scale use of tropical birds' feathers and entire stuffed birds for hats for women[36] and the killing of African elephants for ivory.[37] Reflecting his anthropological research interests, his concept of nature in need of protection also included those human populations that had so far not been exposed to Western civilization. He feared that they might become extinct soon after opening contact.[38]

Facilitated by new technology, Sarasin warned, humans were by now able to hunt animals to extinction even in the remotest corners of the planet, such as mammals and birds on the polar ice of the far-off island of Spitsbergen, for their fat, feathers and fur. The most immediate reason for him to call for international nature protection had been news of Norwegian inventors' most recent innovations to whaling methods, namely the steam-powered factory ship that made it possible to process large whales on the high seas.[39] Sarasin rightly anticipated that this technology would lead to the wholesale destruction of whale populations in both the Arctic and the Antarctic seas.[40]

In order to institutionalize the issue, Sarasin did not just call for an international conference, but demanded the setting up of an IO, a *Weltnaturschutzkommission*, or Commission for International Nature Protection. This commission was initially to consist of individuals as volunteers – based on the model of an INGO[41] – that only included societal actors.[42] Subsequently, it was to be staffed by national delegates, officially representing national governments – more like an intergovernmental IO.[43] Sarasin managed to convince the International Zoological Association's congress to accept his proposal and to authorize him to organize the provisional commission.[44]

During the following years, Sarasin lobbied the mostly academic public within and beyond Switzerland for his goal of establishing global nature protection. Most importantly, he managed to gain the support of the Swiss Federal government for holding an international conference on the issue. On 17–19 November 1913, at the invitation of the Swiss Federal government, thirty-two delegates from seventeen countries gathered in the Swiss Federal building in Bern.[45] At the conference, Sarasin reiterated his agenda, criticizing greed as the main reason for the large-scale exploitation of nature. He appealed to the national delegates to engage in ethically motivated action worldwide instead.[46]

Not all of the delegates were as excited about organizing nature conservation beyond the national level as Sarasin was. Notably, the German delegate, Hugo Conwentz, the internationally influential Prussian Commissioner for Natural Monuments, considered nature protection a national duty. This argument was not only in line with the spirit of the time, but also with his own concept of protecting natural monuments – aesthetically appealing, but often rather small-scale sites of natural beauty – as part of national heritage. Conwentz, as well as the Prussian ministry officials who had sent him with clear instructions, did not consider establishing a centralized office and uniform rules at the international level to be useful for practical reasons, given the vast differences between countries in the kinds of nature to be protected. He also disagreed on more principled political grounds, as uniform rules would have constituted a limitation of national sovereignty. At best, international action was to be limited to the informal exchange of information among experts developing nonbinding recommendations. Representatives from colonial powers also resented Sarasin's inclusion of indigenous populations, and the issue was subsequently dropped.[47]

Aware of these concerns and eager to reach an agreement, the representative of the Swiss Federal government, Ludwig Ferrer, proposed a compromise. Instead of including all concerns, he suggested concentrating on the most urgent issues. He also proposed specifically prioritizing those places in the world that required international action most urgently: places beyond national regulation, namely the 'high seas, the deserts, the steppe'.[48] Regulating and protecting nature in what we today often characterize as 'global commons',[49] where unilateral action only harmed those who refrained from exploitation, required binding international agreements and action of all or at least a group of 'civilized states'. Ferrer also suggested a hierarchy of action and actors, in which intergovernmental cooperation featured as a measure of last resort: first and foremost, nature protection was to be achieved by 'voluntary, private work and forming public opinion within the country', i.e. by domestic NGOs, secondly, by national-level action, and only thirdly by the 'cooperation of civilized nations or some of them'.[50]

On this basis, the conference reached an agreement to which all delegations eventually signed up. It established a Consultative Commission for International Nature Protection (CCINP),[51] consisting of two representatives per member state, from a minimum of nine member states. The seat of the body was to be Sarasin's hometown of Basel. The CCINP's tasks were, however, much closer to Conwentz's concerns than to Sarasin's vision: first, to collect, review and disseminate all available information on international nature protection and, secondly, to propagate international nature conservation – via its members. Member states also made clear that they would not have to provide any funding.[52]

Even if, by the end of 1914, fourteen countries had actually nominated representatives to the committee, Sarasin's efforts did not fall on fertile ground. Already, the first meeting of the CCINP, scheduled for 28 September 1914, was thwarted by the outbreak of the First World War. The committee did not survive the war.

Both events – the conference in 1900 mentioned at the start of the chapter and the 1913 conference – demonstrate key characteristics of international nature protection during the years before the First World War. First, contemporary actors called for international action to solve novel problems that resulted from technological advances and colonialism, and that were beyond the reach of national authorities. The problems were either of a crossborder nature, as in the movement of large animals in Africa as well as the trade of their products or, as in the case of whaling on the high seas or hunting on the remote and yet unclaimed island of Spitsbergen, they resulted from the absence of any national authority. As promoted by Hugo Grotius in the seventeenth century,[53] the high seas were conventionally understood to be free and open to all. When the space of regulation and the scope of the problem did not coincide, reckless behaviour and over-exploitation militated against sustainable use. This led to a situation that has been characterized as a 'tragedy of the commons'[54] or the 'fisherman's problem'.[55] Establishing international rules or IOs seemed the only possible solution.[56]

The first two decades of the twentieth century saw the beginning of such international rule making by conventions. Apart from the London Convention relating to African animals, European governments signed a Convention for the Protection of Birds Useful to Agriculture in 1902, which came into force in 1905.[57] In North America, with the pioneering Migratory Bird Treaty of 1916 between Canada and the United States, governments achieved a remarkable agreement on the crossborder issue of bird migration and set up an International Joint Commission, an IO to deal with issues of contentious crossborder issues more generally, in 1909.[58]

Secondly, the fact that large mammals and birds featured very prominently in prewar international nature conservation directly resulted from the fact that both hunting and bird protection were among the first environmental concerns around which people organized at both the national and international levels. As early as the 1880s, national bird protection NGOs such as the British Royal Society for the Protection of Birds (RSPB) and the National Audubon Society in the United States were founded.[59] The first INGO in nature protection grew out of the concerns raised at the London Conference of 1900. Founded in 1902, the Society for the Preservation of the Wild Fauna of the Empire – today Flora and Fauna International – lobbied to preserve African wildlife, birds and large mammals in particular. Their mostly upper-class sponsors, colonial officials, hunters and naturalists sought to establish game reserves and to restrict hunting. Privileged white men used their international networks and government links to promote these issues. The press soon nicknamed them 'penitent butchers'.[60]

Interwar Years: Nature Conservation and the League of Nations

During the interwar years, the range of issues that a variety of well-organized actors placed on the international agenda broadened substantially. They extended beyond the preservation of wild animals. In institutional terms, however, international nature protection became more concentrated. The establishment of the new all-purpose world organization, the League of Nations,[61] meant that for the first time, societal actors – NGOs and INGOs – had an IO to address their claims and complaints to, and to lobby for international nature protection. Four issues featured most prominently in the League of Nations: oil pollution at sea, animal protection more generally of both domestic and wild animals, marine resources and the protection of natural beauty.[62]

Oil pollution of the sea was the first serious transboundary pollution issue that was of a more global scope.[63] Oil pollution of marine waters was a direct and indirect result of the war. In some ways, to paraphrase John Maynard Keynes, it was an environmental consequence of Mr Churchill.[64] In the 1910s and 1920s, shipping companies followed a trend set by the British navy. In order to improve the speed and range of warships, First Lord of the Admiralty Winston Churchill had decided to change the British navy's fuel from coal to oil during the prewar arms race and to install the necessary infrastructure.[65] Whereas in 1914 only 2.6 per cent of ships at sea had relied on oil, by 1922 the share of those powered with liquid fuel had risen to 22.3 per cent.[66] The unanticipated side-effects of this technological shift resulted from a physical property of oil. Unlike coal, which simply went down to the bottom of the sea, oil floated on water. Oil from sunk battleships and fuel spilled for a variety of reasons, notably the cleaning of tanks, blackened shorelines and covered seabirds' feathers, leading to their slow and agonizing death. Emotionally distressing images of dying birds provided suggestive evidence of the urgency of the issue. For affected seaside communities, fishermen and bird protection activists, they were an important tool to demonstrate that international action was necessary.[67] Hence, despite the novelty of the pollution issue, it made it to the international agenda in a more traditional guise: as an emotionally charged bird protection issue.

Building on their large and well-established national level organizations in numerous countries in Europe and North America, in the interwar years, bird protection groups were already well resourced to take the issue to the international level. Birders had long been excellently connected across borders, drawing on international scientific networks of ornithologists.[68] Meeting in London in 1922 in the context of the campaign against oil pollution, bird protection groups founded their own INGO: the International Council for Bird Protection (ICBP). Rather than creating a strong international body of the type that Sarasin had envisioned for International Nature Protection, their ambition was more modest: mainly to coordinate their efforts transnationally.[69] On the issue of oil spills that mostly threatened seabirds, the large and influential British RSPB pushed the British government hard not to limit their ambition to only domestic legislation to combat oil pollution. In a large-scale letter-writing campaign, the RSPB demanded that the issue be taken to the international level. Bird protection groups on both sides of the English Channel, notably Dutch and British groups, exchanged information and cooperated transnationally, as they tried to lobby their own governments.[70]

Protection of both wild and domestic animals more generally constituted a second broadly environmental issue that was taken to the nascent League of Nations as early as 1919. The range of issues included the international transport and treatment of life animals, for example, an emotionally and ethically charged issue. Referring to national precedents such as British animal protection legislation, animal protection activists lobbied the League for international rules, notably an International Charter for the Prevention of Cruelty to Animals.[71] Animal protection NGOs dated back to the nineteenth century, notably in Protestant Northern Europe and the United States. In Britain, the Society for the Prevention of Cruelty to Animals was founded as early as 1824 (from 1837 'Royal' Society for the Prevention of Cruelty to Animals

[RSPCA]). Its counterpart in the United States, the American Society for the Prevention of Cruelty to Animals (ASPCA), was established in 1866 and quickly gained mass appeal, with local organizations across the Northern federal states in particular. Animal protection groups were internationally well connected, with international conferences before the war in London in 1909 and in Copenhagen in 1911, but also after the war in Paris in 1925. While the concern for African fauna was almost exclusively a preoccupation of men, women made up a large number of the members and activists in bird protection organizations, as well as in animal protection NGOs.[72]

The third major environmental issue that the League of Nations dealt with had already been on Sarasin's agenda. The increasing exploitation of the 'riches of the sea' gave rise to concerns about international solutions to the 'fisherman's paradox'. Managing the global commons of the high seas was considered a conservationist concern about resources. The debate emerged within the expert committee on the 'Progressive Codification of International Law', and arrived from what we today would call the global South in the mid 1920s. The Argentinian international law expert José León Suarez scandalized the exploitation of the Southern oceans by Northern industrial whaling and fishing, and pushed for international legal solutions.[73] When selecting a priority issue for actual treatment by the League, again large mammals mattered more than fish: notably whales and seals.[74]

The issue was shifted to the League's Economic and Financial Section, which was dominated by economists who sought to promote sound management in the spirit of conservation. After all, marine animals were considered a renewable resource. For scientific expertise, they relied on the International Council for the Exploration of the Sea (ICES), a research body that the Western nations engaging in commercial fishing had established in Copenhagen in 1902. This acted as a kind of clearing house for national fisheries interests, facilitating information exchange and research cooperation among its member states, rather than functioning as an independent IO. Its office was only staffed with a secretary general working part-time. However, the ICES subsequently provided a model for further regional fisheries research IOs worldwide, such as the International Commission for the Scientific Exploration of the Mediterranean Sea of 1919, the North American Council on Fisheries Investigation of 1920 and the Indo-Pacific Fisheries Council of 1948. As its name suggests, the ICES's main aim was to facilitate access to marine resources, rather than to be overly concerned with conservation of what still seemed plentiful. Hence, it was hardly surprising that the ICES's experts tended to pursue a policy of 'wait and see' concerning the whaling issue, referring to an apparent lack of evidence.[75]

It was not so much due to the input of experts or societal actors, but due to the pressure of the Norwegian government that the League eventually established a Convention for the International Regulation of Whaling, which was signed in 1931 and entered into force in 1935. Norway, as the most important whaling nation, established an Institute for Whale Research and a Committee for Whaling Statistics to keep track of whaling activities worldwide. From 1930, it published the *International Whaling Statistics*.[76] Worried by the growing number of factory ships from an ever-larger number of countries, the Norwegian government started to push for international regulation in the League's Economic and Financial Section in 1928 in order to safeguard the life of its own industry. The 1931 Convention followed the model of a Norwegian whaling law of 1929, which included the protection of immature animals, cows with calves, of certain whale species, prohibited the killing of whales in the tropics, and was to restrict the number of animals caught in Antarctic waters. It also included an obligation to submit statistical information about the whales killed to the Committee for Whaling Statistics, which for this purpose acted as the relevant IO. What severely limited the effectiveness of the convention was the emergence of Japan, Germany and the Soviet Union as major whaling nations in the 1930s. Germany and Japan left the League in 1933. Japan did not sign up to the convention and Germany failed to ratify it. The Soviet Union refused to sign as long as Japan did not sign.[77]

National parks were a fourth environmental issue discussed within the League of Nations. Despite the transnational spread of the idea and practice of founding national parks across a number of countries worldwide[78] and the establishment of a the first crossborder park in the Tatra Mountains between Poland and Czechoslovakia as a peace project facilitated by the League in 1932,[79] this pioneering issue of nature protection itself proved much harder to institutionalize internationally. The topic lingered in one of the League's technical organizations, the International Committee on Intellectual Cooperation, where it was discussed by prominent scientists, such as Marie Skłodowska Curie, without the debates ever leading to any tangible results.[80]

Likewise, revitalizing and institutionalizing Sarasin's CCINP within or in conjunction with the new world organization proved impossible. As early as 1919, Sarasin had made attempts to that end. He found a partner within the League of Nations, the Japanese deputy secretary general and director of its international office, Inazo Nitobe. This cosmopolitan agronomist, who had studied in the United States and Germany and had taught as a professor in Japan, shared Sarasin's interests in reconciling man and nature and supported his efforts. However, the initial goal to get the CCINP officially recognized with the League as an existing IO failed. So soon after the war, delegates from the former belligerent states refused to cooperate with each other. Nitobe suggested an alternative solution. Sarasin was to convince the Swiss Federal government to lobby for including nature protection as one of the world organization's goals. However, unlike in the prewar period, Swiss foreign policy makers no longer considered nature protection a means to promote Switzerland's international role when peace and reconstruction seemed more urgent tasks. Sarasin, by now an old man suffering from ill health, was no longer able to forge and mobilize the necessary networks to ensure the long-term viability of his project.[81]

In the 1920s, a group of international activists including notably the Dutch promoter of nature and bird conservation Pieter Gerbrand van Tienhoven attempted to revitalize Sarasin's ambition to set up an INGO for nature protection. Tienhoven's global perspective also stemmed from colonial concerns. He had shared Sarasin's worries about the preservation of exotic birds and had also fought against the trade of feathers in the Dutch East Indies. His approach to international action was based on mobilizing scientific networks among biologists and favoured informal structures of cooperation. Following a motion of the International Union of Biological Sciences' General Assembly in support of an international union, the already existing Dutch, Belgian and French committees for the protection of nature established a more limited Central Bureau of Information and Correlation under Tienhoven's presidency in 1928. Seven years later, in 1935, this was turned into the International Office for the Protection of Nature (IOPN) as an INGO with a broader European and North American membership, financed through a separate funding body, the Foundation for International Nature Protection. Based in Brussels, the IOPN set out to collect information, building up a library and editing its own journal, the *International Review of Legislation for the Protection of Nature*.[82] In the latter half of the 1930s, the new organization found little space for international political action as the League's early ambitions were increasingly hollowed out by growing international conflict.[83]

What continued to flourish during the interwar years, however, were the growing number of national NGOs engaged in nature protection. A number of major nature conservation NGOs were founded during this period, notably in the United States, such as the Wilderness Society or the National Wildlife Federation. As in the case of the 'penitent butchers', some of these organizations, such as the Izaak Walton League with more than 100,000 members, were interested in protecting nature, notably wetlands, rivers and lakes, primarily as hunting and fishing grounds. On both sides of the Atlantic, nature protection NGOs became influential players and were increasingly recognized in their role as advisers. Their lobbying focused on some receptive governments, such as the British government, rather than the League of Nations in order to promote wildlife protection and national parks within and across national boundaries.[84]

As a response to such lobbying, in 1933 the British government held a follow-up conference to London of 1900. The resulting 1933 London Convention Relative to the Preservation of Fauna and Flora in their Natural State, which entered into force in 1936, was another nonbinding treaty relating to African wildlife. It reiterated many of the concerns and instruments of the earlier convention, including trade in hunting trophies, and promoted the establishment of national parks and natural reserves. Without regard for traditional indigenous hunting rights and practices, the convention imposed bans on native hunting and hunting methods. While it did not establish an IO, nonetheless it did foresee a privileged advisory and potentially even an agenda-setting role for NGOs and INGOs. Article 5 stipulated that governments were to 'notify' each other of any 'information relevant to the purposes of the present Convention and communicated to them by any national museums or by any societies, national and international'.[85]

By the early 1940s, international nature protection expanded beyond Europe, North America, Africa and the high seas. Following American birds' flyways, US bird protection activists had already lobbied the US Senate in 1920 to extend the Canadian-American Migratory Birds Treaty of 1916 to all of the Americas. However, the United States and Canada only reached an agreement with Mexico in 1936. In 1940 the Pan American Union sponsored a more general Convention on Nature Protection and Wild Life Preservation in the Western Hemisphere, which included migratory birds. The 1933 London Convention provided the model for its contents, including the recognition of the advisory role of NGOs and experts, and the creation of national parks and nature reserves.[86]

Postwar: UNESCO Takes over

The Second World War drastically demonstrated to contemporaries the capacity of modern societies to destroy their livelihoods as well as their natural environments on an unprecedented scale, leading to hunger and destitution. The nuclear bombs of Hiroshima and Nagasaki epitomized the destructive power of scientific advances. Fairfield Osborne's gloomy views of the *Plundered Planet* and William Vogt's *Road to Survival* are probably the most well-known reflections of such views, including neo-Malthusian worries about excessive population growth.[87]

However, the postwar period was also characterized by a revival of internationalism. Already during the war, plans for a new world organization had been advanced. The UN, headquartered in New York, replaced the League of Nations and included an entire family of IOs, such as the Food and Agricultural Organization (FAO) in Rome. The new UN Educational, Scientific and Cultural Organization (UNESCO) took office in Paris in 1947. It drew on the heritage of the League's International Committee on Intellectual Cooperation, where national parks had been discussed during the interwar years. UNESCO was generously staffed, with six hundred officials by the end of 1947. While science and culture seemed clearly related to environmental issues, it was in fact the first Secretary General of UNESCO, Julian Huxley, who promoted nature protection as part of the new organization's agenda. The British zoologist's interest in nature included both aesthetic and scientific concerns. He became a vocal advocate of ecology after his expedition to Spitsbergen in 1921 and was very concerned about the impact of human population growth on natural environments.[88]

Huxley's role was highly instrumental in establishing the IUPN and connecting this hybrid organization – a blend between an IO and an INGO – in which both national government delegates and NGOs were represented, to UNESCO.[89] Sponsored by the French government and UNESCO, the main proponents of international network of nature protection activists, many of them members or partners of the IOPN, met in Fontainebleau, outside of Paris, in September 1948.[90] Founding members of the new organization were eighteen governments, 107 national conservation organizations and seven IOs, who signed the constitutive act on 5 October 1948. The IUPN/IUCN was subsequently mainly active in nature protection, pioneering 'red data books' of species threatened by extinction and the promotion of national parks through publications and conferences, routinely in conjunction with UN bodies.[91]

The original definition of the IUPN's ambition, however, was much more comprehensive. It spoke of 'the preservation of the entire world biotic community, or Man's natural environment, which includes the Earth's renewable natural resources' – a phrasing that reflected a modern global ecological conception of nature and the environment as well as the mid-twentieth-century concern about resource conservation. Nonetheless, the original IUPN also praised the older motivation of nature protection activists by highlighting 'natural beauty [as] one of the higher common denominators of spiritual life'.[92]

Julian Huxley sponsored the establishment of the IUPN as he planned to employ the new body to organize an international conference on nature protection for UNESCO. This conference was to coincide with, but also to counterbalance the more technical United Nations Scientific Conference on the Conservation and Utilisation of Resources (UNSCCUR) in Lake Success, the temporary home of the UN on Long Island, New York, in August 1949. Whereas UNSCCUR, with its more than 500 participants from 49 countries, tried to take stock of the state of natural resources ranging from minerals to fish and game in a more technical, conservationist fashion, the conference on nature protection organized by UNESCO and IUPN had a much more political agenda. Its 150 delegates from 33 countries discussed issues of education (notably environmental education), of ecology, of an international convention on nature protection, as well as the establishment of crossborder national parks.[93]

These issues raised at the conference foreshadowed later controversies and IO action. The Ecology Section's discussion on chemical pesticides predated Rachel Carson's outcry by more than a decade. The issue of environmental education was a central goal promoted by the Stockholm UN Conference.[94] In the immediate postwar period many of these issues however remained unresolved due to different priorities in UNESCO's activities after the departure of Julian Huxley in 1948.[95]

Always strapped for cash, the IUPN/IUCN's activities remained focused on a limited number of projects. In the late 1950s and early 1960s their research and protection efforts again targeted the oldest issue of international nature conservation, namely African wildlife.[96] IUCN members shared the concerns popularized by Bernhard Grzimek's 1959 documentary *Serengeti Shall Not Die*, namely that decolonization and development would inexorably lead to the demise of African wild animal populations by encroaching on their stomping grounds.[97] Moreover, human population growth threatened to put pressure on nature reserves, which colonial authorities had set aside with little concern for native land use practices and needs. Julian Huxley, who undertook a study in Africa for UNESCO in 1960, proposed holding a conference in order to convince Africa's new leaders of the value of nature conservation. Not only relying on the financial support of UNESCO and the FAO, the IUCN also utilized its close ties to the UN system in order to get access to policy makers in the newly established African states. It collaborated with the Commission for Technical Cooperation in Africa South of the Sahara (CCTA), a subsidiary body of the UN's Economic Commission for Africa, to convince African political elites that protecting wildlife also made for an economically viable strategy. The IUCN and CCTA organized a major conference in Arusha in what was then Tanganyika in 1961. Chaired by Tanganyika's leader Julius Nyerere, who was sympathetic to nature protection, about 150 participants from twenty-one African states, six non-African states and five IOs attended the meeting. Framing African wildlife as natural world heritage, promising a source of future income via tourism, and acknowledging the need for expertise and international cooperation, the Arusha conference was to secure the continued existence of nature reserves and national parks, and the ongoing international involvement of wildlife conservation experts and NGOs from Europe and North America. The IUCN's central goal remained 'educating' Africans on nature conservation issues.[98]

In order to improve its finances, the IUCN's leaders developed the idea of establishing a separate fundraising body, a model that the IOPN had already used in the interwar years, with its funding organization, the Foundation for International Nature Protection. This led to the founding of the World Wildlife Fund (WWF, today the Worldwide Fund for Nature) in 1961. Involving celebrities, such as former US President Dwight D. Eisenhower, and European royalty, such as Prince Philip, Duke of Edinburgh, for the first fundraising appeals in the United States and the United Kingdom respectively and Prince Bernhard of the Netherlands as first President, the WWF attracted substantial, and increasingly controversial, corporate funding, for instance from large oil companies like Shell. It soon gained independence and became the first major environmental INGO of a global scope, pre-dating the establishment of Friends of the Earth and Greenpeace by about a decade.[99]

Apart from the protection of nature, in the postwar period specialist IOs were established to deal with more limited crossborder environmental problems. One example of such an IO is the International Commission for the Protection of the Rhine against Pollution. It was founded in 1950 as a complementary IO to the already existing Central Commission on Rhine Navigation. This IO dated back to the Vienna Congress of 1814–15. While the older IO had facilitated turning Western Europe's largest river into a shipping channel in the nineteenth century and a sewer for the chemical industry upstream in the twentieth century for the benefit of economic growth, the new body was to limit the worst side-effects of these developments.[100]

The postwar period also saw the establishment of the International Whaling Commission (IWC) in 1949 to administer the 1946 International Convention for the Regulation of Whaling. This IO was initially a cartel of whaling nations to protect the continued existence of their exploitative business model. It was designed to achieve what the earlier Convention of 1935 in the context of the League had not attained, namely the regulation of whaling in a sustainable manner. The attempt to realize the contradictory objectives of conserving whale populations and developing the industry reflected the postwar hunger for natural resources and an exaggerated belief in scientific management. The IWC originally consisted of sixteen members, among them all the whaling nations and some countries more interested in conservation, like the US, which, however, remained a large consumer of whale products. The IWC was not very effective, due to opt-out clauses and a lack of implementation control. By the time of the 1972 Stockholm Conference, the industry had so thoroughly depleted whale stocks that, apart from the subsidized whalers of the Soviet Union, Japan and Norway, whaling was no longer commercially viable. All of the large whale species had gone commercially extinct and were more or less close to biological extinction. The Stockholm Conference coincided with an IWC meeting and led to demands for a whaling moratorium. Whaling – and the goal to save the last whales – remained a controversial issue.[101] By the 1970s, whaling became a primary target of INGOs, notably Greenpeace.[102]

While international nature protection remained the dominant environmental concern, one important global pollution issue emerged already in the 1950s. As part of the Cold War arms race, the United States, the Soviet Union and the United Kingdom – and subsequently also France and China – tested hundreds of nuclear devices in the atmosphere. Even if these weapons tests were conducted far away from the centres of population, in deserts and on Pacific islands, radioactive isotopes were spread throughout the atmosphere and came back to the ground as so-called fallout. As scientists, including ecologists sponsored by the US military, increasingly learned about the impact of radioactive isotopes like strontium–90 on human health, they started to inform policy makers and the public.[103] Popular protest in North America, Europe, but also around the Pacific, against atmospheric testing led to the Treaty Banning Nuclear Weapon Tests in the Atmosphere, in Outer Space and Under Water, the so-called Limited Test Ban Treaty of 1963, which, however, only covered the United States, the United Kingdom and the Soviet Union.[104] Underground testing, as well as notably French atmospheric testing, continued, as did protest against it. Indeed, in the early 1970s, Greenpeace started out with protests against U.S. and French nuclear testing in the Pacific.[105] The fallout debate thus marks the transition to the new comprehensive and global concept of the environment.

In the Run-up to the Stockholm Conference: From Nature to Pollution

By the 1960s, postwar prosperity fuelled by inexpensive and plentiful oil[106] had for a while pushed aside the resource conservation concerns and the debate about population. These issues only returned during the years immediately prior to the Stockholm Conference, with Paul Ehrlich's 1968 *Population Bomb* and the 1972 Club of Rome Report on the *Limits to Growth*.[107]

What became more pressing instead were the unpleasant side-effects of postwar prosperity, such as noise and waste, as well as the dangers associated with the use of chemical pesticides. Notably, air and water pollution were issues that were of both a local and a transnational nature. Untreated effluents negatively affected people downstream, threatening their health and livelihoods. Moreover, many European rivers, for instance the Rhine, crossed national borders. Similarly, a 'policy of high chimneys' designed to alleviate local pollution turned air pollution into a crossborder issue too.[108]

Confronted with these problems and the growing debate about them, a number of those IOs that had been founded in the aftermath of the Second World War, notably those with an interest in culture and research, as well as economics, started to engage with the pollution concern during the 1960s. Three IOs were pioneers in the run-up to the Stockholm Conference: the Organisation for Economic Co-operation and Development (OECD) founded in this form in 1961–62, the Western European Council of Europe created in 1949 and UNESCO.

In the 1960s the OECD's main goal had been to facilitate economic growth and prosperity. By the end of the 1960s, the OECD sought to carve out a new role for itself by tackling those 'problems of modern society' that resulted from two decades of unprecedented economic expansion. Impressed by the workers and student protests of 1968, and intellectually and personally linked to the Club of Rome, the Director General for Scientific Affairs, Alexander King, developed a political vision of improving the quality of life beyond purely quantitative growth. Both the OECD's outgoing Secretary General, the Danish futurist and economist Thorkil Kristensen, and his successor, the Dutch lawyer Emiel van Lennep, who took office in 1969, strongly supported a new emphasis on 'qualitative aspects' of growth and on environmental problems, drawing on current critical views in economics.[109] It was in this context that the OECD took up the issue of in environmental protection. Within its Committee for Research Cooperation, it set up two working groups to deal with the most pressing pollution issues: a group on water pollution in 1967 and a group on air pollution in 1968. In 1970 the OECD was the first IO to set up an Environmental Committee. Crossborder air and water pollution featured prominently on the Committee's agenda, as did the definition of comparable standards, measures and principles.[110]

The Council of Europe had started discussing issues of crossborder air pollution as early as 1961. In the summer of 1964, the organization invited some 350 experts from its member states and overseas for a conference on the issue to Strasbourg. Debates ranged from the immediate concern of health effects and methods of technically reducing pollution from different sources to legal instruments and plans for a European convention on the issue. In the aftermath of the conference, in 1966, the Council of Europe set up a Committee of Experts on Air Pollution. In close cooperation with other IOs, such as the World Health Organization (WHO), the UN Economic Commission for Europe or the OECD, this committee prepared the basis for what was to be the Declaration of Principles on Air Pollution Control that was accepted by the national ministers in March 1968. Among the principles listed, the resolution stipulated the polluter pays principle for the first time at the international level: 'The cost incurred in preventing or abating pollution should be borne by who-ever causes the pollution. This does not preclude aid from Public Authorities.'[111] This principle was subsequently promoted by the OECD and the European Community (EC), as Jan-Henrik Meyer's second chapter in this volume demonstrates. Concerning the issue of crossborder water pollution, the Council of Europe had already adopted a European Water Charter in 1967.[112]

Apart from pollution, the Council of Europe was also more broadly involved in conservation concerns, declaring 1970 the European Conservation Year. The main conference held in Strasbourg in February 1970, assembling some 330 participants from seventeen member states, dealt with a broad range of issues including urbanization – and the problem of urban sprawl – industrialization, agriculture and forestry, that is, issues classically related to the conservation agenda. At the same time, this conference was important in raising environmental awareness in Europe, in parallel with Earth Day in the United States in April 1970.[113]

Finally, UNESCO's role in raising political awareness for environmental issues is closely associated with an event in Paris in September 1968: the Biosphere Conference. The International Conference of Experts on a Scientific Basis for a Rational Use and Conservation of the Resources of the Biosphere grew out of the international cooperative research efforts during the postwar period. In the wake of the International Geophysical Year of 1957/58,[114] the International Biological Programme (IBP) was designed to enquire into the 'Biological Basis of Productivity and Human Welfare', including not only the 'use and management of natural resources' and 'productivity in terrestrial, freshwater and marine ecosystems', but also the 'conservation of terrestrial ecosystems'. The latter concern also involved cooperation with the IUCN.[115] These unprecedented efforts of international cooperation among scientists and the large-scale application of ecological ideas produced an enormous wealth of data and scientific knowledge, which impressively demonstrated the human impact on the planet for the first time.[116]

Against this backdrop, the Biosphere Conference marked the culmination of the debates of the conferences of Lake Success, combining ecological and conservation agendas with the new crossborder pollution problem. The conference politicized environmental protection, e.g. by adopting twenty recommendations to governments and international organizations for future action.[117] It was in Paris that demands for an international conference within the UN framework first came up. At the time, Sweden pioneered environmental policy at the domestic level, integrating various nature protection and anti-pollution bodies into a comprehensive Swedish Environmental Protection Agency in 1967, and subsequently preparing an Environmental Protection Act that was enacted in 1969.[118] Due to Sweden's geographical location, the country was affected by massive crossborder air pollution that led to the acidification of Scandinavian lakes, which caused major public concern. Thus, in order to raise attention to environmental issues more broadly and upload the crossborder concern to the international level, the Swedish government began to push for such a conference since the spring of 1967. This demand was referred to the UN Economic and Social Commission (ECOSOC). ECOSOC supported these plans in the summer of 1968, referring explicitly to the experts' views from the Biosphere Conference. The UN General Assembly accepted a resolution on 3 December 1968 that called for a world conference on the human environment. This eventually led to the 1972 Stockholm Conference, which was intensely prepared by the different regional UN Economic Commissions around the world, involving experts and scientists, including those from developing countries. Reports were commissioned on the situation of the environment at the national level.[119] All of this research and debate contributed to an unprecedented rise in awareness for environmental problems around the world. It marks the breakthrough of the environmental agenda that Stockholm commonly stands for.[120]

Conclusion

Looking back from the eve of the 1972 Stockholm Conference, what is the record of continuity and change in the role of IOs in environmental protection during the first six decades of the global twentieth century?

First, it has become clear that IOs had to deal with a variety of issues resulting from either crossborder problems or 'global commons' of international anarchy. The focal point of their agendas shifted over time. IO activism started out with the concern for natural beauty and for wild animals in the colonies or other remote places before the First World War. In the interwar years, the League of Nations continued to deal with protecting wild animals, for example those threatened by oil spills. However, the League also included the growing worries of conservationists about managing and sustaining the world's natural resources. In the postwar period, alongside the dominant resource conservation concern, aesthetic and increasingly ecological issues constituted core motivations for action to protect nature within the UN system, especially UNESCO and its partners at the IUPN/IUCN. That the concern for wild animals and particularly African fauna that had started international nature protection in 1900 persisted well into postwar nature protection is a remarkable element of continuity, as was the disregard for indigenous land use, even after decolonization. This resulted from the continuity of the transnationally well-connected international nature protection elites, who were active across a network comprising IOs and INGOs. These elites were constituted mostly of West European and North American white upper-class men or natural scientists whose worldviews were shaped by the age of colonialism.

It was only in the 1960s, in the run-up to the Stockholm Conference, that pollution, which had previously been considered only a local nuisance or reframed as a bird protection issue in the interwar years, became the most prominent issue internationally. This was a consequence of the unprecedented technological and economic development in the so-called 'golden age' of postwar prosperity.[121] Rising mass consumption went hand in hand with rapidly increasing use of natural resources and pollution. Electricity providers' reliance on a 'policy of high chimneys', for example, turned air pollution from coal-fired power stations from a more localized to a crossborder problem.[122] At the same time, the Cold War not only stimulated a nuclear arms race that led to nuclear pollution of the atmosphere – it also contributed to an exponential growth of scientific knowledge and the rise of ecology that framed environmental issues as global in scope.[123]

Secondly, the pendulum seems to have been swinging back and forth between two types of institutional solutions. While the initial attempt of the CCINP was to set up a standalone IO solely for the purpose of nature protection, nature conservation concerns were included in the newly founded all-purpose world organization, the League of Nations, in the 1920s. The period of postwar reconstruction was the heyday of IOs. New IOs were designed to tackle specific problems, such as the IWC, or more comprehensive issues of nature protection, such as the IUPN/IUCN, which drew on predecessors, notably the IOPN, and institutional models from the interwar years. Even if this looked like a revival of the functional, standalone IO solution, it was not. In fact, it reflected the emergence of a network structure of IOs and INGOs over the course of the century. Notably, the IUCN was closely connected with the new world organization, the UN, and its cultural branch UNESCO, but also the FAO, and created the first of the new global environmental INGOs, the WWF.[124] In the run-up to the Stockholm Conference, this network grew, but also competition emerged among IOs: IOs designed for very different purposes started to include the environment among their remit, such as the OECD, the Council for Europe and even NATO.[125]

Thirdly, a broad range of actors – from the white upper-class male elite of wealthy hunters turned penitent butchers to bureaucrats, scientists and female activists, notably in bird and animal protection, promoted and advanced environmental protection efforts within and across IOs. Sympathetic national governments or individual politicians or officials, such as the Swiss Federal government in Sarasin's case or the British government in the case of the London conference, played proactive roles in promoting environmental protection. These and other actors were able to mobilize support and to manufacture a common interest. However, sponsoring governments also benefited from international self-promotion by hosting international events. Some governments tried to solve national problems via the international route: Norway tried to 'upload' its whaling regulation to the international level already in the interwar years, thus creating a level playing field for its whalers and common standards and more 'sustainable' practices on the high seas.

NGOs and INGOs played an important role in placing issues on IO agendas, linking them to societal concerns. Clearly, for most of this period, elite groups of upper-class (former) hunters and well-connected scientists dominated nature protection NGOs and INGOs, and linked them to IOs. Most of these men shared a first-hand encounter with pristine nature in overseas colonies. This clearly had an impact on their ideas about which kinds of nature deserved and required protection and how this could be achieved. This is reflected in the dominant concern for African nature and the establishment of national parks to protect animals and their habitats from local populations. However, nature protection also had a mass appeal, mobilizing those citizens who wrote to various IOs to save the birds,[126] or raising funds for NGOs and INGOs.

In the period before Stockholm, the geographical scope of the problems was not fully reflected in the geographical distribution of actors involved. In the League of Nations, the Argentinian Suarez seemed an early defender of Southern interests in getting a fair share of natural resources of the Southern Atlantic. However, in the interwar years, it was mainly US actors who lobbied for the extension of the Canadian-American Migratory Birds Treaty to Mexico and the entire Western Hemisphere. Moreover, well into the postwar period, Africa, the primary habitat of the most cherished wildlife, firmly remained a space determined and defined by European and North American nature protection elites, even if upper-class hunters were slowly replaced by scientists and experts in nature conservation. While these Northern defenders of nature protection appreciated African nature, African people rather appeared as a potential threat to nature. IOs were making great efforts to convince African leaders to maintain those national parks set aside by the colonial powers and to continue keeping people out. Revoking or adjusting these aspects of colonial rule seemed illegitimate to the Western advocates of nature protection. Even during and after the period of decolonization, in the eyes of IOs and Western nature protection groups, African leaders and people needed to be educated about the importance and value of the natural heritage, which was to be set aside for Northern tourists' appreciation and pleasure. Tourism was promoted as the only legitimate economic activity and source of income. The IUCN also invested substantial resources to establish structures for passing on and thus maintaining their model of nature conservation by training African wildlife wardens and park directors, notably at the College of African Wildlife Management in Tanzania. In many places, the administration of African national parks and nature reserves remained in the hands of white people until the 1970s, as the last remainder of colonialism. Only during the years immediately preceding the Stockholm Conference did African policy makers became more fully involved in setting international rules for Africa, for instance regarding an IUCN-sponsored African Convention on the Conservation of Nature and Natural Resources, accepted by African governments in Algiers in 1968.[127]

Finally, what is the record concerning the success of IOs in building institutions and shaping agendas? Viewed from Stockholm on the eve of the 1972 Conference, the rise of environmental protection seemed remarkable. By this time, a large number of IOs had taken the issue of the environment on board, and plans existed to centralize the UN's efforts concerning the environment in one UN agency – what was to become UNEP.

Nevertheless, while the record of IOs in building institutions, building networks with a broad range of actors and setting agendas is impressive, their real-world achievements were at best mixed. By the early 1970s, national parks had spread widely across the globe and continued to do so thereafter. Although IO efforts to control oil pollution of the seas dated back to the 1920s, they could not prevent the two major oil spills in the 1960s – Torrey Canyon in 1967 and Santa Barbara in 1969 – that had a major impact in terms of placing the environment on the political agenda in the United States and Europe in the late 1960s.[128] The IWC's apparent ineffectiveness to save the last remaining whales after their commercial extinction was the target of much debate and protest on the margins of the Stockholm Conference.[129]

Nonetheless, given that an important part of postwar international policy making was directed at facilitating global economic growth regardless of its side effects,[130] the establishment of such a comprehensive environmental agenda was still remarkable. The institutional infrastructure of the IO world also ensured that the new issue became firmly established across a wide range of IOs, even if this was not their primary concern.[131] However, within less than a year after the Stockholm Conference, the oil crisis of autumn 1973 and its socioeconomic aftermath once more tempted policy makers to prioritize the economy over environmental protection.[132]

Jan-Henrik Meyer is a senior researcher at the Max-Planck-Institute for the History of European Law, Frankfurt, Germany and an associate researcher at the Centre for Contemporary History, Potsdam.

Notes

1. Johannes Paulmann, 'Reformer, Experten und Diplomaten. Grundlagen des Internationalismus im 19. Jahrhundert', in *Akteure der Außenbeziehungen: Netzwerke und Interkulturalität im historischen Wandel*, Hillard von Thiessen and Christian Windler (eds) (Cologne: Böhlau, 2010), 173–98, 180.
2. Kurk Dorsey, *The Dawn of Conservation Diplomacy: U.S.–Canadian Wildlife Protection Treaties in the Progressive Era* (Seattle: University of Washington Press, 1998), 13; Janet M. Davis, 'Bird Day: Promoting the Gospel of Kindness in the Philippines during the American Occupation', in *Nation-States and the Global Environment: New Approaches to International Environmental History*, Erika Marie Bsumek, David Kinkela and Mark Atwood Lawrence (eds) (Oxford: Oxford University Press, 2013), 181–206, 186.
3. Mark Cioc, *The Game of Conservation. International Treaties to Protect the World's Migratory Species* (Athens, OH: Ohio University Press, 2009), 20 f.
4. E.g. Brad Scriber, '100,000 Elephants Killed by Poachers in Just Three Years, Landmark Analysis Finds. Central Africa Has Lost 64 Percent of its Elephants in a Decade', *National Geographic*, 18 August 2014, http://news.nationalgeographic.com/news/2014/08/140818-elephants-africa-poaching-cites-census, accessed 18 May 2016.
5. Cioc, *Game of Conservation*, 20–28 (including ivory trade statistics).
6. Joachim Radkau, *Die Ära der Ökologie. Eine Weltgeschichte* (Munich: Beck, 2011), 76 f.
7. Bernhard Gissibl, 'German Colonialism and the Beginnings of International Wildlife Preservation in Africa', *GHI Bulletin*, Supplement 3 (2006), 121–43, http://www.ghi-dc.org/publications/ghi-bulletin/bulletin-supplements/bulletin-supplement-3-2006.html?L=0, accessed 30 June 2016, 130–37.

8. Davis, 'Bird Day', 187.
9. Madeleine Herren, *Hintertüren zur Macht: Internationalismus und modernisierungsorientierte Außenpolitik in Belgien, der Schweiz und den USA, 1865–1914* (Munich: Oldenbourg, 2000), 300–62.
10. This chapter has been developed in the context of a larger project on the origins of the European Communities' environmental policy, supported by the German Science Foundation-funded KFG 'The Transformative Power of Europe' at FU Berlin, a Marie Curie Intra-European Fellowship and a Marie Curie Reintegration Grant within the 7th European Framework Programme, by the Danish Research Council for Culture and Communication (FKK) within the 'Transnational History' project at Aarhus University, and by a Rachel Carson Fellowship of the Rachel Carson Center, Munich.
11. Donald Worster, *Nature's Economy. A History of Ecological Ideas* (Cambridge: Cambridge University Press, 1994 [1977]), 350–73.
12. Cf. Barry Commoner, *The Closing Circle. Nature, Man and Technology* (New York: Knopf, 1971).
13. Jacob Darwin Hamblin, *Poison in the Well: Radioactive Waste in the Oceans at the Dawn of the Nuclear Age* (New Brunswick, NJ: Rutgers University Press, 2008).
14. Rachel Carson, *Silent Spring* (Greenwich, CT: Fawcett, 1962); Sarah L. Thomas, 'A Call to Action: Silent Spring, Public Discourse and the Rise of Modern Environmentalism', in *Natural Protest. Essays on the History of American Environmentalism*, Michael Egan and Jeff Crane (eds) (New York: Routledge, 2009), 185–204.
15. Michael Egan, *Barry Commoner and the Science of Survival. The Remaking of American Environmentalism* (Cambridge, MA: MIT Press, 2007), 126.
16. Ibid., 6 f.
17. Jens Ivo Engels, 'Modern Environmentalism', in *The Turning Points of Environmental History*, Frank Uekötter (ed.) (Pittsburgh: University of Pittsburgh Press, 2010), 119–131, 124f.
18. Claes Bernes and Lars J. Lundgren, *Use and Misuse of Nature's Resources. An Environmental History of Sweden* (Varnamo: Falth & Hassler, 2009).
19. Micheal R. Vickery, 'Conservative Politics and the Politics of Conservation. Richard Nixon and the Environmental Protection Agency', in *Green Talk in the White House. The Rhetorical Presidency encounters Ecology*, Tarla Rai Peterson (ed.) (College Station: Texas A&M University Press, 2004), 113–33.
20. Paul R. Josephson et al., *An Environmental History of Russia* (Cambridge: Cambridge University Press, 2013).
21. Peter Menke-Glückert, 'Der Umweltpolitiker Genscher', in *In der Verantwortung. Hans Dietrich Genscher zum Siebzigsten*, Klaus Kinkel (ed.) (Berlin: Siedler, 1997), 155–68.
22. Dorsey, *Dawn of Conservation Diplomacy*, 11; Mark Stoll, *Inherit the Holy Mountain: Religion and the Rise of American Environmentalism* (Oxford: Oxford University Press, 2015).

23. John McCormick, *The Global Environmental Movement* (Chichester: John Wiley, 1995), 16 f., 22 f.; Steven Stoll, *US Environmentalism since 1945: A Brief History with Documents* (New York: Palgrave Macmillan, 2007), 8–11.
24. A famous example is the conflict over the Hetch Hetchy valley in early twentieth century in the United States. See Dorsey, *Dawn of Conservation Diplomacy*, 11; John M. Meyer, 'Gifford Pinchot, John Muir, and the Boundaries of Politics in American Thought', *Polity* 30(2) (1997): 267–84; Robert W. Righter, 'The Hetch Hetchy Controversy', in *Natural Protest*, Michael Egan and Jeff Crane (eds) (New York: Routledge, 2009), 117–35.
25. William T. Markham and C.S.A. van Koppen, 'Nature Protection in Nine Countries. A Framework for Analysis', in *Protecting Nature. Organizations and Networks in Europe and the USA*, William T. Markham and C.S.A. van Koppen (eds) (Cheltenham: Edward Elgar, 2008), 1–33; Karl Ditt, 'Nature Conservation in England and Germany 1900–70: Forerunner of Environmental Protection?', *Contemporary European History* 5(1) (1996): 1–28.
26. Lee M. Talbot, 'IUCN in Retrospect and Prospect', *Environmental Conservation* 10(1) (1983): 5–11, 8.
27. McCormick, *Global Environmental Movement*, 45; Martin Holdgate, *The Green Web: A Union for World Conservation* (Abingdon: Earthscan, 2013), 64.
28. Byron W. Daynes and Glen Sussman, *White House Politics and the Environment: Franklin D. Roosevelt to George W. Bush* (College Station: Texas A&M University Press, 2010), 77. The choice of the term protection seems in line with the Nixon administration's conception of the environment as relevant for security: Thorsten Schulz, 'Transatlantic Environmental Security in the 1970s? NATO's "Third Dimension" as an Early Environmental and Human Security Approach', *Historical Social Research* 35(4) (2010): 309–28.
29. This is conventionally translated into 'international nature protection'. A more literal translation would be 'world nature protection' or 'global nature protection'.
30. Bernhard C. Schär, *Tropenliebe. Schweizer Naturforscher und niederländischer Imperialismus in Südostasien um 1900* (Frankfurt: Campus, 2015).
31. Bernhard C. Schär, 'Earth Scientists as Time Travelers and Agents of Colonial Conquest: Swiss Naturalists in the Dutch East Indies', *Historical Social Research* 40(2) (2015): 67–80; Christian Simon, *Reisen, Sammeln und Forschen. Die Basler Naturhistoriker Paul und Fritz Sarasin* (Basel: Schwabe, 2015).
32. Andreas Kley, 'Die Weltnaturschutzkonferenz 1913 in Bern', *Umweltrecht in der Praxis* 7(7) (2007): 685–705, 688–90; Anna-Katharina Wöbse, *Weltnaturschutz: Umweltdiplomatie in Völkerbund und Vereinten Nationen 1920–1950* (Frankfurt: Campus, 2011), 39–45.
33. Patrick Kupper, 'Translating Yellowstone: Early European National Parks, Weltnaturschutz and the Swiss Model', in *Civilizing Nature. National Parks in Global Historical Perspective*, Bernhard Gissibl, Sabine Höhler and Patrick Kupper (eds) (New York: Berghahn Books, 2012), 123–39.
34. McCormick, *Global Environmental Movement*, 27; Wöbse, *Weltnaturschutz*, 39.

35. Paul Sarasin, 'Weltnaturschutz. Vortrag gehalten auf dem VIII. Internationalen Zoologenkongress in Graz, am 16. August und an der 93. Versammlung der Schweizerischen Naturforschenden Versammlung in Basel am 5. September 1910', *Verhandlungen der Schweizerischen Naturforschenden Gesellschaft* 93 (1910): 50–73.
36. On this practice and its scandalization, see Anna-Katharina Wöbse, 'Als eine Mode untragbar wurde. Die Kampagnen gegen den Federschmuck im Deutschen Kaiserreich', in *Federn kitzeln die Sinne*, Dorothea Deterts (ed.) (Bremen: Überseemuseum, 2004), 43–50; Davis, 'Bird Day', 185.
37. Sarasin, 'Weltnaturschutz', 66–68 (here and subsequently, my translations from the German).
38. Ibid., 71 f.; Raf de Bont, '"Primitives" and Protected Areas: International Conservation and the "Naturalization" of Indigenous People, ca. 1910–1975', *Journal of the History of Ideas* 76(2) (2015): 215–36, 219–22.
39. On these inventions, see e.g. Kurk Dorsey, *Whales and Nations: Environmental Diplomacy on the High Seas* (Seattle: University of Washington Press, 2014), 6 f.
40. Sarasin, 'Weltnaturschutz', 62–64.
41. Pierre-Yves Saunier, 'International Non-governmental Organizations (INGOs)', in *The Palgrave Dictionary of Transnational History*, Akira Iriye and Pierre-Yves Saunier (eds) (Basingstoke: Palgrave Macmillan, 2009), 573–80.
42. For the terminological debate, cf. Wolfram Kaiser and Jan-Henrik Meyer, 'Beyond Governments and Supranational Institutions. Societal Actors in European Integration', in *Societal Actors in European Integration: Polity-Building and Policy-Making 1958–1992*, Wolfram Kaiser and Jan-Henrik Meyer (eds) (Basingstoke: Palgrave Macmillan, 2013), 1–14, 1–6.
43. Laura Elisabeth Wong, 'Intergovernmental Organizations (IGOs)', in *The Palgrave Dictionary of Transnational History*, Akira Iriye and Pierre-Yves Saunier (eds) (Basingstoke: Palgrave Macmillan, 2009), 555–61.
44. Sarasin, 'Weltnaturschutz', 72 f.
45. Kley, 'Weltnaturschutzkonferenz', 691 f. The countries represented were: Argentina, Austria, Belgium, Denmark, France, Germany, Great Britain, Hungary, Italy, Netherlands, Norway, Portugal, Russia, Spain, Sweden, Switzerland and the United States.
46. Ibid., 693–95.
47. Anna-Katharina Wöbse, 'Naturschutz global – oder: Hilfe von außen : internationale Beziehungen des amtlichen Naturschutzes im 20. Jahrhundert', in *Natur und Staat 1906–2006*, Hans-Werner Frohn (ed.) (Münster: Landwirtschaftsverlag, 2006), 625–727, 635–39.
48. Kley, 'Weltnaturschutzkonferenz', 703 f. (Anhang II. Eröffnungsrede an der internationalen Weltnaturschutzkonferenz von Bundesrat Ferrer vom 17. November 1913 = Annex II. Opening speech by Federal Council Ferrer at the international World Nature Protection Conference, 17 November 1913).
49. For this debate, see Elinor Ostrom, et al., 'Revisiting the Commons: Local Lessons, Global Challenges', *Science* 284(5412) (1999): 278–82; Isabella Löhr and Andrea Rehling, '"Governing the Commons": Die global commons und

das Erbe der Menschheit im 20. Jahrhundert', in *Global Commons im 20. Jahrhundert. Entwürfe für eine globale Welt*, Isabella Löhr and Andrea Rehling (eds) (Berlin: De Gruyter, 2014), 3–32.

50. Kley, 'Weltnaturschutzkonferenz', 703 f.
51. Translations of the title into English vary: 'Consultative Commission for the International Protection of Nature' (McCormick) or 'Consultative Commission of International Nature Protection (CCINP)' (de Bont). McCormick, *Global Environmental Movement*, 27; de Bont, '"Primitives" and Protected Areas', 218.
52. Kley, 'Weltnaturschutzkonferenz', 698.
53. Hugo Grotius, *The Free Sea* (Indianapolis: Liberty Fund, 2004 [1609]).
54. Garrett Hardin, 'The Tragedy of the Commons', *Science* 162(3859) (1968): 1243–48.
55. Arthur F. McEvoy, *The Fisherman's Problem. Ecology and Law in the California Fisheries, 1850–1980* (Cambridge: Cambridge University Press, 1986), 10.
56. Dorsey, *Dawn of Conservation Diplomacy*, 12 f.
57. McCormick, *Global Environmental Movement*, 19; Convention for the Protection of Birds Useful to Agriculture, signed in Paris, 19 March 1902, http://www.ecolex.org/ecolex/ledge/view/RecordDetails?id=TRE-000067&index=treaties, accessed 18 May 2016.
58. Dorsey, *Dawn of Conservation Diplomacy*, 165–237.
59. McCormick, *Global Environmental Movement*, 18.
60. Joachim Radkau, *Nature and Power: A Global History of the Environment* (Cambridge: Cambridge University Press, 2008), 52; Richard S.R. Fitter and Peter Scott, *The Penitent Butchers: The Fauna Preservation Society, 1903–1978* (London: Collins, 1978).
61. On the League of Nations context, see Patricia Clavin, *Securing the World Economy: The Reinvention of the League of Nations, 1920–1946* (Oxford: Oxford University Press, 2013).
62. Wöbse, *Weltnaturschutz*.
63. Another prominent cross-border pollution issue at the time was the Trail Smelter case. With its tall chimney built in 1926 the lead and zinc smelting plant in British Columbia impacted on U.S. territory, however, only on a more local or regional scale. The issue was resolved in 1941 through the existing structures of the bilateral U.S.-Canadian International Joint Commission. See James R. Allum, '"An Outcrop of Hell": History, Environment, and the Politics of the Trail Smelter Case', in *Transboundary Harm in International Law: Lessons from the Trail Smelter Arbitration*, Rebecca M. Bratspies and Russell A. Miller (eds) (Cambridge: Cambridge University Press, 2006), 13–26, 14–16.
64. John Maynard Keynes, *The Economic Consequences of Mr. Churchill* (London: L. and V. Woolf, 1925).
65. Erik J. Dahl, 'Naval Innovation: From Coal to Oil', *Joint Force Quarterly* 27 (2001): 50.

66. Anna-Katharina Wöbse, 'Der Schutz der Natur im Völkerbund – Anfänge einer Weltumweltpolitik', *Archiv für Sozialgeschichte* 43(1) (2003): 177–91, 185.
67. Anna-Katharina Wöbse, 'Oil on Troubled Waters? Environmental Diplomacy in the League of Nations', *Diplomatic History* 32(4) (2008): 519–37.
68. The first ornithologists' international congress was held in Vienna in 1884. 'Sitzungsprotokolle des 1. Ornithologen–Congresses, 7.–11. April 1884, Wien', *Proceedings of the International Ornithological Congress*, 1 (1884), http://www.biodiversitylibrary.org/item/110498, accessed 28 June 2016.
69. McCormick, *Global Environmental Movement*, 28.
70. Wöbse, *Weltnaturschutz*, 79 f., 86 f.
71. Ibid., 133–70.
72. Davis, 'Bird Day', 185; Wöbse, *Weltnaturschutz*, 135 f.; McCormick, *Global Environmental Movement*, 4.
73. Wöbse, *Weltnaturschutz*, 174–76.
74. Ibid., 171–235.
75. Ibid., 188 f., 200–4; Lynton Keith Caldwell, *International Environmental Policy: From the Twentieth to the Twenty-First Century* (Durham, NC: Duke University Press, 1996), 45.
76. Charlotte Epstein, *The Power of Words in International Relations: Birth of an Anti-whaling Discourse* (Cambridge, MA: MIT Press, 2005), 73–77; Johan Nicolai Tønnessen and Arne Odd Johnsen, *The History of Modern Whaling* (London: C. Hurst & Company, 1982), 365, 399 f.
77. Wöbse, *Weltnaturschutz*, 220–37; Cioc, *Game of Conservation*, 127–32; Dorsey, *Whales and Nations*, 42–48, 72–82.
78. Bernhard Gissibl, Sabine Höhler and Patrick Kupper (eds) *Civilizing Nature. National Parks in Global Historical Perspective* (New York: Berghahn Books, 2012).
79. Bianca Hönig, 'Profoundly National Yet Transboundary: The Tatra National Parks', *Environment & Society Portal: Arcadia* 16 (2014), http://www.environmentandsociety.org/arcadia/profoundly-national-yet-transboundary-tatra-national-parks, accessed 30 June 2016.
80. Anna-Katharina Wöbse, 'Framing the Heritage of Mankind: National Parks on the International Agenda', in *Civilizing Nature. National Parks in Global Historical Perspective*, Bernhard Gissibl, Sabine Höhler and Patrick Kupper (eds) (New York: Berghahn Books, 2012), 140–56, 144–47.
81. Wöbse, *Weltnaturschutz*, 54–60.
82. Holdgate, *Green Web*, 13; E. Pelzers, Entry 'Tienhoven, Pieter Gerbrand van (1875-1953)', *Biografisch Woordenboek van Nederland* (2013), http://resources.huygens.knaw.nl/bwn1880-2000/lemmata/bwn4/tienhoven, accessed 28 June 2016.
83. McCormick, *Global Environmental Movement*, 28.
84. Holdgate, *Green Web*, 14.
85. McCormick, *Global Environmental Movement*, 22; Convention Relative to the Preservation of Fauna and Flora in their Natural State, 1933, http://www.jus.

uio.no/english/services/library/treaties/06/6-02/preservation-fauna-natural.
xml, accessed 18 May 2016.

86. Caldwell, *International Environmental Policy*, 39; M.J. Bowman, 'International
    Treaties and the Global Protection of Birds: Part I', *Journal of Environmental
    Law* 11(1) (1999): 87–119, 92.
87. Fairfield Osborn, *Our Plundered Planet* (Boston, MA: Little Brown, 1948);
    William Vogt, *Road to Survival* (New York: Sloane Associates, 1948); Thomas
    Robertson, 'Total War and the Total Environment: Fairfield Osborn, William
    Vogt, and the Birth of Global Ecology', *Environmental History* 17(2) (2012):
    336–64.
88. Wöbse, *Weltnaturschutz*, 274–81; Peder Anker, *Imperial Ecology.
    Environmental Order in the British Empire, 1895–1945* (Cambridge, MA:
    Harvard University Press, 2001), 86–96.
89. For a more extensive treatment, see Stephen Macekura, *Of Limits and Growth:
    The Rise of Global Sustainable Development in the Twentieth Century*
    (Cambridge: Cambridge University Press, 2015), 17–53.
90. Anna-Katharina Wöbse, 'L'Unesco et l'Union internationale pour la protection
    de la nature. Une impossible transmission de valeurs?', *Relations internationales*
    152(4) (2012): 29–38; Wöbse, *Weltnaturschutz*, 298.
91. Holdgate, *Green Web*, 90 f.
92. Ibid., 31 f; Talbot, 'IUCN in Retrospect and Prospect', 5 f.
93. Wöbse, *Weltnaturschutz*, 301–15.
94. Ibid., 302–7; Jan-Henrik Meyer, Appropriating the Environment: How the
    European Institutions Received the Novel Idea of the Environment and Made
    it Their Own, *KFG Working Paper* 31 (2011), 1–33, http://edocs.fu-berlin.de/
    docs/receive/FUDOCS_document_000000012522, accessed 18 May 2016,
    21f.
95. Chloé Maurel, *Histoire de l'UNESCO. Les trente premières années. 1945–1974*
    (Paris: L'Harmattan, 2010).
96. Macekura, *Of Limits and Growth*, 42–90.
97. Thomas Lekan, 'Serengeti Shall Not Die: Bernhard Grzimek, Wildlife Film,
    and the Making of a Tourist Landscape in East Africa', *German History* 29(2)
    (2011): 224–64.
98. Holdgate, *Green Web*, 71–74.
99. Ibid., 61–87; Alexis Schwarzenbach, *Saving the World's Wildlife. WWF – The
    First 50 Years* (London: Profile Books, 2011); John Vidal, 'WWF International
    Accused of "Selling its Soul" to Corporations', *The Observer*, 4 October 2014,
    http://www.theguardian.com/environment/2014/oct/04/wwf-international-
    selling-its-soul-corporations, accessed 28 June 2015. On other global INGOs,
    see Frank Zelko, *Make it a Green Peace!: The Rise of Countercultural
    Environmentalism* (Oxford: Oxford University Press, 2013); Brian Doherty
    and Timothy Doyle, *Environmentalism, Resistance and Solidarity: The Politics
    of Friends of the Earth International* (Basingstoke: Palgrave Macmillan, 2013).
100. Mark Cioc, 'Europe's River: The Rhine as a Prelude to Transnational
    Cooperation and the Common Market', in *Nation-States and the Global*

Environment, Erika Marie Bsumek, David Kinkela and Mark Atwood Lawrence (eds) (Oxford: Oxford University Press, 2013), 25–42, 25–27.

101. Cioc, *Game of Conservation*, 139–45; Kurk Dorsey, 'National Sovereignty, the International Whaling Commission, and the Save the Whales Movement', in *Nation-States and the Global Environment*, Erika Marie Bsumek, David Kinkela and Mark Atwood Lawrence (eds) (Oxford: Oxford University Press, 2013), 43–61.

102. Zelko, *Make it a Green Peace*, 161–230.

103. Barry Commoner, 'The Fallout Problem', *Science* 127(3305) (1958): 1023–26; Rachel Rothschild, 'Environmental Awareness in the Atomic Age: Radioecologists and Nuclear Technology', *Historical Studies in the Natural Sciences* 43(4) (2013): 492–530.

104. Holger Nehring, *Politics of Security: The British and West German Protests against Nuclear Weapons and the Early Cold War, 1945–1970* (Oxford: Oxford University Press, 2013), 47–50, 68, 232; Dario Fazzi, 'The Blame and the Shame: Kennedy's Choice to Resume Nuclear Tests in 1962', *Peace & Change* 39(1) (2014): 1–22, 3.

105. Zelko, *Make it a Green Peace*, 78–160.

106. Christian Pfister, 'The "1950s Syndrome" and the Transition from a Slow-Going to a Rapid Loss of Global Sustainability', in *The Turning Points of Environmental History*, Frank Uekötter (ed.) (Pittsburgh: University of Pittsburgh Press, 2010), 90–118.

107. Dennis Meadows et al., *The Limits to Growth* (New York: Universe Books, 1972); Paul Ehrlich, *The Population Bomb* (New York: Ballentine Books, 1968).

108. Frank Uekötter, *The Age of Smoke. Environmental Policy in Germany and the United States, 1880–1970* (Pittsburgh: University of Pittsburgh Press, 2009).

109. Matthias Schmelzer, 'The Crisis before the Crisis: The "Problems of Modern Society" and the OECD, 1968–74', *European Review of History* 19(6) (2012): 999–1020, 1002–8.

110. Rachel Rothschild, 'Burning Rain: The Long-Range Transboundary Air Pollution Project', in *Toxic Airs: Chemical and Environmental Histories of the Atmosphere*, James Rodger Fleming and Ann Johnson (eds) (Pittsburgh: University of Pittsburgh Press, 2014), 181–207, 185.

111. Gaetano Adinolfi, 'First Steps toward European Cooperation in Reducing Air Pollution. Activities of the Council of Europe', *Law and Contemporary Problems: A Quarterly* 33(2) (1968): 421–26; Council of Europe, *Resolution (68)4, Adopted by the Ministers' Deputies on 8 March 1968, Approving the 'Declaration of Principles' on Air Pollution Control* (Strasbourg, 1968), Part II § 6.

112. Council of Europe, 'European Water Charter, adopted by the Consultative Assembly on 22 April 1967 (Recommendation 493 (1967) and by the Committee of Minsters on 26 May 1967 (Resolution (67) 10), proclaimed in Strasbourg on 6 May 1968', *Yearbook of the International Law Commission* 26(II) part 2 (1974): 342–43.

113. Thorsten Schulz, 'Das "Europäische Naturschutzjahr 1970" – Versuch einer europaweiten Umweltkampagne', *WZB-Discussion Papers*, P 2006: 7 (2006): 1–34, http://bibliothek.wzb.eu/pdf/2006/p06-007.pdf, accessed 18 May 2016; Adam Rome, *The Genius of Earth Day: How a 1970 Teach-in Unexpectedly Made the First Green Generation* (New York: Hill & Wang, 2013).

114. Christy Collis and Klaus Dodds, 'Assault on the Unknown: The Historical and Political Geographies of the International Geophysical Year (1957–8)', *Journal of Historical Geography* 34(4) (2008): 555–73.

115. Holdgate, *Green Web*, 93–96.

116. Elena Aronova, Karen S. Baker and Naomi Oreskes, 'Big Science and Big Data in Biology: From the International Geophysical Year through the International Biological Program to the Long Term Ecological Research (LTER) Network, 1957–Present', *Historical Studies in the Natural Sciences* 40(2) (2010): 183–224.

117. Nicholas Polunin, 'The Biosphere Conference', *Biological Conservation* 1(2) (1969): 186–87; Caldwell, *International Environmental Policy*, 53 f.

118. Claes Bernes and Lars J. Lundgren, *Use and Misuse of Nature's Resources: An Environmental History of Sweden* (Varnamo: Falth & Hassler, 2009), 88–89.

119. E.g. Jack Davis (Minister of the Environment for Canada), *Canada and the Human Environment. A Contribution of the Government of Canada to the United Nations Conference on the Human Environment, Stockholm, Sweden, June 1972* (Ottawa: Information Canada, 1972).

120. Caldwell, *International Environmental Policy*, 57–62.

121. Eric J. Hobsbawm, *Age of Extremes: The Short Twentieth Century 1914–1991* (London: Joseph, 1994).

122. Uekötter, *The Age of Smoke*; Birgit Metzger, *'Erst stirbt der Wald, dann du!'. Das Waldsterben als westdeutsches Politikum (1978–1986)* (Frankfurt: Campus, 2015).

123. Jacob Darwin Hamblin, *Arming Mother Nature. The Birth of Catastrophic Environmentalism* (Oxford: Oxford University Press, 2013).

124. Anna-Katharina Wöbse, '"The World after All was One": The International Environmental Network of UNESCO and IUPN, 1945–1950', *Contemporary European History* 20(3) (2011): 331–48.

125. Schulz, 'Transatlantic Environmental Security in the 1970s?'; Jacob Darwin Hamblin, 'Environmentalism for the Atlantic Alliance: NATO's Experiment with the "Challenges of Modern Society"', *Environmental History* 15(1) (2010): 54–75.

126. Such letter writing continued well into the 1970s. See e.g. Jan-Henrik Meyer, 'Saving Migrants: A Transnational Network Supporting Supranational Bird Protection Policy in the 1970s', in *Transnational Networks in Regional Integration. Governing Europe 1945–83*, Wolfram Kaiser, Brigitte Leucht and Michael Gehler (eds) (Basingstoke: Palgrave Macmillan, 2010), 176–98.

127. Holdgate, *Green Web*, 73.

128. Stanley P. Johnson, *The Politics of the Environment: The British Experience* (London: Tom Stacey, 1973), 82–85.

129. 'Eco-Barbarians or New Leaders', *Stockholm Conference Eco*, 6 June 1972, 1, 8.
130. E.g. Matthias Schmelzer, *The Hegemony of Growth: The OECD and the Making of the Economic Growth Paradigm* (Cambridge: Cambridge University Press, 2016).
131. John W. Meyer, et al., 'The Structuring of a World Environmental Regime, 1870–1990', *International Organization* 51(4) (1997): 623–51.
132. Thorsten Schulz-Walden, *Anfänge globaler Umweltpolitik. Umweltsicherheit in der internationalen Politik (1969–1975)* (Munich: Oldenbourg, 2013), 279 f.; Matthias Schmelzer, 'The Growth Paradigm: History, Hegemony, and the Contested Making of Economic Growthmanship', *Ecological Economics* 118 (2015), 262–71.

Bibliography

Adinolfi, Gaetano, 'First Steps toward European Cooperation in Reducing Air Pollution. Activities of the Council of Europe', *Law and Contemporary Problems: A Quarterly* 33(2) (1968): 421–26.
Allum, James R., '"An Outcrop of Hell": History, Environment, and the Politics of the Trail Smelter Case', in *Transboundary Harm in International Law: Lessons from the Trail Smelter Arbitration*, Rebecca M. Bratspies and Russell A. Miller (eds) (Cambridge: Cambridge University Press, 2006), 13–26.
Anker, Peder, *Imperial Ecology. Environmental Order in the British Empire, 1895–1945* (Cambridge, MA: Harvard University Press, 2001), 86–96.
Aronova, Elena, Karen S. Baker and Naomi Oreskes, 'Big Science and Big Data in Biology: From the International Geophysical Year through the International Biological Program to the Long Term Ecological Research (LTER) Network, 1957–Present', *Historical Studies in the Natural Sciences* 40(2) (2010): 183–224.
Bernes, Claes and Lars J. Lundgren, *Use and Misuse of Nature's Resources: An Environmental History of Sweden* (Varnamo: Falth & Hassler, 2009).
Bont, Raf de, '"Primitives" and Protected Areas: International Conservation and the 'Naturalization' of Indigenous People, ca. 1910–1975', *Journal of the History of Ideas* 76(2) (2015): 215–36.
Bowman, M.J., 'International Treaties and the Global Protection of Birds: Part I', *Journal of Environmental Law* 11(1) (1999): 87–119.
Caldwell, Lynton Keith, *International Environmental Policy. From the Twentieth to the Twenty-First Century* (Durham, NC: Duke University Press, 1996), 45.
Carson, Rachel, *Silent Spring* (Greenwich, CT: Fawcett, 1962).
Cioc, Mark, *The Game of Conservation. International Treaties to Protect the World's Migratory Species* (Athens, OH: Ohio University Press, 2009).
———. 'Europe's River. The Rhine as a Prelude to Transnational Cooperation and the Common Market', in *Nation-States and the Global Environment*, Erika Marie Bsumek, David Kinkela and Mark Atwood Lawrence (eds) (Oxford: Oxford University Press, 2013), 25–42.

Clavin, Patricia, *Securing the World Economy. The Reinvention of the League of Nations, 1920–1946* (Oxford: Oxford University Press, 2013).

Collis, Christy and Klaus Dodds, 'Assault on the Unknown. The Historical and Political Geographies of the International Geophysical Year (1957–8)', *Journal of Historical Geography* 34(4) (2008): 555–73.

Commoner, Barry, 'The Fallout Problem', *Science* 127(3305) (1958): 1023–26.

———. *The Closing Circle. Nature, Man and Technology* (New York: Knopf, 1971).

Convention for the Protection of Birds Useful to Agriculture, signed in Paris, 19 March 1902,http://www.ecolex.org/ecolex/ledge/view/RecordDetails?id=TRE-000067&index=treaties, accessed 18 May 2016.

Convention Relative to the Preservation of Fauna and Flora in Their Natural State, 1933,http://www.jus.uio.no/english/services/library/treaties/06/6-02/preservation-fauna-natural.xml, accessed 18 May 2016.

Council of Europe, *Resolution (68)4, adopted by the Ministers' Deputies on 8 March 1968, Approving the 'Declaration of Principles' on Air Pollution Control* (Strasbourg, 1968).

———. 'European Water Charter, Adopted by the Consultative Assembly on 22 April 1967 (Recommendation 493 (1967) and by the Committee of Minsters on 26 May 1967 (Resolution (67) 10), proclaimed in Strasbourg on 6 May 1968', *Yearbook of the International Law Commission* 26(II), part 2 (1974): 342–43.

Dahl, Erik J., 'Naval Innovation: From Coal to Oil', *Joint Force Quarterly* 27 (2001): 50.

Davis, Jack (Minister of the Environment for Canada), *Canada and the Human Environment: A Contribution of the Government of Canada to the United Nations Conference on the Human Environment, Stockholm, Sweden, June 1972* (Ottawa: Information Canada, 1972).

Davis, Janet M., 'Bird Day. Promoting the Gospel of Kindness in the Philippines during the American Occupation', in *Nation-States and the Global Environment: New Approaches to International Environmental History*, Erika Marie Bsumek, David Kinkela and Mark Atwood Lawrence (eds) (Oxford: Oxford University Press, 2013), 181–206.

Daynes, Byron W. and Glen Sussman, *White House Politics and the Environment. Franklin D. Roosevelt to George W. Bush* (College Station: Texas A&M University Press, 2010).

Ditt, Karl, 'Nature Conservation in England and Germany 1900–70: Forerunner of Environmental Protection?', *Contemporary European History* 5(1) (1996): 1–28.

Doherty, Brian and Timothy Doyle, *Environmentalism, Resistance and Solidarity: The Politics of Friends of the Earth International* (Basingstoke: Palgrave Macmillan, 2013).

Dorsey, Kurk, *The Dawn of Conservation Diplomacy: U.S.–Canadian Wildlife Protection Treaties in the Progressive Era* (Seattle: University of Washington Press, 1998).

———. 'National Sovereignty, the International Whaling Commission, and the Save the Whales Movement', in *Nation-States and the Global Environment*, Erika Marie Bsumek, David Kinkela and Mark Atwood Lawrence (eds) (Oxford: Oxford University Press, 2013), 43–61.

———. *Whales and Nations: Environmental Diplomacy on the High Seas* (Seattle: University of Washington Press, 2014).

'Eco-Barbarians or New Leaders', *Stockholm Conference Eco*, 6 June 1972, 1, 8.

Egan, Michael, *Barry Commoner and the Science of Survival: The Remaking of American Environmentalism* (Cambridge, MA: MIT Press, 2007).

Ehrlich, Paul, *The Population Bomb* (New York: Ballentine Books, 1968).

Engels, Jens Ivo, 'Modern Environmentalism', in *The Turning Points of Environmental History*, Frank Uekötter (ed.) (Pittsburgh: University of Pittsburgh Press, 2010), 119–31.

Epstein, Charlotte, *The Power of Words in International Relations: Birth of an Anti-whaling Discourse* (Cambridge, MA: MIT Press, 2005).

Fazzi, Dario, 'The Blame and the Shame: Kennedy's Choice to Resume Nuclear Tests in 1962', *Peace & Change* 39(1) (2014): 1–22.

Fitter, Richard S.R. and Peter Scott, *The Penitent Butchers: The Fauna Preservation Society, 1903–1978* (London: Collins, 1978).

Gissibl, Bernhard, 'German Colonialism and the Beginnings of International Wildlife Preservation in Africa', *GHI Bulletin*, Supplement 3 (2006), 121–43, http://www.ghi-dc.org/publications/ghi-bulletin/bulletin-supplements/bulletin-supplement-3-2006.html?L=0, accessed 30 June 2016.

Gissibl, Bernhard, Sabine Höhler and Patrick Kupper (eds), *Civilizing Nature. National Parks in Global Historical Perspective* (New York: Berghahn Books, 2012).

Grotius, Hugo, *The Free Sea* (Indianapolis: Liberty Fund, 2004 [1609]).

Hamblin, Jacob Darwin, *Poison in the Well. Radioactive Waste in the Oceans at the Dawn of the Nuclear Age* (New Brunswick, NJ: Rutgers University Press, 2008).

———. 'Environmentalism for the Atlantic Alliance: NATO's Experiment with the "Challenges of Modern Society"', *Environmental History* 15(1) (2010): 54–75.

———. *Arming Mother Nature: The Birth of Catastrophic Environmentalism* (Oxford: Oxford University Press, 2013).

Hardin, Garrett, 'The Tragedy of the Commons', *Science* 162(3859) (1968): 1243–48.

Herren, Madeleine, *Hintertüren zur Macht: Internationalismus und modernisierungsorientierte Außenpolitik in Belgien, der Schweiz und den USA, 1865–1914* (Munich: Oldenbourg, 2000), 300–62.

Hobsbawm, Eric J., *Age of Extremes: The Short Twentieth Century 1914–1991* (London: Joseph, 1994).

Holdgate, Martin, *The Green Web: A Union for World Conservation* (Abingdon: Earthscan, 2013).

Hönig, Bianca, 'Profoundly National Yet Transboundary: The Tatra National Parks', *Environment & Society Portal: Arcadia* 16 (2014), http://www.

environmentandsociety.org/arcadia/profoundly-national-yet-transboundary-tatra-national-parks, accessed 30 June 2016.

Johnson, Stanley P., *The Politics of the Environment. The British Experience* (London: Tom Stacey, 1973).

Josephson, Paul R. et al., *An Environmental History of Russia* (Cambridge: Cambridge University Press, 2013).

Kaiser, Wolfram and Jan-Henrik Meyer, 'Beyond Governments and Supranational Institutions. Societal Actors in European Integration', in *Societal Actors in European Integration. Polity-Building and Policy-Making 1958–1992*, Wolfram Kaiser and Jan-Henrik Meyer (eds) (Basingstoke: Palgrave Macmillan, 2013), 1–14.

Keynes, John Maynard, *The Economic Consequences of Mr. Churchill* (London: L. and V. Woolf, 1925).

Kley, Andreas, 'Die Weltnaturschutzkonferenz 1913 in Bern', *Umweltrecht in der Praxis* 7(7) (2007): 685–705.

Kupper, Patrick, 'Translating Yellowstone: Early European National Parks, Weltnaturschutz and the Swiss Model', in *Civilizing Nature. National Parks in Global Historical Perspective*, Bernhard Gissibl, Sabine Höhler and Patrick Kupper (eds) (New York: Berghahn Books, 2012), 123–39.

Lekan, Thomas, 'Serengeti Shall Not Die: Bernhard Grzimek, Wildlife Film, and the Making of a Tourist Landscape in East Africa', *German History* 29(2) (2011): 224–64.

Löhr, Isabella and Andrea Rehling, '"Governing the Commons": Die global commons und das Erbe der Menschheit im 20. Jahrhundert', in *Global Commons im 20. Jahrhundert. Entwürfe für eine globale Welt*, Isabella Löhr and Andrea Rehling (eds) (Berlin: De Gruyter, 2014), 3–32.

Macekura, Stephen, *Of Limits and Growth. The Rise of Global Sustainable Development in the Twentieth Century* (Cambridge: Cambridge University Press, 2015).

Markham, William T. and C.S.A. van Koppen, 'Nature Protection in Nine Countries. A Framework for Analysis', in *Protecting Nature. Organizations and Networks in Europe and the USA*, William T. Markham and C.S.A. van Koppen (eds) (Cheltenham: Edward Elgar, 2008), 1–33.

Maurel, Chloé, *Histoire de l'UNESCO. Les trente premières années. 1945–1974* (Paris: L'Harmattan, 2010).

McCormick, John, *The Global Environmental Movement* (Chichester: John Wiley, 1995).

McEvoy, Arthur F., *The Fisherman's Problem. Ecology and Law in the California Fisheries, 1850–1980* (Cambridge: Cambridge University Press, 1986).

Meadows, Dennis et al., *The Limits to Growth* (New York: Universe Books, 1972).

Menke-Glückert, Peter, 'Der Umweltpolitiker Genscher', in *In der Verantwortung. Hans Dietrich Genscher zum Siebzigsten*, Klaus Kinkel (ed.) (Berlin: Siedler, 1997), 155–68.

Metzger, Birgit, *'Erst stirbt der Wald, dann du!'. Das Waldsterben als westdeutsches Politikum (1978–1986)* (Frankfurt: Campus, 2015).

Meyer, Jan-Henrik, 'Saving Migrants: A Transnational Network Supporting Supranational Bird Protection Policy in the 1970s', in *Transnational Networks in Regional Integration: Governing Europe 1945–83*, Wolfram Kaiser, Brigitte Leucht and Michael Gehler (eds) (Basingstoke: Palgrave Macmillan, 2010), 176–98.

———. 'Appropriating the Environment: How the European Institutions Received the Novel Idea of the Environment and Made it Their Own', *KFG Working Paper* 31 (2011): 1–33, http://edocs.fu-berlin.de/docs/receive/FUDOCS_document_000000012522, accessed 18 May 2016.

Meyer, John M., 'Gifford Pinchot, John Muir, and the Boundaries of Politics in American Thought', *Polity* 30(2) (1997): 267–84.

Meyer, John W. et al., 'The Structuring of a World Environmental Regime, 1870–1990', *International Organization* 51(4) (1997): 623–51.

Nehring, Holger, *Politics of Security: The British and West German Protests against Nuclear Weapons and the Early Cold War, 1945–1970* (Oxford: Oxford University Press, 2013).

Osborn, Fairfield, *Our Plundered Planet* (Boston, MA: Little Brown, 1948)

Ostrom, Elinor et al., 'Revisiting the Commons: Local Lessons, Global Challenges', *Science* 284(5412) (1999): 278–82.

Paulmann, Johannes, 'Reformer, Experten und Diplomaten. Grundlagen des Internationalismus im 19. Jahrhundert', in *Akteure der Außenbeziehungen: Netzwerke und Interkulturalität im historischen Wandel*, Hillard von Thiessen and Christian Windler (eds) (Cologne: Böhlau, 2010), 173–98.

Pelzers, E., Entry 'Tienhoven, Pieter Gerbrand van (1875–1953)', *Biografisch Woordenboek van Nederland* (2013), http://resources.huygens.knaw.nl/bwn1880-2000/lemmata/bwn4/tienhoven, accessed 30 June 2016.

Pfister, Christian, 'The "1950s Syndrome" and the Transition from a Slow-Going to a Rapid Loss of Global Sustainability', in *The Turning Points of Environmental History*, Frank Uekötter (ed.) (Pittsburgh: University of Pittsburgh Press, 2010), 90–118.

Polunin, Nicholas, 'The Biosphere Conference', *Biological Conservation* 1(2) (1969): 186–87.

Radkau, Joachim, *Nature and Power. A Global History of the Environment* (Cambridge: Cambridge University Press, 2008).

———. *Die Ära der Ökologie. Eine Weltgeschichte* (Munich: Beck, 2011).

Righter, Robert W., 'The Hetch Hetchy Controversy', in *Natural Protest*, Michael Egan and Jeff Crane (eds) (New York: Routledge, 2009), 117–35.

Robertson, Thomas, 'Total War and the Total Environment: Fairfield Osborn, William Vogt, and the Birth of Global Ecology', *Environmental History* 17(2) (2012): 336–64.

Rome, Adam, *The Genius of Earth Day: How a 1970 Teach-in Unexpectedly Made the First Green Generation* (New York: Hill & Wang, 2013).

Rothschild, Rachel, 'Environmental Awareness in the Atomic Age: Radioecologists and Nuclear Technology', *Historical Studies in the Natural Sciences* 43(4) (2013): 492–530.

————. 'Burning Rain: The Long-Range Transboundary Air Pollution Project', in *Toxic Airs: Chemical and Environmental Histories of the Atmosphere*, James Rodger Fleming and Ann Johnson (eds) (Pittsburgh: University of Pittsburgh Press, 2014), 181–207.

Sarasin, Paul, 'Weltnaturschutz. Vortrag gehalten auf dem VIII. Internationalen Zoologenkongress in Graz, am 16. August und an der 93. Versammlung der Schweizerischen Naturforschenden Versammlung in Basel am 5. September 1910', *Verhandlungen der Schweizerischen Naturforschenden Gesellschaft* 93 (1910): 50–73.

Saunier, Pierre-Yves, 'International Non-governmental Organizations (INGOs)', in *The Palgrave Dictionary of Transnational History*, Akira Iriye and Pierre-Yves Saunier (eds) (Basingstoke: Palgrave Macmillan, 2009), 573–80.

Schär, Bernhard C., 'Earth Scientists as Time Travelers and Agents of Colonial Conquest: Swiss Naturalists in the Dutch East Indies', *Historical Social Research* 40(2) (2015): 67–80.

————. *Tropenliebe. Schweizer Naturforscher und niederländischer Imperialismus in Südostasien um 1900* (Frankfurt: Campus, 2015).

Schmelzer, Matthias, 'The Crisis before the Crisis: The "Problems of Modern Society" and the OECD, 1968–74', *European Review of History* 19(6) (2012): 999–1020.

————. 'The Growth Paradigm: History, Hegemony, and the Contested Making of Economic Growthmanship', *Ecological Economics* 118 (2015): 262–71.

Schmelzer, Matthias, *The Hegemony of Growth. The OECD and the Making of the Economic Growth Paradigm* (Cambridge: Cambridge University Press, 2016).

Schulz, Thorsten, 'Das "Europäische Naturschutzjahr 1970" – Versuch einer europaweiten Umweltkampagne, *WZB-Discussion Papers*, P 2006: 7 (2006): 1–34, http://bibliothek.wzb.eu/pdf/2006/p06-007.pdf, accessed 18 May 2016.

————. 'Transatlantic Environmental Security in the 1970s? NATO's 'Third Dimension' as an Early Environmental and Human Security Approach', *Historical Social Research* 35(4) (2010): 309–28.

Schulz-Walden, Thorsten, *Anfänge globaler Umweltpolitik. Umweltsicherheit in der internationalen Politik (1969–1975)* (Munich: Oldenbourg, 2013).

Schwarzenbach, Alexis, *Saving the World's Wildlife. WWF – The First 50 Years* (London: Profile Books, 2011).

Scriber, Brad, '100,000 Elephants Killed by Poachers in Just Three Years, Landmark Analysis Finds. Central Africa Has Lost 64 Percent of its Elephants in a Decade', *National Geographic* 18 August 2014, http://news.nationalgeographic.com/news/2014/08/140818-elephants-africa-poaching-cites-census, accessed 18 May 2016.

Simon, Christian, *Reisen, Sammeln und Forschen. Die Basler Naturhistoriker Paul und Fritz Sarasin* (Basel: Schwabe, 2015).

'Sitzungsprotokolle des 1. Ornithologen-Congresses, 7.–11. April 1884, Wien', *Proceedings of the International Ornithological Congress*, 1 (1884), http://www.biodiversitylibrary.org/item/110498, accessed 28 June 2016.

Stoll, Mark, *Inherit the Holy Mountain. Religion and the Rise of American Environmentalism* (Oxford: Oxford University Press, 2015).

Stoll, Steven, *US Environmentalism since 1945: A Brief History with Documents* (New York: Palgrave Macmillan, 2007).

Talbot, Lee M., 'IUCN in Retrospect and Prospect', *Environmental Conservation* 10(1) (1983): 5–11.

Thomas, Sarah L., 'A Call to Action: Silent Spring, Public Discourse and the Rise of Modern Environmentalism', in *Natural Protest. Essays on the History of American Environmentalism*, Michael Egan and Jeff Crane (eds) (New York: Routledge, 2009), 185–204.

Tønnessen, Johan Nicolai and Arne Odd Johnsen, *The History of Modern Whaling* (London: C. Hurst & Company, 1982).

Uekötter, Frank, *The Age of Smoke: Environmental Policy in Germany and the United States, 1880–1970* (Pittsburgh: University of Pittsburgh Press, 2009 [German 2003]).

Vickery, Micheal R., 'Conservative Politics and the Politics of Conservation. Richard Nixon and the Environmental Protection Agency', in *Green Talk in the White House: The Rhetorical Presidency Encounters Ecology*, Tarla Rai Peterson (ed.) (College Station: Texas A&M University Press, 2004), 113–33.

Vidal, John, 'WWF International Accused of "Selling its Soul" to Corporations', *The Observer*, 4 October 2014, http://www.theguardian.com/environment/2014/oct/04/wwf-international-selling-its-soul-corporations, accessed 28 June 2015.

Vogt, William, *Road to Survival* (New York: Sloane Associates, 1948).

Wöbse, Anna-Katharina, 'Als eine Mode untragbar wurde. Die Kampagnen gegen den Federschmuck im Deutschen Kaiserreich', in *Federn kitzeln die Sinne*, Dorothea Deterts (ed.) (Bremen: Überseemuseum, 2004), 43–50.

———. 'Der Schutz der Natur im Völkerbund – Anfänge einer Weltumweltpolitik', *Archiv für Sozialgeschichte* 43(1) (2003): 177–91.

———. 'Framing the Heritage of Mankind: National Parks on the International Agenda', in *Civilizing Nature. National Parks in Global Historical Perspective*, Bernhard Gissibl, Sabine Höhler and Patrick Kupper (eds) (New York: Berghahn Books, 2012), 140–56.

———. 'L'Unesco et l'Union internationale pour la protection de la nature. Une impossible transmission de valeurs?', *Relations internationales* 152(4) (2012): 29–38.

———. 'Naturschutz global – oder: Hilfe von außen : internationale Beziehungen des amtlichen Naturschutzes im 20. Jahrhundert', in *Natur und Staat 1906–2006*, Hans-Werner Frohn (ed.) (Münster: Landwirtschaftsverlag, 2006), 625–727.

———. 'Oil on Troubled Waters? Environmental Diplomacy in the League of Nations', *Diplomatic History* 32(4) (2008): 519–37.

———. *Weltnaturschutz: Umweltdiplomatie in Völkerbund und Vereinten Nationen 1920–1950* (Frankfurt: Campus, 2011), 39–45.

———. "'The World after All was One": The International Environmental Network of UNESCO and IUPN, 1945–1950', *Contemporary European History* 20(3) (2011): 331–48.

Wong, Laura Elisabeth, 'Intergovernmental Organizations (IGOs)', in *The Palgrave Dictionary of Transnational History*, Akira Iriye and Pierre-Yves Saunier (eds) (Basingstoke: Palgrave Macmillan, 2009), 555–61.

Worster, Donald, *Nature's Economy. A History of Ecological Ideas* (Cambridge: Cambridge University Press, 1994 [1977]).

Zelko, Frank, *Make it a Green Peace!: The Rise of Countercultural Environmentalism* (Oxford: Oxford University Press, 2013).

CHAPTER 2

Environmental Problem-Solvers?

Scientists and the Stockholm Conference

Enora Javaudin

From global warming to biotechnologies, scientific data has turned into a powerful tool and legitimizing capital in the development of national and international environmental policies.[1] The case of the 1972 Stockholm Conference and its preparations from 1968 onwards constitute a key entry point to understanding the relationship between science and policy making in the early 1970s.

From the 1940s to the 1960s, the military requirements and industrial production of the Second World War and the Cold War helped to transform scientists into professional experts, 'binding' scientific authority to the decision makers.[2] While scientists became more involved in policy making, movements of 'critical science' intensified throughout the 1960s and 1970s in the United States and Western Europe. They condemned the science-military ties, the idea that 'science was now associated with destructive rather than constructive research.'[3] They also questioned the role of science and scientists in society and expressed doubts about the benefits of technology.[4] A new figure of the scientist as a 'citizen-professional' emerged, of scientists participating 'not only as citizens but as professionals, in the great work of improving and maintaining a viable environment on the earth' and using their 'knowledge to help right the ecological wrongs that have been made and are being committed in the name of scientific and technological progress.'[5]

The Stockholm Conference marked a radical change from previous United Nations (UN) expert conferences. Not only was it the first international conference centred on the issue of the 'environment,'[6] it also involved multiple actors including governments, scientists, international non-governmental organizations (INGOs) and individual citizens. Additionally, it was prominently reported in the media. When the UN decided to hold the conference in 1968 and began preparations, the definition of what constituted the 'environment' was still fluid. Moreover, global scientific data on the state of the world environment was not yet centrally available. The issue of the human environment differed from its previous natural resources framing. It involved new and global interrelated problems. As Maurice Strong, the Stockholm Conference secretary observed: 'the new environmental issues ... are matters in which the need for joint efforts between scientists and politicians is apparent'.[7]

This chapter thus explores to what extent the Stockholm Conference actually was such a 'joint effort between scientists and politicians'. Did the Stockholm Conference and its preparatory process mark a turning point in the way scientists interacted with international organizations (IOs) in international environmental policy making? Who were these scientists and how did they participate in the preparations for and at the Stockholm Conference?

This chapter builds on historical studies on how science and technology have been significant factors in foreign policy and international relations since the Second World War.[8] It also draws on the new political sociology of science, exploring the relations between scientists and other social groups, and the links between expert assessments and politics – especially the influence of scientific knowledge on social movements, public debates and political processes.[9] It contributes to studying such links at the level of IOs and global environmental protection by analysing how the UN and the Stockholm Conference secretariat interacted with international and interdisciplinary scientists. Based on the analysis of a variety of archival sources, this chapter argues that the Stockholm Conference and its preparatory process can be understood as a significant event in the long-term process of increasing the authority of scientific knowledge in international environmental policy making.[10]

The first section explores how by the mid 1960s, scientists – mostly earth scientists and biologists from Western countries – enhanced awareness of what they characterized as the 'environmental crisis'. They shared and expressed critical views and concerns about the state of the 'human environment' and advocated the need for scientists and the international community to take action. Through widely disseminated publications and collective statements, they helped transform the environment from a scientific into an international political issue. The second section investigates why the conference secretariat claimed to act as 'an interface' between the larger UN organization and the scientific milieu. It explores how the UN and the conference secretariat worked with international interdisciplinary scientists and INGOs. Appointed as advisors and consultants, these scientists helped the conference secretariat gather global data and develop an intellectual and conceptual framework for international environmental policies. Finally, the last section analyses the Stockholm intergovernmental conference and the meetings, events and so-called counter-conferences on its fringes. This conference not only created historical momentum for formulating international environmental policies but also a space for debate among scientists.

Scientists and the Rise of the Environment as a Public Concern

After its creation, the UN rapidly organized several international scientific conferences, including the Scientific Conference on the Conservation and Utilization of Resources in 1949.[11] This conference gathered five hundred scientists.[12] However, the issues were not framed in terms of environmental problems, but of natural resources, focusing on development and food issues. The discussions mostly focused on soil conservation, forests, wildlife and fisheries.[13]

Two decades later, scientists still had a central position in the proceedings of world conferences regarding nature. In this vein, the United Nations Educational, Scientific and Cultural Organization (UNESCO) organized an Intergovernmental Conference of Experts on the Scientific Basis for Rational Use and Conservation of Resources of the Biosphere in Paris in 1968.[14] While this event marked a crucial step in conceiving 'nature' in a global and interactive perspective, as a 'biosphere',[15] it was organized for the exchange of scientific data and study, not to devise international policies. Later that year, however, as Jan-Henrik Meyer explains in his introductory chapter in this book, the UN General Assembly created a new field of action on their agenda: the 'problems of the human environment'.[16]

According to the then Swedish Prime Minister Olof Palme, 'the environment/ecological question was not regarded as political, neither nationally nor internationally, but was indeed a very relevant question in the scientific world'.[17] Peter S. Thacher, an American space scientist and Programme Director of the Stockholm Conference, declared that the 'role of the scientific community in alerting mankind to the inter-relatedness between man and his environment was a key factor in leading governments at the U.N. to decide to have this conference'.[18] But who were these scientists? What were their concerns? Why and how did they express them? This section argues that some scientists already advocated the environment as a global issue before the Stockholm Conference, and alerted the UN through publications as well as international collective statements.

In the mid 1960s, some scientists began expressing concerns about 'the complexity and fragility of nature'[19] and environmental destruction. These scientists, who were mostly American and from other English-speaking countries with a background in the natural sciences, shared the belief that the world was facing an unprecedented crisis. Even if they focused on different issues, they all highlighted global changes that were threatening life on earth. They also shared the conviction that, as scientists, they had the responsibility to alert the public and governments. They published books and articles, participated in research projects and contributed towards collective public statements. In 1962, *Silent Spring*, written by the American biologist Rachel Carson, was one of the first publications denouncing the intensive use of pesticides and the contamination of the environment by toxic substances, highlighting the risks for future generations.[20] In 1965, French biologist and ornithologist Jean Dorst published *Avant que nature ne meure*, denouncing the impact of human activities on the environment.[21] In 1968, in the disturbingly titled book *The Population Bomb*, biologist, entomologist and demographer Paul Ehrlich denounced the global pillaging of natural resources due to the demographic explosion.[22] In the same year, the international thinktank Club of Rome was founded publishing its first report, *The Limits to Growth*, just weeks before the Stockholm Conference. The American biologist Barry Commoner, founder of the Committee for Nuclear Information, published *The Closing Circle: Nature, Man, and Technology* in 1971. Positioning himself as a scientist with 'social responsibility', he sought to alert the public to the links between technological factors and the environmental crisis.[23] All of these publications aimed at bringing knowledge on specific environmental issues to a larger non-academic audience. They thus utilized a nonscientific discourse, often writing in an alarmist tone.

At the turn of the 1970s, international scientists mobilized in other ways, too. The first example is the international manifesto *A Blueprint for Survival*, which was published six months before the Stockholm Conference in *The Ecologist* magazine.[24] This manifesto was signed by more than twenty British scientists, among them one Nobel Prize winner.[25] It claimed that 'governments … are either refusing to face the relevant facts, or are briefing their scientists in such a way that their seriousness is played down'.[26] The authors highlighted several key issues contributing to the ongoing environmental crisis, such as industrial life and its 'ethos of expansion', population growth and growing consumption. According to them, these factors were upsetting the balance of the ecosystem, leading to lack of food, exhaustion of resources and, ultimately, a breakdown of society. Furthermore, the authors demanded the formulation of a new philosophy to work towards the creation of a 'new society'. The manifesto called for minimizing disruption to ecological processes, a conversion to a new economy, a stabilization of the population and the creation of a new social system. In all, *A Blueprint for Survival* gave rise to numerous debates and was a topic of discussion at so-called teach-ins in the United Kingdom.[27]

Other scientists drafted the *Menton Statement, A Message to Our 3 1/2 Billion Neighbours on Planet Earth*, a petition published on 28 February 1971. It originated at a meeting organized in the French town of Menton on 4–6 May 1970. This meeting was organized by a peace organization called Fellowship of Reconciliation (FOR)[28] with the participants of Dai Dong, a FOR transnational project launched in 1969 aimed at linking both peace and the environment. Gathering for three days, environmental scientists from Western European countries, the United States and Asia expressed their concerns and discussed endangered nature and the problems facing humanity, as well as potential solutions.[29] Their petition intended to 'dramatize the urgency and interrelatedness of the problems'.[30] It defined the environment broadly as 'the habitat of all living things'[31] and sketched 'relationships between pollution, poverty, population, politics and war'.[32] Nearly two thousand natural and social scientists signed the petition. Some of them also came to Stockholm in June 1972, such as Paul Ehrlich and Margaret Mead, an American anthropologist and member of the Scientists' Institute for Public Information (SIPI).

These scientists' publications and collective public statements, widely shared and publicized outside the scientific milieu, constituted landmark documents, shaping new and transnational ideas on the environmental crisis, shared and circulated before the Stockholm Conference. The Western natural and social sciences scientists involved in these initiatives shared the idea that man was responsible for the environmental crisis threatening life on earth. They mainly focused on issues such as pesticides, pollution, population, growth and the impact of technology and industrial production. While they framed these issues in a global perspective, some scientists from the Global South, as we will see later, argued that they were only environmental problems for Western countries and were not necessarily relevant to the developing countries.

To what extent were these scientists' discourses about the environmental crisis and the Stockholm Conference preparations interrelated? The scientists used their publications and collective statements to grab the UN's attention for the environmental crisis. The *Menton Statement* was in fact addressed to the UN Secretary General, Sithu U Thant. On 4 March 1971, the FOR Director Alfred Hassler requested a meeting with U Thant to submit the *Statement* to him. A delegation of scientists representing its signatories met with U Thant on 11 May 1971.[33] During this get-together, U Thant declared: 'Your urgent message to all the inhabitants of our small planet Earth must be heeded – and acted upon – without delay if our human habitat is to be preserved for future generations.'[34] Following this, Maurice Strong suggested that the delegation of scientists present the *Statement* at the Stockholm Conference plenary meeting, which they did.[35] He also offered to create a delegation of scientists who would participate in the conference as observers.[36] In addition, in July 1971, the *Menton Statement* was published in the *UNESCO Courier*.[37]

Moreover, these scientists' actions were known, read and quoted by the Stockholm Conference secretariat. In an article published in the journal *Ambio*[38] in June 1972, Strong wrote that 'the proposals before [the conference] draw heavily on the wide variety of contributions made by the world's scientific community and a number of specifically scientific meetings which took place in preparing for the conference.'[39] He specifically cited *The Limits to Growth* and the *Blueprint for Survival*. Strong declared that the conference secretariat worked 'at the interface between governments and the scientific community ... to synthesize the concerns and advice ... and to prepare recommendations on which governments could act.'[40] But how exactly did this cooperation between the conference secretariat and scientists play out? With which scientists did the secretariat work and how?

'Our Inspiration Came from Science': The Conference Secretariat

As Jan-Henrik Meyer sets out in his introductory chapter in this book, the UN and the conference secretariat prepared the Stockholm Conference between 1968 and 1972. In order to prepare the topics to be discussed, the conference secretariat had to gather and analyse scientific data on the global environment and coordinate information, before rephrasing the data in accessible language geared towards formulating public policy. The preparation process was divided into three phases. First, the conference secretariat developed an 'intellectual and conceptual framework' in order to define the known problems, identify new ones, establish areas of consensus, produce priority frameworks and report on the state of the environment. Second, from this first intellectual and conceptual framework, the secretariat produced an 'Action Plan'. It aimed to define solutions, establish consensus on necessary action, suggest institutional alternatives and outline action programmes, decisions and recommendations. The third and last step was to agree and get governments and IOs to take specific action.

Building an intellectual and conceptual framework required scientific information and data that had never been centrally collated before. The secretariat relied on different sources for this. Initially, it received numerous national reports on environmental issues from governments. In addition, the UN specialized agencies, UNESCO, the World Health Organization (WHO), the International Atomic Energy Agency(IAEA) and the Food and Agriculture Organization (FAO) all sent documents on their environmental activities. The secretariat also relied on the report resulting from the UN Economic Commission for Europe (UNECE) symposium on problems relating to the environment, held in May 1971 in Prague, where scientists discussed environmental issues.[41] Building on these contributions, the conference secretariat created a conceptual framework for outlining the wide range of complex issues constituting the human environment.

In addition, Maurice Strong and the conference secretariat worked closely with several national and international scientific INGOs on the environmental issues to be considered during the conference. Initially, the secretariat relied upon the work of the International Council for Scientific Unions (ICSU). In a speech given at the ICSU 1972 General Assembly, Peter Thacher mentioned that the conference secretariat 'received numerous proposals, advice, and assessments from the scientific community, including many from members of [their] scientific unions, as well as associated national academies and research councils, who played an active role in an almost continuous dialogue throughout 1971 and the early part of this [1972] year'.[42] Founded in 1931, the ICSU worked as an 'interdisciplinary platform' and linked members of national scientific organizations with international scientific unions. This INGO aimed at promoting international scientific activity in the various scientific fields and its application to the wellbeing of mankind, as well as providing a source of scientific advice to governmental agencies. Connected to the UN system, the ICSU partly contributed to the financing of the programmes of the International Geophysical Year 1957–58, to the international biological programme from 1964 to 1974 and to the Global Atmospheric Research Programme from 1967 to 1980.

The Scientific Committee on Problems of the Environment (SCOPE),[43] an ICSU ad hoc committee set up in 1969, which gathered interdisciplinary scientists, was especially important in the Stockholm Conference preparations. In December 1970, Strong requested the SCOPE Commission on Monitoring to prepare a report recommending 'the design, the parameters and technical organisation needed for a coherent global environmental monitoring system'.[44] SCOPE accepted this request and began this preparatory work in Stockholm in March 1971.[45] The SCOPE Commission on Monitoring gathered and analysed information on the regional and global environment. Its report resulting from this study was written by a group of twenty-four scientists, coming from the Global North and South.[46] The intergovernmental group formed in 1971 by the conference secretariat to oversee and police global environmental issues relied significantly on the commission's report.[47] The Earthwatch programme, the global monitoring system of the Stockholm Conference Action Plan designed for evaluating the impact of human activity on the global environment, was a direct contribution by SCOPE scientists assisting the conference secretariat.[48]

The conference secretariat also relied on studies undertaken by Carroll L. Wilson, a professor of problems of contemporary technology at the Massachusetts Institute of Technology (MIT) in Boston. Wilson was keen to create a network of influential scientists, conscious of global problems and the environment, with a vision for the future. He chaired two panels of scientists in July 1970: the 'Study of Critical Environmental Problems' (SCEP) and the 'Study of Man's Impact on Climate' (SMIC). Meteorologists, oceanographers, ecologists, chemists, physicists, biologists, geologists, engineers, economists, social scientists and lawyers examined the new environmental issue of the global climatic and ecological effects of human activities. They published two interdisciplinary scientific reports funded by the MIT: *Man's Impact on the Global Environment: Assessment and Recommendations for Action* (1970) and *Inadvertent Climate Modification* (1971).[49] These two reports were interdisciplinary and their tone was radically different from conventional scientific publications. Both were meant to contribute directly to the

Stockholm Conference.[50] Of the reports' priorities, some were taken up and accepted during the Stockholm Conference, including the establishment of an atmospheric monitoring network.[51]

In addition to the work of SCOPE, the ICSU and the MIT studies, the conference secretariat was invited to take part in scientific events organized by national and international NGOs and research organizations. These events allowed UN officials, the conference secretariat and scientists to gather and exchange ideas on the environmental crisis. Thus, a meeting on 'International Organization and the Human Environment' was held on 21–23 May in Rensselaerville in the United States in order to assist preparations for the Stockholm Conference.[52] The meeting was co-sponsored by the Institute on Man and Science and the Aspen Institute for Humanistic Studies, an educational and policy studies organization founded in 1950. Richard Gardner, a law professor at Columbia University who also represented the International Union for the Conservation of Nature (IUCN) at the UN, organized this meeting. It gathered U Thant and Strong as well as twenty representatives from IOs.[53] UN specialized and affiliated agencies (e.g. UNESCO, FAO) and non-UN IOs like the Organisation for Economic Co-operation and Development (OECD) also participated in this event, along with environmental and scientific INGOs like the IUCN and SCOPE.

The meeting's central focus was to devise 'new or improved international arrangements to protect the global environment [and to] explore alternative means of organising international action'.[54] Special emphasis was placed on how to 'achieve a more effective coordination of activities within the UN system and between the UN system and private scientific organizations'[55] and on the new international structure that could be set up after the Stockholm Conference. The report issued after this meeting directly contributed to the Stockholm Conference. It was prepared as a so-called basic paper and addressed to conference agenda item VI, 'International organisational implications of action proposals'.

The Stockholm Conference preparation process created a momentum during which the UN and the conference secretariat relied on scientists and scientific NGOs in order to prepare the issues and actions to be handled during the conference. At the same time, scientific NGOs benefited from this intense context during which the environment became an international top priority issue and invited the UN officials to events in order to map 'needed areas of study in the international environmental area'.[56]

As Jan-Henrik Meyer analyses in greater detail in his chapter, during the preparations of the Stockholm Conference, developing countries criticized the conference's apparent focus on the environmental problems of the developed countries. In the light of this criticism, Strong convened the Founex meeting in Switzerland in June 1971, which brought together development economists, as discussed by Michael W. Manulak in his chapter in this book. The Founex group's report stated that environmental issues had to be considered as an integral part of the development process. Science and technology were considered as a response to the needs of the new 'environmental management'.[57] However, the role of science and technology in the new environmental management was not only debated at the Founex meeting, but also framed the rest of the conference preparations, the Stockholm Conference itself and its results. Preoccupied with the North–South divide on environmental questions, Strong also organized a meeting sponsored by SCOPE in cooperation with the UN Advisory Committee on the Application of Science and Technology to Development (ACAST) in Canberra in Australia from 24 August to 3 September 1971.[58] It was designed 'especially to involve scientists from developing or Third World countries and to provide advice to the conference secretariat'.[59] Although Strong worked closely with scientific INGOs and with scientists during the Founex and Canberra meetings, he also needed a permanent team of scientists to coherently synthesize global data on the environment.

A 'Committee of Corresponding Consultants'

In May 1971 Strong appointed René Dubos, a French-born microbiologist and experimental pathologist who had spent his career at Rockefeller University, as head of the so-called Committee of Corresponding Consultants. This committee was to evaluate the existing knowledge on the state of the world environment, collate the data in a report, and highlight the main agreements and differences of view among scientists and intellectuals, so that governments could then act upon it.[60] The British economist Barbara Ward was to draft the report.[61]

The committee was supposed to be representative of a global scientific and intellectual milieu. Strong described it as a 'unique experiment in international collaboration – one that engaged many of the world's leading authorities as consultants in the multiple branches of environmental affairs'.[62] It was composed of 152 scientists from more than sixty countries.[63]

It included natural scientists (biologists, ecologists, biophysicists, marine and soil science specialists, and agriculture development specialists) and social scientists (sociologists, anthropologists, economists and international relations specialists) and humanities scholars as well as businessmen, architects and lawyers. These 'consultants' held influential professional positions as professors, directors and presidents in national and international universities, research centres and academies of science. Some of them had affiliations with IOs like the OECD, UNESCO and the UN, and others with scientific INGOs like the ICSU, SCOPE, the IUCN and the International Social Sciences Council (ISSC). Some were authors of books and studies about the environmental crisis mentioned earlier, like Barry Commoner, Margaret Mead, Aurelio Peccei, Roger Revelle and Carroll Wilson.

The Committee of Corresponding Consultants was to give advice and not to make concrete proposals for government action or international agreements: 'the main goal of the report is not to bridge the differences, but to expose them objectively. It is not a question of making recommendations, but rather to provide an overall picture for the people making decisions'.[64] However, setting out global scientific data coherently and synthetizing it in a report without making any recommendations proved to be difficult.

First, the committee had to compile data and correspond with scientists from around the world. A first draft of the state of the world's environmental problems was sent to all members of the committee so that it could be criticized, corrected and revised. Seventy of the 152 committee members sent their comments on the first draft. In total, more than 350 pages of critical commentary were received and analysed in the preparation of the final report.[65] This enabled Ward and Dubos to draft the final report, which became the *World Report on the Human Environment*.

Secondly, the committee also had to bridge the gap between different scientists' views of the environmental problems and how to act on them. Dubos and Ward mentioned in the report's introduction that the members of the committee all agreed on the gravity and global dimensions of the environmental crisis that required global solutions.[66] According to Dubos and Ward, however, the corresponding consultants disagreed on how to interpret the issues and data.[67] Corresponding consultants from the Global North and South equally disagreed over the impact of industrial production and technologies. For some, industry was primarily a source of pollution, while for others, it was necessary for economic development and improved living conditions. The same debate concerned the technology issue – for example, the use of pesticides. While some scientists claimed that infectious diseases and food hygiene problems would spread if DDT use were restricted, others were convinced that pesticides profoundly disrupted natural ecological systems.[68] The introduction to the report also highlighted the different viewpoints among scientists concerning the advantages or dangers of nuclear technology.[69]

Committee scientists were consulted on an ad hoc basis to contribute their knowledge and advise the UN and national governments.[70] Despite the limits of their cooperation, it nonetheless constituted an unprecedented way of producing global knowledge on the state of the global environment. *Only One Earth: The Care and Maintenance of a Small Planet* constituted the world's first state-of-the-environment report.[71] It was circulated within the UN and its member states before the Stockholm Conference, and was published in several languages on its first day. *Only One Earth* became the conference slogan.

The Stockholm Conference: Translating Scientific Facts into Political Action?

To what extent did the Stockholm Conference create a public space for discussion on the environment among scientists, governments and the UN? And how did scientists take part in forms of social mobilization in Stockholm outside of the intergovernmental conference? What topics were debated?

UN Secretary General U Thant declared from the start that the Stockholm Conference 'should not be a scientific meeting, but a meeting aimed at making recommendations to Governments and international agencies regarding planning, rational management and control of the human environment'.[72] Its goal was to 'provide guidelines for action by national governments and international organisations in their attempts to achieve solutions for the problems of the human environment'.[73] The conference should foster public interest in the issue. In this sense, the Stockholm Conference was first and foremost intended to engage governments and motivate them to take action. The objective was to establish 'intensified action at the national, regional, and international level to limit and, where possible, to eliminate the impairment of the human environment'.[74] Nonetheless, scientists also took part in the Stockholm proceedings.

The Stockholm Conference included so-called official scientists. Alongside Ward and Dubos, other scientists from the Committee of Corresponding Consultants were invited. Among them were Thor Heyerdahl, a Norwegian ethnographer; Solly Zuckerman, a British zoologist; Aurelio Peccei of the Club of Rome; and Gunnar Myrdal, a Swedish sociologist and politician, Nobel Prize winner in Economics and former director of the UN Economic Commission for Europe.[75] At the same time, however, scientists also contributed to the so-called counterconferences. Various meetings took place in parallel with the negotiation process: the Peoples' Forum, the Environment Forum and the Dai Dong Conference.[76]

The Peoples' Forum was organized by a coalition of Scandinavian political and environmental groups. The Environment Forum was set up 'to satisfy the express desires of a growing number of pressure groups, lobbyists and field workers ... [and] participants from the scientific and technological community [who] might wish to mount a small number of informative demonstrations'.[77] It was financed by the Swedish government and planned by the UN Association of Sweden and the National Council of Swedish Youth. It gathered INGOs such as the World Wildlife Fund, the National Audubon Society, the Sierra Club, the IUCN, Friends of the Earth and many others.

Among the roots of the environmental crisis, the issue of population was one of the most debated at the Environmental Forum.[78] Two influential and mediatized scientists, Paul Ehrlich and Barry Commoner, were present and contributed to the debate.[79] Ehrlich focused on the demographic issue and the pressure population growth put on the planet's resources. In contrast, Commoner argued that what he called capitalist technologies were chiefly responsible for environmental degradation. Thus, in these counterconferences, too, scientists frequently disagreed on the roots, problems and appropriate solutions to environmental problems.

The Dai Dong counterconference finally met in Stockholm for two weeks beginning in the week before the UN conference opened. It gathered INGOs and some thirty scientists from around the world, except from the Soviet Bloc. The Dai Dong Conference set up by the FOR in line with the *Menton Statement* protested that the UN conference was not going to discuss chemical and biological warfare, overpopulation and population control or 'ecocidal American activities in Indochina'.[80] It was also meant to enable speakers to present and share their knowledge and debate their views on environmental priorities, with the aim of negotiating an alternative environmental declaration that would take into account the positions of the countries of the Global South.[81]

Scientists who came to Stockholm also debated their own role in national and international environmental policy making. An important meeting took place in Harpsund on 11 June 1972. The meeting brought together politicians, notably Strong, Olof Palme, the Swedish Prime Minister, and the Stockholm Conference President Ingemund Bengtsson and scientists, including Commoner, Dubos, Mead, Peccei, Ward and Wilson. Scientists from the Global South also participated in this meeting. Among them were Letitia Obeng, a zoologist and one of the first female scientists from Ghana; the Indian physicist Ashok Parthasarathi; and Mohamed El-Kassas, an Egyptian botanist and environmentalist who also worked for UNESCO and was particularly concerned with the desertification issue.

During this meeting in Harpsund, scientists were united in demanding a stronger role for themselves and their expertise in international environmental policy making. However, controversially, they discussed how they should organize their relationship with IOs and governments, as well as establish exchange among members of the scientific community? How could critical and independent scientists become involved in international environmental policy making without jeopardising their own ideas and research? Should they create an international scientific network or an INGO like the ICSU? Or should their priority be to more effectively include the scientific community in governments' policy making? Some scientists like Barry Commoner strongly denounced the consensual scientific institutions such as the US National Academy of Sciences, which 'not only failed to detect the environmental issue but fought against the few independent scientists who raised the issue, like Rachel Carson'.[82] Ward stressed the need to enable scientists to meet more systematically and create an institutional structure where they could participate as advisors in international environmental decision making. This idea was directly inspired by the paper 'The Human Environment: Science and International Decision-Making'[83] prepared for the conference secretariat by the International Institute of Environmental Affairs (IIEA), an international development and environment policy research organization founded by Ward in 1971. It recommended the creation of an interdisciplinary scientific structure for environmental research and development, composed of 'social scientists, engineers and humanists along with natural scientists on an unprecedented scale'.[84] Scientists could develop research on the environment and would be 'the counterpart of the UN environmental unit and its principal partner in the non-governmental world of science and technology'. They would 'perform ongoing advisory services for the United Nations system on environmental problems' by 'identifying gaps in present knowledge, advice with respect to research priorities, and help in the analysis of global environmental data'.[85]

The relationship between scientists and policy makers was not the only issue discussed during the Harpsund meeting. As Ashok Parthasarathi pointed out, 'it is not just the question of a dialogue between the scientists and the political leaders but a dialogue between scientists in developed countries and scientists in developing countries'.[86] Scientists from Asian, African and Latin American countries argued that most of the information and analysis of the environmental crisis came from scientists from developed countries. This information and analysis did not necessarily fit the needs of developing countries. This issue was raised during the Stockholm governmental meetings as well. Well aware of environmental deterioration, they expressed criticism of the way in which environmental issues were framed by Western scientists' views. According to the Italian ecologist Francesco Di Castri, who taught in Chile and worked as Secretary General for the UNESCO Man and the Biosphere (MAB) programme: 'There is no alternative because all the scientists are inspired – and the research – from the US or from Europe ... An enormous amount of the information comes from the developed countries. This recent Club of Rome initiative, the Blueprint for Survival, and so on had very controversial effects on the public opinion in developing countries ... For the public opinion [there] it seems to be just a kind of propaganda from the developed countries.'[87] Denouncing the Stockholm Conference's organization and proceedings as a 'rotten international tradition of running an international conference', Mohamed El-Kassas claimed that 'we came to Stockholm as scientists, not as delegates of our country ... but in the organization of the conference we could not free ourselves from the long standing traditions of international conferences that are apparently meant to the white people'.[88] This particular debate among scientists from the Global North and South thus reflected the broader debates at the Stockholm Conference itself.

Conclusion

As this chapter has highlighted, at the turn of the 1970s, scientists actively took part, along with other social actors, in fostering the rise of the environment as a global political issue. According to Strong, 'for the first time, scientific truth about the condition of man and society engaged the attention of political leaders on a world scale'.[89] Not all scientists participated and contributed in the same way, however, and they did not share the same ideas on the environmental crisis, its roots and its possible resolution. Some scientists acted as advocates, raising and sharing their environmental concerns through widely disseminated publications and collective international statements. As this chapter has highlighted, these actions were very important and influential with the Stockholm Conference secretariat. In addition, some of the same scientists – Barry Commoner, Paul Ehrlich and Margaret Mead, for example – went to Stockholm to debate in the counterconferences and other meetings. According to some scientists from developing countries, most of the issues discussed there, as in the main intergovernmental conference, reflected Western scientists' conceptions of the environmental crisis. Other scientists participated more directly in the conference preparatory process as advisors and consultants, helping translate their scientific knowledge into international environmental policies. In all, the conference secretariat relied on the work of a very heteronomous group of scientists – from developed as well as developing countries, from different academic fields, and with different views concerning the environmental crisis and its roots.

While the intergovernmental negotiations incorporated a few scientists with observer status, the counterconferences and meetings constituted platforms for discussions among scientists from the Global North and South. Although they shared the same basic view that the world was facing a global crisis, they did not share a common approach. In particular, the North–South split also divided scientists, who interpreted the scientific data differently depending on their professional background and ideological preferences. Moreover, while they agreed that scientific expertise should play a key role in policy making, they were not united on the issue of how to organize their relationship with IOs and national governments.

Scientific knowledge was crucial in the conference's conceptual framework, structure and process, both in the collation of scientific data on a global scale and for devising environmental policies. The *Declaration on the Human Environment* adopted at the end of the Stockholm Conference confirmed the key contribution of scientific knowledge and research for the 'identifications, avoidance, and control of environmental risks and a solution of environmental problems'.[90] In addition, the *Declaration* asked for a 'greater flow of scientific information and transfer of experience',[91] especially to developing countries.

Scientists debated not only environmental issues, but also their own role in environmental policy making and their relationships with IOs. The preparatory process, the conference itself and its results embedded the idea in the work of IOs and intergovernmental cooperation that effective communication and cooperation between scientists and political decision makers was 'indispensable to successful environmental management'. The action proposals added that 'carrying out the Action Plan effectively [requires] the full cooperation and involvement of the scientific community'.[92] At the end of the conference, UN Resolution 2997 (XXVII) created the Governing Council of the United Nations Environment Programme (UNEP). One of its tasks was to 'promote the contribution of the relevant international scientific and other professional communities to the acquisition, assessment and exchange of environmental knowledge and information and, as appropriate, to the technical aspects of the formulation and implementation of environmental programmes within the United Nations system'.[93]

Although the newly created intergovernmental body and the secretariat had to 'work in close liaison with the scientific community ... [which] should embrace all branches of the scientific community including medicine and the social sciences',[94] cooperation among scientists and between scientists, the UN and governments was not easy.[95] According to Peter S. Thacher, one of the deficiencies of the prevailing institutional arrangements was 'the inadequate relationship between the scientific community and social and political decision-making processes'.[96] According to him, one of the key problems was the lack of consensus among the scientific community at large and the lack of closer cooperation between natural and social scientists. This in turn made it difficult for scientists to provide 'clear guidance' to political decision makers.[97]

According to Strong, 'substantial changes within the scientific community itself and its relationship with society'[98] were required to achieve better international cooperation. He thus suggested that 'it would seem better to proceed by convening expert groups on an ad hoc basis. In this way, the intergovernmental body and secretariat would be able to seek advice from those persons whose expertise was most relevant to the particular problem under review'.[99] In line with similar processes of professionalization of expertise in policy making processes in other sectors,[100] the Stockholm Conference and the creation of UNEP thus provided a stimulus for strengthening relations between IOs and a new international figure: the 'environment expert'. Frequently appointed by IOs on an ad hoc basis, according to his or her relevant expertise regarding a particular issue, the role of the new 'environment expert' was to give advice, 'solve' specific issues and help implement new forms of 'environmental management' through international public policies.

Enora Javaudin is a Technical advisor for the NGO Apiflordev.

Notes

1. Ulrich Beck, *Risk Society: Towards a New Modernity* (London: Sage, 1992); Pierre Lascoumes, *L'éco-pouvoir, Environnement et politiques* (Paris: La Découverte, 1994).
2. Ronald E. Doel, 'Scientists as Policymakers, Advisors, and Intelligence Agents: Linking Diplomatic History with the History of Science', in *The Historiography of The History of Contemporary Science, Technology, and Medicine*, Thomas Söderqvist (ed.) (London: Harwood Academic Press, 1997), 33–62.
3. Kelly Moore, *Disrupting Science: Social Movements, American Scientists, and the Politics of the Military, 1945–1975* (Princeton, NJ: Princeton University Press, 2008), 12.
4. Martin Brown (ed.), *The Social Responsibility of the Scientist* (New York: Free Press, 1971); Joseph Ben-David, *The Scientist's Role in the Society* (Englewood Cliffs, NJ: Prentice Hall, 1971); Barry Commoner, 'Social Aspects of Science', *Science*, 127 (1957), 25 January 1957, 143–47.
5. Jack M. Hollander, 'Scientists and the Environment: New Responsibilities', *Ambio* 1(3) (1972): 73–120, 116.
6. John Boli and George M. Thomas, *Constructing World Culture, International Non-governmental Organizations since 1875* (Stanford: Stanford University Press, 1999), 84.
7. Maurice Strong, 'The Stockholm Conference – Where Science and Politics Meet', *Ambio* 1(3) (1972): 73–120, 73.
8. John Krige and Kai-Henrik Barth, *Global Knowledge Power: Science and Technology in International Affairs* (Chicago: University of Chicago Press, 2006), 3–4.
9. Francis Chateauraynaud and Didier Torny, *Les sombres précurseurs: une sociologie pragmatique de l'alerte et du risque* (Paris: EHESS, 1999).
10. Moore, *Disrupting Science*, 2–3.

11. 'Questions Pertaining to Natural Resources', *United Nations Yearbook* 1966, 321–34.
12. *United Nations Yearbook* 1948–1949, 481.
13. Rowland Wade, *The Plot to Save the World: The Life and Time of the Stockholm Conference* (Toronto: Clarke, Irwin & Company, 1973), 31.
14. UNESCO, *Utilisation et conservation de la Biosphère*, Actes de la Conférence intergouvernementale d'experts sur les bases scientifiques de l'utilisation rationnelle et de la conservation des ressources de la biosphère, Paris, 4–13 September 1968
15. The term 'biosphere' is a concept invented by the Austrian geologist Eduard Suess and taken over by the Russian scientist Wladimir Vernadsky (1863–1945) to refer to terrestrial life as an independent whole. Wladimir Vernadsky, *La Biosphère* (Paris: Librairie Felix Alcan, 1929).
16. United Nations Economic and Social Council, 23rd Session, 3 December 1968, 'Problems of the Human Environment', *United Nations Yearbook* 1968, 475.
17. Swedish Prime Minister Olof Palme, 'Address', Meeting at Harpsund, 11 June 1972, in connection with the UN Conference on the Human Environment. Maurice F. Strong Papers: 1948–2000, Environmental Science and Public Policy Archives, Harvard College Library. Maurice F. Strong, 1929–2012, box 29, file 291.
18. 'The International Council of Scientific Unions (ICSU)'s Role in the Light of the Stockholm Conference', Statement by Peter S. Thacher, Programme Director, UN Conference on the Human Environment to the 14th General Assembly of the International Council of Scientific Unions, Helsinki, Finland, 16 September 1972. Peter S. Thacher Environment Collection, 1960–1996. Environmental Science and Public Policy Archives, Harvard College Library, box 44, file 388.
19. Stephen Bocking, *Nature Experts: Science, Politics and the Environment* (Piscataway, NJ: Rutgers University Press, 2004), 26.
20. Rachel Carson, *Silent Spring* (Boston, MA: Houghton Mifflin, 1962).
21. Jean Dorst, *Avant que nature ne meure...* (Neuchâtel: Delachaux et Niestlé, 1965).
22. Paul Ehrlich, *The Population Bomb* (New York: Ballantine Books, 1968).
23. Barry Commoner, *The Closing Circle: Nature, Man, and Technology* (New York: Knopf, 1971). See also Michael Egan, *Barry Commoner and the Science of Survival: The Remaking of American Environmentalism* (Cambridge, MA: MIT Press, 2007).
24. 'A Blueprint for Survival', *The Ecologist* 2(1) (1972): 1–43.
25. The list is published in ibid., preface.
26. Wade, *The Plot to Save the World*, 20
27. Leading to Michael Schwab (ed.), *Teach-in for Survival* (London: Robinson and Watkins, 1972).
28. Founded by a group of Christians in Cambridge, England in December 1914, with the aim of working for world peace. The organization's first members were mostly Protestants (the movement started with the British Quaker

Henry Hodgkin (1877–1933) and the German Lutheran Friedrich Siegmund-Schultze).

29. Menton Conference of Scientists, 1970, Series H, box 3, in the Fellowship of Reconciliation Records (DG 013), Swarthmore College Peace Collection.

30. 13 October 1970, Series H, box 3, in the Fellowship of Reconciliation Records (DG 013), Swarthmore College Peace Collection.

31. Fred H. Knelman, 'What Happened at Stockholm', *International Journal* 28(1) (1972/1973): 28–49, 32.

32. Ibid.

33. 'Text of Statements at Visit of Environmental Scientists to Secretary General', U.N. Press release, 11 May 1971. Series H, box 2, in the Fellowship of Reconciliation Records (DG 013), Swarthmore College Peace Collection.

34. UN Secretary-General to the scientists of the *Menton Statement*, introduced by the secretary of the Fellowship of Reconciliation organization, during their meeting at the UN, 11 May 1971, in file Environmental Scientist Visit, 11 May 1971, Speeches, Messages, Statements, and Addresses, S-0885-0002-32, United Nations Archives.

35. Knelman, 'What Happened at Stockholm', 32

36. Extracts from the report following this meeting between the UN Secretary-General and the delegation of scientists representing all the signatories of the Menton Declaration. Series H, box 2, in the Fellowship of Reconciliation Records (DG 013), Swarthmore College Peace Collection.

37. 'S.O.S Environnement: Un message de 2200 savants aux 3 milliards et demi d'habitants de notre planète', *Courrier de l'UNESCO*, 4–5 July 1971.

38. *Ambio* was a bimonthly international journal published by the Royal Academy of Sciences, introduced in 1972.

39. Strong, 'The Stockholm Conference', 76

40. Peter S. Thacher, 'ICSU's Role in the Light of the Stockholm Conference', 16 September 1972.

41. Report by the Secretary-General during the third session of the preparatory committee of the conference, *ECE Symposium on Problems Relating to Environment, Proceedings and Documentation of the Symposium Organized by the United Nations Economic Commission for Europe and Held from 2 to 15 May 1971 in Prague (Czechoslovakia): With a Study Tour to the Region of Ostrava (Czechoslovakia) and Katowice (Poland)* (New York: United Nations, 1971).

42. Thacher, 'ICSU's Role in the Light of the Stockholm Conference'.

43. Maurice Strong, 'Relationship with the Scientific Community', in 'United Nations Conference on the Human Environment, International Organizational Implications of Action Proposals', Speech delivered during the 11th General Assembly of the International Union for Conservation of Nature, Banff, Canada, 11 September 1972, 23, United Nations Conference on the Human Environment correspondence, Economic and Social Affairs, S-0288-0020-17, United Nations Archives; Maurice F. Strong Papers: 1948–2000, Environmental

Science and Public Policy Archives, Harvard College Library. Maurice F. Strong, 1929–2012, box 29, file 295.

44. Commission on Monitoring of the International Council of Scientific Unions (ICSU) Scientific Committee on Problems of the Environment (SCOPE), 'Request to SCOPE from the U.N. Conference on the Human Environment', in *Global Environmental Monitoring, A Report Submitted to the United Nations Conference on the Human Environment, Stockholm 1972* (Stockholm: ICSU SCOPE, 1971), 14.

45. Abridged Report on Global Environmental Monitoring, Selected Sections on the complete draft report (submitted by the SCOPE Commission on Monitoring to the Intergovernmental Working Group on Monitoring and Surveillance at its First Meeting, Geneva, 16 August 1971). Maurice F. Strong Papers: 1948–2000, Environmental Science and Public Policy Archives, Harvard College Library. Maurice F. Strong, 1929–2012, box 36, file 359.

46. The United States, Canada, the United Kingdom, Australia, France, the Netherlands, Germany, Sweden, Spain, Bulgaria, Hungary, India, Indonesia, Japan, the USSR and Kenya. SCOPE, *Global Environmental Monitoring*, 1971.

47. Thacher, 'ICSU's Role in the Light of the Stockholm Conference'.

48. United Nations System-wide Earthwatch, 'Earthwatch 1972–1992. A Review of the Development of Earthwatch Prepared for the UNEP Annual Report 1992', http://www.un.org/earthwatch/about/docs/annrpt92.htm, accessed 2 July 2016.

49. MIT Study of Critical Environmental Problems (SCEP), *Man's Impact on the Global Environment: Assessment and Recommendations for Action* (Cambridge, MA: MIT Press, 1970); MIT Study of Man's Impact on Climate (SMIC), *Inadvertent Climate Modification* (Cambridge, MA: MIT Press, 1971).

50. 'Activist on the World Stage: Carroll Wilson Remembered', *MIT Tech Review*, February/March 1984, http://web.mit.edu/idi/carroll_wilson_award/CLW-Activist.pdf, accessed: 20 May 2016.

51. Hart, David and David Victor, 'Scientific Elites and the Making of US Policy for Climate Change Research, 1957–74', *Social Studies of Science* 23(4) (1993): 643–680.

52. IIEA, *International Organization and the Human Environment: The Proceedings of an International Conference Held in May, 1971, in Rensselaerville, New York in Preparation for the United Nations Stockholm Conference of 1972*, Rensselaerville, NY, 1971.

53. Institute on Man and Science to U Thant, 13 January 1971, International Organization on the Human Environment's Conference opening, 20 May 1971, Speeches, Messages, Statements, and Addresses, S-0885-0003-42, United Nations Archives.

54. Institute on Man and Science to U Thant, 13 January 1971.

55. Ibid.

56. Benjamin H. Read to Maurice Strong, 14 September 1970. Maurice F. Strong Papers: 1948–2000, Environmental Science and Public Policy Archives, Harvard College Library. Maurice F. Strong, 1929–2012, box 16, file 161.

57. *Development and Environment: Report and Working Papers of a Panel of Experts Convened by the Secretary-General of the United Nations Conference on the Human Environment, Founex, Switzerland, June 4–12, 1971* (Paris: Mouton, 1972).

58. Report by the Secretary-General, Third Session of the Conference Preparatory Committee, § 31; 32; 33; 34. French Diplomatic Archives, La Courneuve. Direction of political affairs, international non-governmental organizations, 'United Nations and International Organizations (1954–1973)', box 1360.

59. Lynton Keith Caldwell, *International Environmental Policy from the Twentieth to the Twenty-First Century* (Durham, NC: Duke University Press, 1996), 61.

60. Report of the Second Session Preparatory Committee (8–19 February 1971 Geneva). Direction of political affairs, international non-governmental organizations, 'United Nations and International Organizations (1954–1973)', French Diplomatic Archives, La Courneuve, box 1360.

61. Maurice Strong's speech during the National Foreign Trade Convention, New York, 17 November 1971, Conference on Human Environment, United Nations Commissions, Committees and Conferences, S-0858-0005-06, United Nations Archives.

62. Press release, 18 April 1972. René Dubos Papers, record group 450D851, box 51, folder 13, Rockefeller Foundation Archives, Rockefeller Archive Center, Sleepy Hollow, New York, USA.

63. Argentina (2); Australia (3); Austria; Barbados; Belgium (3); Brazil (3); Bulgaria; Cameroun; Canada (5); Ceylon; Chile (2); Colombia; Cyprus; Denmark (2); East and West Germany (8); Egypt (3); Ethiopia; Finland; France (7); Ghana (2); Greece; Guatemala; Hungary (2); India (3); Indonesia (3); Iraq; Iran; Israel; Italy (4); Japan (7); Kenya; Lebanon; Malaysia; Malta; Mexico (2); the Netherlands (4); New Zealand; Niger; Nigeria; Norway; Pakistan; the Philippines (2); Poland (2); Romania (2); Senegal; Singapore; Sweden (7); Switzerland (2); Tanzania; Thailand; Trinidad; United Kingdom (12); Uruguay; United States (21); USSR (6); Venezuela; Yugoslavia (2).

64. Report by the Secretary-General, Third Session of the Conference Preparatory Committee, 12–14. French Diplomatic Archives, La Courneuve., 'United Nations and International Organizations (1954–1973)', Direction of political affairs, international non-governmental organizations, box 1360.

65. Maurice F. Strong Papers: 1948–2000, Environmental Science and Public Policy Archives, Harvard College Library. Maurice F. Strong, 1929–2012, box 40, file 402.

66. Barbara Ward and René Dubos, *Only One Earth: The Care and Maintenance of a Small Planet: An Unofficial Report Commissioned by the Secretary-General of the United Nations Conference on the Human Environment. Prepared with the Assistance of a 152-Member Committee of Corresponding Consultants in 58 Countries* (London: Penguin Books, 1972), xii–xiii, xvii–xviii.

67. Ibid., xiv.

68. Ibid, xiv–xvi.

69. Ibid., xvi.

70. Maurice F. Strong, 'Relationship with the Scientific Community', in 'United Nations Conference on the Human Environment, International Organizational Implications of Action Proposals', Speech delivered during the 11th General Assembly of the International Union for Conservation of Nature, Banff, Canada, 11 September 1972, 23, United Nations Conference on the Human Environment Correspondence, Economic and Social Affairs, S-0288-0020-17, United Nations Archives.
71. Ward and Dubos, *Only One Earth*.
72. 'Report to the Secretary General by the Advisory Committee on the Application of Science and Technology to Development (ACAST), May 1, 1969'. Peter S. Thacher Environment Collection, 1960–1996. Environmental Science and Public Policy Archives, Harvard College Library, box 11, file 99; United Nations Secretariat, 'Problems of the Human Environment: report of the Secretary General', E/4667, 26 March 1969.
73. 'Report to the Secretary General by the ACAST', 1 May 1969.
74. 'Problems of the Human Environment', 3 December 1968.
75. Wade, *The Plot to Save the World*, 38.
76. Memo on conversation with Dr Barry Commoner on alternative environmental conference in Stockholm, 28 July 1971. Series H, box 4, in the Fellowship of Reconciliation Records (DG 013), Swarthmore College Peace Collection.
77. 'Environment Forum Proposal: The Reasons Why it is Needed'. Maurice F. Strong Papers: 1948–2000, Environmental Science and Public Policy Archives, Harvard College Library. Maurice F. Strong, 1929–2012, box 28, file 285 001.
78. Lars Emmelin, 'The Stockholm Conferences', *Ambio* 1(4) (1972): 123–152, 139.
79. 'Environment Forum Proposal: The Reasons Why it is Needed'; Peter Nilsson, 'NGOs Involvement in the U.N. Conference on the Human Environment in Stockholm 1972, Interrelations between Intergovernmental Discourse Framing and Activist Influence', Thesis, Linköping University, Department of Management and Economics, 27.
80. 'Open Options: A Guide to Stockholm's Alternative Environmental Conferences'. Series H, box 4, file 1 and 2, Fellowship of Reconciliation Records (DG 013), Swarthmore College Peace Collection.
81. Anne E. Egelston, *Sustainable Development: A History* (Dordrecht: Springer, 2013), 68–69.
82. Barry Commoner, Meeting at Harpsund, 11 June 1972.
83. IIEA; 'The Human Environment: Science and International Decision-Making', A Basic Paper prepared for the Secretariat of the United Nations Conference on the Human Environment, by the International Institute of Environmental Affairs (IIEA). Peter S. Thacher Environment Collection, 1960–1996. Environmental Science and Public Policy Archives, Harvard College Library, Box 12, file 102.
84. Ibid.
85. Ibid.

86. Ashok Parthasarathi, Meeting at Harpsund, 11 June 1972. Maurice F. Strong Papers: 1948–2000, Environmental Science and Public Policy Archives, Harvard College Library. Maurice F. Strong, 1929–2012, box 29, file 291.
87. Francesco Di Castri, Meeting at Harpsund, 11 June 1972.
88. Mohamed Kassas, Meeting at Harpsund, 11 June 1972.
89. 'Science and Society in the Environmental Age', Speech delivered by Maurice Strong, in Montreal, Canada, on 21 August 1971, Maurice F. Strong Papers: 1948–2000, Environmental Science and Public Policy Archives, Harvard College Library. Maurice F. Strong, 1929–2012, box 29, file 296.
90. *Declaration of the United Nations Conference on the Human Environment*, Principle 18.
91. Ibid., Principle 20.
92. Strong, 'Relationship with the Scientific Community'.
93. Resolution 2997 (XXVII) of the UN General Assembly, 'Institutional and Financial Arrangements for International Environmental Co-operation', http://www.un.org/french/documents/view_doc.asp?symbol=A/RES/2997(XXVII)&Lang=F, accessed 20 May 2016.
94. *International Organizational Implications of Action Proposals* (A/CONF.48/11), 23.
95. Thacher, 'ICSU's Role in the Light of the Stockholm Conference'.
96. Strong, 'Relationship with the Scientific Community'.
97. Ibid., 23.
98. Strong, 'Science and Society in the Environmental Age'.
99. *International Organizational Implications of Action Proposals* (A/CONF.48/11), 23.
100. For Europe, see Martin Kohlrausch and Helmuth Trischler, *Building Europe on Expertise. Innovators, Organizers, Networkers* (Basingstoke: Palgrave Macmillan, 2014); Wolfram Kaiser and Johan Schot, *Making the Rules for Europe: Experts, Cartels, International Organizations* (Basingstoke: Palgrave Macmillan, 2014).

Bibliography

'A Blueprint for Survival', *The Ecologist* 2(1) (1972), 1–43.
'Activist on the World Stage: Carroll Wilson Remembered', *MIT Tech Review*, February/March 1984, http://web.mit.edu/idi/carroll_wilson_award/CLW-Activist.pdf, accessed 4 April 2016.
Beck, Ulrich, *Risk Society: Towards a New Modernity* (London: Sage, 1992).
Ben-David, Joseph, *The Scientist's Role in Society* (Englewood Cliffs, NJ: Prentice Hall, 1971).
Bocking, Stephen, *Nature Experts: Science, Politics and the Environment* (Piscataway, NJ: Rutgers University Press, 2004).
Boli, John and George M. Thomas, *Constructing World Culture: International Non-governmental Organizations since 1875* (Stanford: Stanford University Press, 1999).

Brown, Martin (ed.), *The Social Responsibility of the Scientist* (New York: Free Press, 1971).

Caldwell, Lynton Keith, *International Environmental Policy from the Twentieth to the Twenty-First Century* (Durham, NC: Duke University Press, 1996).

Carson, Rachel, *Silent Spring* (Boston, MA: Houghton Mifflin, 1962).

Chateauraynaud, Francis and Didier Torny, *Les sombres précurseurs: une sociologie pragmatique de l'alerte et du risque* (Paris: L'École des hautes études en sciences sociales, 1999).

Commission on Monitoring of the International Council of Scientific Unions' (ICSU) Scientific Committee on Problems of the Environment (SCOPE), 'Request to SCOPE from the U.N. Conference on the Human Environment', in *Global Environmental Monitoring, A Report submitted to the United Nations Conference on the Human Environment, Stockholm 1972* (ICSU SCOPE: Stockholm 1971), 14.

Commoner, Barry, 'Social Aspects of Science', *Science* 127 (1957), 25 January 1957, 143–47.

———. *The Closing Circle: Nature, Man, and Technology* (New York: Knopf, 1971).

Doel, Ronald E., 'Scientists as Policymakers, Advisors, and Intelligence Agents: Linking Diplomatic History with the History of Science', in *The Historiography of The History of Contemporary Science, Technology, and Medicine*, Thomas Söderqvist (ed.) (London: Harwood Academic Press, 1997), 33–62.

Dorst, Dorst, *Avant que nature ne meure...* (Neuchâtel: Delachaux et Niestlé, 1965).

Egan, Michael, *Barry Commoner and the Science of Survival: The Remaking of American Environmentalism* (Cambridge, MA: MIT Press, 2007).

Egelston, Anne E. *Sustainable Development: A History* (Dordrecht: Springer, 2013).

Ehrlich, Paul, *The Population Bomb* (New York: Ballantine Books, 1968).

Emmelin, Lars 'The Stockholm Conferences', *Ambio* 1(4) (1972): 123–52.

Environment Forum, 'Environment Forum Proposal: The Reasons Why it is Needed'. Maurice F. Strong Papers: 1948–2000, Environmental Science and Public Policy Archives, Harvard College Library. Maurice F. Strong, 1929–2012, Box 28, file 285 001.

Extracts from the report following the meeting between the UN General-Secretary and the delegation of scientists representing all the signatories of the Menton Declaration. Series H, box 2, in the Fellowship of Reconciliation Records (DG 013), Swarthmore College Peace Collection.

Hart, David and David Victor, 'Scientific Elites and the Making of US Policy for Climate Change Research, 1957–74', *Social Studies of Science* 23 (4) (1993), 643–680.

Hollander, Jack M. 'Scientists and the Environment: New Responsibilities', *Ambio* 1(3) (1972): 73–120.

IIEA. 'The Human Environment: Science and International Decision-Making', A basic paper prepared for the secretariat of the United Nations Conference on the Human Environment, by the International Institute of Environmental Affairs (IIEA). Peter S. Thacher Environment Collection, 1960–1996.

Environmental Science and Public Policy Archives, Harvard College Library, Box 12, file 102.

———. *International Organization and the Human Environment: The Proceedings of an International Conference Held in May, 1971, in Rensselaerville, New York in Preparation for the United Nations Stockholm Conference of 1972*, Rensselaerville, NY, 1971.

Institute on Man and Science to U Thant, 13 January 1971, International Organization on the Human Environment's Conference Opening, 20 May 1971, Speeches, Messages, Statements, and Addresses, S-0885-0003-42, United Nations Archives.

Kaiser, Wolfram and Johan Schot, *Making the Rules for Europe: Experts, Cartels, International Organizations* (Basingstoke: Palgrave Macmillan, 2014).

Knelman, Fred H., 'What Happened at Stockholm', *International Journal* 28(1) (1972/1973): 28–49.

Kohlrausch, Martin and Helmuth Trischler, *Building Europe on Expertise: Innovators, Organizers, Networkers* (Basingstoke: Palgrave Macmillan, 2014).

Krige, John and Kai-Henrik Barth, *Global Knowledge Power: Science and Technology in International Affairs* (Chicago: University of Chicago Press, 2006).

Lascoumes, Pierre, *L'éco-pouvoir, Environnement et politiques* (Paris: La Découverte, 1994).

Memo on conversation with Dr Barry Commoner on Alternative Environmental Conference in Stockholm, 28 July 1971. Series H, box 4, in the Fellowship of Reconciliation Records (DG 013), Swarthmore College Peace Collection.

Menton Conference of Scientists, 1970, Series H, box 3, in the Fellowship of Reconciliation Records (DG 013), Swarthmore College Peace Collection.

MIT, *Study of Critical Environmental Problems (SCEP): Man's Impact on the Global Environment. Assessment and Recommendations for Action* (Cambridge, MA: MIT Press, 1970).

———. *Study of Man's Impact on Climate (SMIC): Inadvertent Climate Modification* (Cambridge, MA: MIT Press, 1971).

Moore, Kelly, *Disrupting Science: Social Movements, American Scientists, and the Politics of the Military, 1945–1975* (Princeton, NJ: Princeton University Press, 2008).

Nilsson, Peter, 'NGOs Involvement in the U.N. Conference on the Human Environment in Stockholm 1972, Interrelations Between Intergovernmental Discourse Framing and Activist Influence', Thesis, Linköping University, Department of Management and Economics.

'Open Options: A Guide to Stockholm's Alternative Environmental Conferences'. Series H, box 4, file 1 and 2, in the Fellowship of Reconciliation Records (DG 013), Swarthmore College Peace Collection.

Palme, Olof, Swedish Prime Minister, 'Address', Meeting at Harpsund, 11 June 1972, in connection with the UN Conference on the Human Environment. Maurice F. Strong Papers: 1948–2000, Environmental Science and Public Policy Archives, Harvard College Library. Maurice F. Strong, 1929–2012, box 29, file 291.

Parthasarathi, Ashok, Meeting at Harpsund, 11 June 1972. Maurice F. Strong Papers: 1948–2000, Environmental Science and Public Policy Archives, Harvard College Library. Maurice F. Strong, 1929–2012, box 29, file 291.

Press release, 18 April 1972. René Dubos Papers, record group 450D851, box 51, folder 13, Rockefeller Foundation Archives, Rockefeller Archive Center, Sleepy Hollow, New York, USA.

Read, Benjamin H. to Maurice Strong, 14 September 1970. Maurice F. Strong Papers: 1948–2000, Environmental Science and Public Policy Archives, Harvard College Library. Maurice F. Strong, 1929–2012, box 16, file 161.

Schwab, Michael (ed.), *Teach-in for Survival* (London: Robinson and Watkins, 1972).

SCOPE, Abridged Report on Global Environmental Monitoring, Selected Sections on the complete draft report, submitted by the SCOPE Commission on Monitoring to the Intergovernmental Working Group on Monitoring and Surveillance at its First meeting, Geneva, 16 August 1971. Maurice F. Strong Papers: 1948–2000, Environmental Science and Public Policy Archives, Harvard College Library. Maurice F. Strong, 1929–2012, box 36, file 359.

Strong, Maurice F., Papers: 1948–2000, Environmental Science and Public Policy Archives, Harvard College Library. Maurice F. Strong, 1929–2012, box 29, file 295.

———. Papers: 1948–2000, Environmental Science and Public Policy Archives, Harvard College Library. Maurice F. Strong, 1929–2012, box 40, file 402.

———. Speech during the National Foreign Trade Convention, New York, 17 November 1971, Conference on Human Environment, United Nations Commissions, Committees and Conferences, S-0858-0005-06, United Nations Archives.

———. 'Relationship with the Scientific Community', in 'United Nations Conference on the Human Environment, International Organizational Implications of Action Proposals', Speech delivered during the 11th General Assembly of the International Union for Conservation of Nature, Banff, Canada, 11 September 1972, 23, United Nations Conference on the Human Environment Correspondence, Economic and Social Affairs, S-0288-0020-17, United Nations Archives.

———. 'Science and Society in the Environmental Age', Speech in Montreal, Canada, on 21 August 1971, Maurice F. Strong Papers: 1948–2000, Environmental Science and Public Policy Archives, Harvard College Library. Maurice F. Strong, 1929–2012, box 29, file 296.

———. 'The Stockholm Conference – Where Science and Politics Meet', *Ambio* 1(3) (1972): 73–120.

Thacher, Peter S., 'The International Council of Scientific Unions (ICSU)'s Role in the Light of the Stockholm Conference', Statement by Programme Director, UN Conference on the Human Environment to the 14th General Assembly of the International Council of Scientific Unions, Helsinki, Finland, 16 September 1972, Peter S. Thacher Environment Collection, 1960–1996. Environmental Science and Public Policy Archives, Harvard College Library, box 44, file 388.

United Nations System-wide Earthwatch, 'Earthwatch 1972–1992. A Review of the Development of Earthwatch Prepared for the UNEP Annual Report 1992', http://www.un.org/earthwatch/about/docs/annrpt92.htm, accessed 2 July 2016.

UN, *United Nations Yearbook* 1948–1949.

UN, 'Questions Pertaining to Natural Resources', *United Nations Yearbook* 1966, 321–34.

———. 'Report to the Secretary General by the Advisory Committee on the Application of Science and Technology to Development (ACAST), May 1, 1969'. Peter S. Thacher Environment Collection, 1960–1996. Environmental Science and Public Policy Archives, Harvard College Library, box 11, file 99.

———. 'Text of Statements at Visit of Environmental Scientists to Secretary General', UN press release, 11 May 1971, Series H, box 2, in the Fellowship of Reconciliation Records (DG 013), Swarthmore College Peace Collection.

———. Report of the Second Session Preparatory Committee (8–19 February 1971, Geneva). Direction of political affairs, international non-governmental organisations, 'United Nations and international organisations (1954–1973)' French Diplomatic Archives, La Courneuve, box 1360.

———. Report by the Secretary-General during the third session of the preparatory committee of the conference. *ECE Symposium on Problems Relating to Environment, proceedings and documentation of the symposium organized by the United Nations Economic Commission for Europe and held from 2 to 15 May 1971 in Prague (Czechoslovakia): with a study tour to the region of Ostrava (Czechoslovakia) and Katowice (Poland)* (United Nations: New York, 1971).

———. *Development and Environment: Report and Working Papers of a Panel of Experts Convened by the Secretary-General of the United Nations Conference on the Human Environment, Founex, Switzerland, June 4–12, 1971* (Paris: Mouton, 1972).

———. *International Organizational Implications of Action Proposals* (A/CONF.48/11).

———. Report by the Secretary-General, Third Session of the Conference Preparatory Committee. Direction of political affairs, international non-governmental organisations, 'United Nations and international organisations (1954–1973)', French Diplomatic Archives, La Courneuve, box 1360.

———. Resolution 2997 (XXVII) of the UN General Assembly, 'Institutional and Financial Arrangements for International Environmental Co-operation', http://www.un.org/french/documents/view_doc.asp?symbol=A/RES/2997(XXVII)&Lang=F, accessed 16 May 2016.

———. UN Secretary-General to the scientists of the *Menton Statement*, introduced by the secretary of the Fellowship of Reconciliation organisation, during their meeting at the UN, 11 May 1971, Environmental Scientist Visit, 11 May 1971, Speeches, Messages, Statements, and Addresses, S-0885-0002-32, United Nations Archives.

———. United Nations Economic and Social Council, 23rd Session, 3 December 1968, 'Problems of the Human Environment', *United Nations Yearbook* 1968, 475.

———. United Nations Secretariat, 'Problems of the Human Environment: Report of the Secretary General', E/4667, 26 March 1969.

UNESCO, *Utilisation et conservation de la Biosphère*, Actes de la Conférence intergouvernementale d'experts sur les bases scientifiques de l'utilisation rationnelle et de la conservation des ressources de la biosphère, Paris, 4–13 September 1968.

———. 'S.O.S Environnement: Un message de 2200 savants aux 3 milliards et demi d'habitants de notre planète', *Courrier de l'UNESCO*, 4–5 July 1971.

Vernadsky, Wladimir, *La Biosphère* (Paris: Librairie Felix Alcan, 1929).

Wade, Rowland, *The Plot to Save the World: The Life and Time of the Stockholm Conference* (Toronto: Clarke, Irwin & Company, 1973).

Ward, Barbara and René Dubos, *Only One Earth: The Care and Maintenance of a Small Planet: An Unofficial Report Commissioned by the Secretary-General of the United Nations Conference on the Human Environment. Prepared with the Assistance of a 152-Member Committee of Corresponding Consultants in 58 Countries* (London: Penguin Books, 1972).

CHAPTER 3

Developing World Environmental Cooperation
The Founex Seminar and the Stockholm Conference

Michael W. Manulak

Debates about the relationship between international development goals and environmental policies have long been at the centre of global efforts to address the world's mounting environmental crisis. The tenor and structure of these debates were shaped and given institutional legitimacy at the 1972 Stockholm Conference. Stockholm institutionalized a vision of North–South environmental cooperation that, to a significant extent, prevails today. Principled and normative ideas about global environmental cooperation shaped the conference declaration and action plan, and heavily influenced the institutional design of the United Nations Environment Programme (UNEP).

Where did these ideas come from? What explains their influence on institution building in the United Nations (UN) context? Firsthand accounts suggest that a group of development economists, meeting in Founex, Switzerland in June 1971, had a leading influence on the institutional outcomes of the Stockholm Conference. Lars-Göran Engfeldt, a Swedish diplomat, for instance, has argued that without the Founex initiative, 'an open North-South confrontation at the [United Nations General Assembly] session [preceding the conference] would have been almost unavoidable and would have had a devastating effect on the Conference'.[1] Christian Herter Jr., who served on the United States' conference delegation, reflected that 'in some ways [the Founex Report] was as important as anything subsequently produced at the conference itself'.[2] According to Peter B. Stone, a member of the conference secretariat, the Founex panel's report 'began life as a barely comprehensible diplomatic-economic paper, but eventually took on an aura like that of the Authorized version of the Bible'.[3]

These assertions present an important historical and theoretical puzzle. Though it is widely mentioned in the global environmental politics literature, no detailed investigations of the Founex meeting exist outside of the enthusiastic recollections of Stockholm Conference participants. Because of the transnational influence of the Founex Report, appreciating its impact demands a broad analytical lens that is sensitive to the role of ideas and intellectual leadership in world politics. The study of the Founex seminar also raises important theoretical questions. What power do knowledge-based networks wield in world politics? What means do they use to influence events? What historical tools can analysts use to assess their impact?

Employing the epistemic communities framework, this chapter analyses the activities of a group of interventionist development experts during the Stockholm Conference preparatory process. Epistemic communities are defined by Peter M. Haas as networks of 'professionals with recognized expertise and competence in a particular domain and an authoritative claim to policy-relevant knowledge within that domain or issue area'.[4] These professionals are united by a shared set of normative or principled convictions, common causal beliefs, shared notions of validity and a common policy enterprise. They influence patterns of international cooperation through policy innovation, diffusion and selection.[5]

Based on a multi-archival investigation and oral history interviews, this chapter contributes to understanding the role of transnational academic networks in institution building at the 1972 Stockholm Conference. It finds that a knowledge community of interventionist development economists used the Founex seminar and report as a means of advancing their wider intellectual agenda. Their success in seizing the opportunity presented by the conference to diffuse social development ideas is explained by their authoritativeness *and* their political acumen. Their impact on policy selection is explained by their ability to produce a conspicuous focal point in a negotiating context that was in urgent need of one. The chapter first discusses the North–South divide on environmental questions that emerged during the Stockholm Conference preparatory process. It then proceeds in a roughly chronological fashion, analysing the convening of the Founex seminar and the resulting report, as well as assessing the impact of the Founex process for institution building at Stockholm. The concluding section discusses the chapter's implications for theorizing the role of epistemic communities in global environmental politics and affairs.

The Contested Development–Environment Linkage

Ominous scientific projections of global pollution crises, natural resource shortages, and population explosions formed the starting point for the development-environment debate during the Stockholm preparatory process. These concerns are discussed in greater detail in Enora Javaudin's chapter in this volume. Widely read books, such as *The Limits to Growth*, *The Population Bomb*, and *A Blueprint for Survival* captured the Western public's attention and reinforced the sense of ecological crisis first stimulated by Rachel Carson's *Silent Spring* – published in 1962 – and by the foundering of the Torrey Canyon super tanker in 1967.[6] In 1970, prominent Stanford biologist Paul Ehrlich wrote that world resource scarcities dictated that developing countries should content themselves with semi-development and that rich countries should aim to de-develop.[7] Stimulated by these assessments, policy insiders, such as George Kennan, Leroy Wehrle and Kingman Brewster Jr., engaged in 'big ideas' debates in major academic journals espousing radical policy reorientations of global environmental management regimes. Their proposals argued for vigorous Western leadership in devising global strategies for preventing 'a world wasteland'.[8]

Less developed countries regarded this emerging Western environmental discourse with scepticism and hostility. Developing country observers argued that environmental concern was a trendy, bourgeois fad.[9] A pollution conference had little relevance for poor countries and could provide an obstacle to development, as U.S. presidential scientific advisor Lee Talbot recalled: 'Pollution was seen by many nations as external evidence of industrial development, not as a threat. "What we need is more pollution" was frequently reiterated by representatives of developing countries. Therefore, efforts to control pollution were looked on by some as efforts to control development.'[10] Talk of limits to growth was heard in a period where development assistance flows were ebbing, leading to fears that the 'liberal hour' of generous development assistance had passed.[11] Many worried that Western environmental concern could lead to additional trade barriers, miserly development assistance allocations and the consequent entrenchment of extant global inequalities.

Discussion of restricting world population growth necessitated by global resource scarcities generated further resentment in the developing world. After all, at the time, developing countries consumed only one-tenth of the earth's natural resources.[12] To the extent that they existed, resource scarcities were clearly a product of voracious Western lifestyles, they argued. Prominent developing world leaders doubted the validity of the conclusions of Western ecologists, derisively labelling them 'pseudoscientific extrapolations of the doomsday variety'.[13] Acting on these extrapolations was at best premature and at worst a neocolonial plot to retard necessary economic development.

This debate was not limited to academics and policy communities. In late 1970, the planned Stockholm Conference 'showed signs of becoming a spirited confrontation pitting the north against its neighbours to the south'.[14] In December of that year, less developed countries, increasingly concerned with the direction of the conference, voted in the UN General Assembly to recommend that the conference secretariat include agenda items 'relating to economic and social aspects … with a view to reconciling the national environmental policies with their national development plans and priorities'.[15] Confidentially, the secretariat received word from an official from Yugoslavia, a leader of the non-aligned movement, that there was deep dissatisfaction among developing countries. From their perspective, the Stockholm Conference was oriented disproportionately towards the concerns of the industrialized countries. The official alluded to a potential developing country boycott of the conference.[16] These warnings were corroborated by persistent rumours that Latin American countries were among those considering such a boycott.[17]

A chief virtue of the UN process was its capacity to bring together leaders of the Global North and South.[18] If developing states boycotted or remained hostile to the process, the UN's fledgling environmental mandate would have been badly discredited. Thus, when Maurice Strong was appointed the conference's secretary-general in the autumn of 1970, one of his main objectives was to engage the developing countries.[19] At the conference's second preparatory committee meeting in February 1971, he proposed a 'radical remake' of the conference agenda that had been approved at the first preparatory committee meeting. This new agenda linked environmental matters directly to the development process and to the interests of the developing countries. At the meeting, Strong spoke of the need to integrate environmental considerations within development goals and announced a series of regional seminars on the environment and development to be held in Asia, Africa, Latin America and the Middle East later that year.[20]

According to Strong, his efforts to involve the developing countries were underpinned by a 'basic thesis' that 'environmental and economic priorities are intrinsically two sides of the same coin ... The key was to insist that the needs of developing countries would best be served by treating the environment as an integral dimension of development rather than an impediment.[21] This appreciation, coloured by Strong's experience as executive director of the Canadian International Development Agency, was more tactical thinking than firm analytical conclusion. As Strong recalled, it 'lacked the benefit of rigorous analysis and a well-developed policy thesis'.[22] He convened an informal meeting in New York to flesh out this idea with a group of eminent development experts, including the well-known British economist Barbara Ward, Gamani Corea of Ceylon, Pakistan's Mahbub ul Haq, the St Lucian economist Arthur Lewis and Enrique Iglesias of Uruguay. In his memoirs, Strong recalled:

> Pakistan's Mahbub ul Haq, one of the most brilliant and provocative development economists, made a spirited attack on the whole concept of 'the environment'. His position was devastating and simple: industrialization had given developed countries disproportionate benefits and huge reservoirs of wealth and at the same time caused the very environmental problems we were now asking the developing countries to join in resolving. The cost of cleaning up the mess, therefore, should be borne by the countries that had caused it in the first place. If they wanted developing countries to go along, they'd have to provide the financial resources to do so.[23]

In response, Strong proposed that an expert working group be established to consider the environment-development topic in more depth.[24] Despite their scepticism, Haq, Corea and Iglesias accepted Strong's proposal and agreed to participate in coordinating the meeting.

Convening the Founex Seminar

After the New York meeting, the Stockholm Conference secretariat appointed a planning group to coordinate the invitations and programme for the Founex meeting. This group, chaired by Corea, included Haq, Ward and Iglesias.[25] Corea had served as deputy governor of Ceylon's central bank and was 'in great demand to guide the trickier international working groups' because of his international experience and prestige.[26] The primary aim of the Founex meetings was to provide a forum for developing world environmental interests to be 'articulated clearly [and] analyzed objectively' by a regionally representative grouping.[27]

The twenty-seven Founex panel participants were chosen based on reputational, personal and intellectual considerations. Many of them seem to have been connected personally to Ward and Strong. For example, numerous experts who attended the Founex meeting, such as Samir Amin, Pitamber Pant, Hans Singer and Jan Tinbergen, had collaborated with Corea, Iglesias, Haq, Lewis and Ward as recently as a February 1970 Columbia University conference on International Economic Development, an event set up by Ward. She also reached out to Ignacy Sachs from France and Miguel Ozorio de Almeida from Brazil.[28] The eminent Argentine economist Raúl Prebisch, who did not attend the Founex seminar, was another well-connected backer of Strong's work. It was Prebisch who had originally encouraged Corea to support Strong's efforts and who appears to have connected Iglesias to the secretary-general.[29] University of Sussex professor Hans Singer, another Founex participant, had collaborated closely with Prebisch in the past. The economists recruited for the Founex seminar therefore had longstanding personal and professional associations.

The list of participants reads as a 'who's who' in the new wave of interventionist development thinking. Leading participants of the Founex meeting, such as Corea, Haq, Iglesias, Sachs and Singer, were disillusioned with conventional, growth-oriented approaches to development economics. They shared causal and normative conclusions within the field of international development. These economists rejected the notion that economic development was best achieved through building up the physical infrastructure of countries. Instead of viewing social expenditures as the fruits of development rather than its seeds, economic interventionists advocated proactive expenditures to stimulate development and reduce poverty. The assumption that wealth would trickle down to alleviate poverty was, they maintained, suspect.[30]

Haq, who would later produce the Human Development Index, was a prominent voice for this new wave of thinking. Reflecting on his experience in the Pakistani planning commission at the May 1971 Society for International Development Conference in Ottawa, Haq advocated a 'direct attack on poverty'. The singular pursuit of gross national product (GNP) growth had, stressed Haq, been an obstacle to poverty reduction.[31] Corea shared this viewpoint, advocating 'a style of development in which economic and social goals reinforce each other'.[32] Similarly, Sachs, a professor at the Parisian École Pratique des Hautes

Études, advocated extensive planning and a move from 'narrow concepts of economic growth to the more comprehensive, although less easily quantifiable concept of development'.[33]

Many of these actors – especially Ward and Haq – had been highly influential in precipitating a shift in World Bank lending policy from the more laissez-faire economic philosophy that had predominated through the Bank's first two decades. In the early 1970s, World Bank President Robert McNamara embraced lending policies that targeted an increase in the productivity of the poor. Under McNamara, the institution became more interventionist in orientation, seeking to improve literacy, nutrition and health.[34]

Two points were essential for the interventionists. First, GNP growth was regarded as an incomplete and sometimes misleading measure of economic progress. Overreliance on per capita GNP figures distracted policy makers from serious market distortions, including extremely uneven wealth distributions and the living conditions of the poor. Accordingly, policy makers should seek to balance national income indicators with qualitative measures of wellbeing. Secondly, institutional reforms and government interventions were the best means of poverty reduction. Ambitious government development planning would provide a sound basis for improving the socioeconomic conditions that fuel economic progress. Unregulated markets could not underpin sensible development efforts.[35]

This interventionist orientation was controversial among development economists and practitioners in 1971. Several Founex participants, such as Wilfred Beckerman and Miguel Ozorio de Almeida, argued strenuously that development schemes should be predicated on market mechanisms and measured in large part by GNP growth indicators. But while the participants of the Founex meeting held diverse views, it is crucial to note that those oriented towards robust government social intervention in development planning were numerous and likely to dominate the proceedings. When Corea and Haq were, respectively, elected chair and rapporteur of the Founex panel, the influence of this perspective was further augmented.

The Founex Seminar: An Inconclusive Debate

The Founex seminar was convened in the unattractive Motel de Founex, located along a highway outside Geneva, from 4 to 12 June 1971. A heatwave, the lack of air conditioning and the motel's tiny meeting room made for an uncomfortable working environment.[36] Archival sources, including the papers that seminar participants prepared for discussion, suggest that competing development outlooks contributed to a strong debate among economists that prescribed fundamentally different approaches to the subject at hand.

Three main viewpoints on the development–environment relationship were present in policy debates at the time. Of these, only two perspectives were seriously considered by the participants in Founex. The approach that was rejected out-of-hand was the semi-development option of Paul Ehrlich. Perhaps because the Founex meeting was, as Peter Stone of the Stockholm Conference secretariat has observed, 'long on economists and short on ecologists', this concept was never considered as a viable policy alternative.[37] Most participants regarded the limited growth approach with 'detached amusement' because its conclusions invoked 'Malthusian ghosts' through 'careless and casual' assumptions and 'mere intellectual fantasy'.[38]

The two main perspectives on the development–environment subject entertained at Founex mirrored broader disciplinary divisions in the field of development economics. The first line was articulated by Almeida, Beckerman and representatives of the secretariat of the General Agreement on Tariffs and Trade (GATT). This argument suggested that environmental regulations were a potential millstone that threatened economic development. It highlighted inherent contradictions between ecologically and economically ideal conditions. The issue, Almeida wrote in his seminar working paper, resembled a zero-sum game. He argued that there was 'a fundamental contradiction between the process of economic development most specifically in its initial and intermediate stages, and the environmental conditions hoped for by the idealistic sociologist'.[39] The main responsibility of the nonindustrialized countries, Almeida stressed, was to accelerate their economic growth. Once a satisfactory level of development was reached, these states could afford to consider environmental initiatives.[40] Environmental benefits would, in effect, trickle down as wealth accumulated.

Beckerman, who argued that rapid growth was the best means of environmental protection, shared this point of view.[41] Since the environment was, he argued at Founex, solely a middle-class concern of industrialized countries, it should not be a major priority for the developing countries.[42] Summing up his arguments after Founex, he described the dramatic anti-pollution progress that had been made in Britain since the mid 1950s in spite of a rise in GNP. He hypothesized that rapid economic growth was the key to accelerating the shift away from polluting production techniques.[43] Similarly, the GATT seminar paper advocated minimal environmental regulation among the developing countries because future technological progress would reduce the costs of pollution abatement. Years of national income growth would render the costs of environmental standards less onerous.[44]

By emphasizing GNP growth and satisfactory levels of industrialization as prerequisites to environmental action, proponents of this view implied that rigorous pollution prevention and natural resource management policies in the developing world were decades away. Almeida categorized the benefits of environmental preservation with consumptive expenditures, such as nutrition, clothing and housing. Consequently, just as social expenditures must be regarded as consumption and therefore stoically forsaken in the short and medium term, environmental amenities for underdeveloped countries should also be sacrificed in the immediate term.[45] Beckerman shared this economic calculus, arguing that pollution should be reduced only when the costs of doing so did not exceed the benefits. In developing countries, pollution abatement should not be pursued at the cost of slowing economic growth.[46]

Corea, Haq, Iglesias and Sachs articulated the alternative, interventionist perspective. Critical to their position was the notion that environmental problems in the Global South were of a different kind from those in the developed world. Issues such as pollution or natural resource depletion were, for the time being, less urgent than problems like polluted water, poor sanitation, inadequate housing, disease, soil erosion and vulnerability to natural disasters. In his seminar paper and at Founex, Iglesias highlighted the socioeconomic dimensions of environmental issues: 'the destruction of the environment is primarily a result of the conditions of poverty that prevail in vast regions of the world'.[47] Poverty was a principal cause of pollution. For proponents of this approach, particularly Sachs, environmental matters in nonindustrialized countries were fundamentally questions of human wellbeing. Human-centred development was the solution to the environmental dilemmas of the Global South.

In the same way that their attitudes diverged from Almeida and Beckerman in the socioeconomic realm, Corea and others differed in their environmental prescriptions. In 1972, Corea situated his environmental approach in the context of the 'virtual dethronement of

GNP and a new sensitivity to social and other goals. He stressed that environmental factors 'are so much a part of social and economic conditions in developing countries that their treatment is but an aspect of the whole approach to economic and social development.'[48] Similarly, Sachs asserted that the only economic measure that environmental protection could hinder was the narrow concept of economic growth.[49]

Just as social problems needed to be managed by government action, environmental imperatives should be addressed through development planning. The environment, it was argued, represented a new dimension to the problem of development and although each country had the sovereign right to exploit its natural resources, no society could avoid the costs involved in the destruction of its environment.[50] While it would be extremely difficult for developing countries to foreclose avenues for economic progress, Haq conceded, 'the choice is not so much between present and future, between development and environment, but it is essentially the choice of a certain pattern or style of development.'[51] This pattern of development supported government engagement in the planning process to minimize environmental disruption and safeguard human wellbeing. Environmental regulations, suggested Sachs, should be formulated and enforced by government agencies.[52]

This strategy for reconciling environment and development was decidedly anthropocentric. In fact, seminar papers and debate implicitly refuted the universalist approaches to global environmental management that were prevalent in the developed world. The environment mattered to Corea, Haq, Iglesias and Sachs, mainly to the extent that it impinged on human welfare.[53] Their approach to the development–environment question was informed by their attitude towards social matters in development. This belief, and the ideas that underlay it, were based on a different school of economic thinking than those advocating that developing countries should avoid environmental policies until they reached an acceptable level of development. Their rejection of laissez-faire, trickle-down theories of economic development in the social realm had been broadened to encompass an environment dimension.

While diverging economic philosophies informed two different appreciations of the development–environment relationship, several issues were subject to broad agreement at the Founex seminar. First, all participants believed that the industrialized countries must refrain from resorting to non-tariff trade barriers that could present new obstacles to developing country access to lucrative Western export markets. There was particular concern that additional production costs resulting from environmental regulation in the West would precipitate industrial pressure for remedial government actions, such as import quotas or subsidies.[54] The idea that polluting industries might become a source of foreign direct investment in developing countries was also discussed favourably in several seminar papers.[55] A second point of agreement was that developed countries should increase international development assistance. Many feared that environmental considerations could increase the costs of development, slow the pace of project implementation and become a formidable competitor for scarce resources that might have otherwise been earmarked for aid.[56] Third, all the working papers agreed that sovereignty over resources and environmental policies must not be compromised.

The Founex Report: A Manufactured Consensus

The main output of the Founex seminar was a report on the topic of development and environment.[57] Analysing the working papers and the seminar discussions that informed the Founex Report sheds light on the origins of its recommendations. The report argued that poor countries had a vital stake in environmental questions. However, it claimed that it was important to distinguish between the environmental problems of the developed and the developing world. While many environmental problems arose from industrial activities, important environmental problems were also caused by a lack of development.[58] The report stated that 'the major environmental problems of the developing countries are essentially of a different kind. They are problems, in other words, of both rural and urban poverty. In both the towns and in countryside, not merely the "quality of life", but life itself is endangered by poor water, housing, sanitation and nutrition, by sickness and disease and by natural disasters.'[59] The report argued that, because 'environmental quality is virtually synonymous with social welfare', international environmental cooperation should be structured with the dilemmas of developing countries in mind.

The environmental problems of nonindustrialized countries, the report maintained, were best resolved by the development process itself. Only rapid development could lift developing states out of poverty and enable them to tackle environmental despoliation. Although the report vigorously affirmed the right of developing countries to exploit their natural resources, it recommended that environmental objectives should be incorporated into development plans. Sound, farsighted planning was the key to addressing the environmental challenges facing the Global South. Development policies ought to embrace a wider set of priorities than GNP growth. Planning should be related directly to the goals and objectives of the UN's Second Development Decade.

The consensus among participants on issues of development assistance and trade was reflected in the Founex Report. The panel argued that if 'the concern for human environment reinforces the commitment to development, it must also reinforce the commitment to international aid' because resource limitations were a principal barrier to environmental protection in developing regions. The panel also endorsed the 'additionality' principle that increased costs of environmental measures in development should be met by the rich countries. On trade issues, it stressed that the potential neoprotectionist temptation to favour domestic industries must be resisted. If environmental policies led to changes in the distribution of global industries in industrialized counties and to new competitive advantages for nonindustrialized countries, this potentiality should be embraced. GATT, the report maintained, must be used to combat any undue trade discrimination or protectionism.

The extent to which the report privileged and adopted the causal beliefs of social interventionists demonstrates the control that this small clique of interventionist development economists had on the Founex process. The distinction made between developed and developing world environmental problems was inspired, to a considerable extent, by the arguments used by Iglesias. The report's attack on growth-oriented policies bore a close relationship to the ideas Haq advanced at his May 1971 presentation to the Society for International Development.[60] Sachs' enthusiasm for the integration of environmental considerations into the planning process was also reflected in the Founex panel's report.[61] Their views were so unambiguously integrated into the report because the final version was only loosely based on the actual seminar conclusions.

Notes from the meeting reflect a meandering and inconclusive debate that exposed the participants' divergent economic philosophies.[62] Those who did not agree with the conclusions of social interventionists did not change their point of view following the Founex seminar discussions. There were, as one participant recalled, 'few epiphanies' at Founex.[63] Beckerman, for example, the most vocal proponent of growth-focused development, left the meeting before its conclusion. Despite having his name on the published report, he claims not to have read the report until it was in its final form.[64] David Runnalls, another participant in the meeting, confirms this assessment: 'They had not really come to a proper consensus at the meeting. No amount of arm-twisting really brought a true consensus ... Eventually they just gave it to Mahbub [ul Haq] and he took it to New York and produced the Founex Report.'[65]

The Founex Report was therefore a manufactured consensus. Recognizing the urgent need for a decisive result, Haq did not attempt to closely reflect the meandering Swiss debate or the diverse contents of the Founex seminar papers in the final report. Instead, he composed a document that gave unambiguous expression to the views of interventionist development economists. The contents of the Founex Report suited the pre-existing intellectual agenda of Haq's professional and intellectual network. The environmental virtues of development planning were extolled, while laissez-faire prescriptions were treated with suspicion.

The Impact of the Founex Report on Institution Building

The Founex Report had a major impact on Stockholm institution building through processes of policy innovation, diffusion and selection. While it did not necessarily break new conceptual ground, the Founex Report contributed to policy innovation at the Stockholm Conference in three ways. First, by framing considerations of the development–environment link as a national choice reflecting different paths or approaches to development, it embedded the environment issue in broader theoretical debates on the role of social factors in development. The extent to which countries elected to address ecological exigencies was one for countries to evaluate in relation to their development priorities. Rather than an adherence to global environmental standards, there would be a two-speed vision of UN environmental cooperation. Secondly, by highlighting the array of environmental problems typically found in developing countries, the Founex Report affirmed the relevance of the Stockholm Conference for the Global South. Conceptions of 'the environment' were broadened to encompass issues often treated in the context of international development strategies, such as sanitation and soil erosion. Environment despoliation should, the report maintained, no longer be treated solely as a 'disease of the rich'. Third, by widening the international environmental policy agenda to explicitly encompass developing country concerns, Founex provided a rationale for a prominent role for the UN in multilateral environmental affairs.

The development economists also influenced events through policy diffusion. The Founex Report was circulated by the secretariat to all countries at the 1971 UN regional seminars on development and environment, held in Mexico City, Addis Ababa, Bangkok and Beirut.⁶⁶ At these meetings, the report served as a basis for discussion and helped developing country diplomats to familiarize themselves with its arguments. ⁶⁷ Members of the interventionist epistemic community, such as Corea and Iglesias, attended several of the regional meetings and discussed the contents of the report in person. Furthermore, after Founex, Maurice Strong 'went around to developing world leaders and stuck the report on their desks. He would point to it and say "these leading Third World economists said this"'.⁶⁸ Considering the fact that Strong travelled to over thirty developing countries during the Stockholm Conference preparatory process, this did much to spread the Founex ideas to developing country leaders. The report was formally included in its entirety as an appendix in the Stockholm Conference secretariat's discussion paper 'Development and Environment' circulated to all governments participating in the conference. It served as the basis of the conference secretariat's discussion paper on the topic and informed many of the recommendations that were considered by governments at Stockholm. Thus, the Founex Report's conclusions fed into the work and policies of governments through multiple channels.

The Founex seminar and report also influenced policy selection by creating a 'focal point' in the Stockholm action plan negotiations.⁶⁹ Since the institutional and conceptual framework for dealing with the development–environment issue was still relatively immature, the negotiating compromise embodied in the report represented a 'conspicuous' formula for the resolution of the cardinal dilemma of the Stockholm Conference preparatory process.⁷⁰ The divergent preferences of the parties, the paucity of mutually satisfactory solutions and the urgency imposed by the fast-approaching conference all combined to give the Founex formula great resonance in North–South talks. Leading developed countries immediately recognized the importance of the Founex Report for North–South bargaining. Canada's permanent UN mission, for instance, noted that it 'strikes us as being extremely important since it may form the main basis [for negotiations]'.⁷¹ Even leading negotiators from the most unreceptive and implacable of the developed countries, the United Kingdom, quietly regarded it as a 'reasonable' document, with recommendations that were 'much less contentious than most people had apparently expected'.⁷²

Perhaps most importantly, the Founex Report was welcomed by developing countries. Brazilian Ambassador Miguel Ozorio de Almeida described the conclusions of the Founex Report as 'well-balanced, very objective and in general endorsed by Brazil'. Bernardo de Azevedo Brito, Brazil's first secretary, argued that by following the recommendations of these 'world-renowned economists', countries could move beyond the 'initial period of over-reaction' to environmental questions.[73] Due to a change in the policy stance of Brazil, which had been a principal and vociferous opponent of the Stockholm Conference, the UN's Economic Commission for Latin America endorsed the Founex conclusions. At the September 1971 regional meeting of the UN's Economic Commission for Africa in Addis Ababa, participating countries unanimously supported the report.[74] As one observer reflected, Founex was 'regarded [by both developed and developing countries] as the political compromise necessary to bridge the development/environment question'.[75]

Once established as a basis for negotiations, offering value to all sides, the Founex compromise acquired a pull that made reformulation problematic, lest diplomats reopen messy, time-consuming bargaining. Despite having significant sway among developing countries, later Brazilian efforts at the third preparatory meeting to stretch developing country claims beyond the terms of the Founex paper received 'no support' from either developed or developing countries.[76] At Stockholm, Peter Stone recalls, 'diplomats eventually began to say things like, "I think we should stick closely to the text of the Founex Report here"' or, "We should not go beyond the position of the Founex Report".'[77] Countries from the North and South saw Founex as a suitable basis for agreement. Altering its recommendations would have risked an impasse.

At Stockholm, governments approved several key principles and recommendations based on the Founex Report. First, the Stockholm declaration and recommendations treated the environment as a component of wider development strategies, as stressed in the Founex Report. Environmental problems in the developing world were considered to be of a different kind from those found in industrialized countries and often could be solved by the development process itself. The Founex Report's interventionist thrust, grounded in socially motivated development approaches, is captured in point 4 of the Stockholm declaration's preamble. It stated that: 'In the developing countries most of the environmental problems are caused by under-development ... Therefore, the developing countries must direct their efforts to development, bearing in mind their priorities and the need to safeguard and improve the environment.' Principle 8 contained a similar point of emphasis. It noted that: 'Economic and Social development is essential for ensuring a favourable living and working environment for man and for creating conditions on earth that are necessary for the improvement of the quality of life.' The Founex Report was the crucial reference point for these formulations.[78]

Principle 9 of the declaration reflected the Founex Report's emphasis on the relationship between environmental deficiencies, under-development and vulnerability to natural disasters. It also stressed the importance of development assistance in meeting these challenges.[79] Principle 11 captured Founex's warnings concerning the potential rise of 'neo-protectionism' or other possible barriers to development.[80] The concept of additionality, given further weight by the Founex findings, was included in principle 12.[81] Just as in the Founex Report, the Stockholm declaration reflected the economic interventionists' great confidence in the ability of development planners to manage the delicate environment–development interrelationship in specific contexts.[82] Finally, the origins of principle 23, which provided a potential basis for developing country exemptions from international environmental standards, can also be traced to Founex.[83]

The Stockholm action plan was also heavily influenced by the Founex formula. Recommendations 103(a & b), 107 and 109 were particularly important to specifying the nature of North–South environmental relations. Recommendation 103(a) mirrored principle 11 of the Stockholm declaration in its condemnation of environmentally motivated trade restrictions. The corollary, controversial recommendation 103(b), suggested that states establish appropriate compensatory means of dealing with the potential detrimental effects of environmental policies on developing country exports. Recommendations 107 and 109 concerned the provision of additional international financing. The former recommendation proposed a study of suitable mechanisms for financing international environmental action. Recommendation 109 argued that environmental measures should not affect the flow of development assistance and that industrialized countries should meet any additional costs associated with integrating environmental questions in development schemes.

The fingerprints of interventionist development experts were therefore all over the Stockholm declaration and action plan. Institutional form, as Maurice Strong insisted during the Stockholm Conference process, followed function. UNEP's design closely reflected the

Stockholm action agenda. Many of the principles, rules and norms that formed the basis for UN environment cooperation after 1972 were thus an outgrowth of the Founex process. Furthermore, developing countries secured predominant representation on the UNEP Governing Council and the environmental agenda of the Global South had a prominent place in the UN environmental regime. UNEP's organizational form was a concrete result of the North–South environmental bargain of the Stockholm Conference. The position of the developing world was bolstered further when states voted in the UN General Assembly to locate UNEP's headquarters in Nairobi, Kenya.

Much of the Founex Report's power came from its unambiguous policy prescriptions. The report's arrival came at a pivotal moment in the Stockholm Conference's preparatory process, when governments were urgently seeking a solution to the main bargaining problem of the conference. A dearth of competing concepts – a condition deliberately fostered by the economic interventionists – ensured the salience of their proposed institutional alternative. Thus, in addition to the economists' perceived authoritativeness, their shrewd use of the opportunity provided by the fast-approaching conference is essential to accounting for their influence.

Conclusion

This chapter has demonstrated that a group of interventionist development economists heavily influenced institution building at the 1972 Stockholm Conference. With Barbara Ward's active intellectual and organizational support, Gamani Corea, Mahbub ul Haq, Enrique Iglesias and Ignacy Sachs collaborated to articulate a unified view about the nature and scale of environmental problems facing developing countries. Their strategic positioning within the Stockholm Conference preparatory process and their keen political acumen allowed them to shape institution building at Stockholm. Their approach to the problem was underpinned by an economic philosophy that called for robust government intervention to ensure that social factors, such as nutrition and housing, were not neglected in development. Because of the Founex meeting, this economic view was broadened to integrate an environmental dimension.

These economists' shared consensual knowledge was based on common analytical and methodological approaches. Their professional training and similar development experiences led them to validate knowledge in a similar fashion. As with their economic conclusions, the ideas expressed in the Founex Report had a strong normative dimension. For economic interventionists, environmental policies were an instrument for avoiding unnecessary environmental disruption and for improving the human condition.

The group's conclusions and diplomacy affected the Stockholm Conference's action recommendations and declaration through policy innovation, diffusion and selection. By integrating environmental issues within a wider social development framework, the Founex Report provided an intellectual and conceptual foundation for agreement at Stockholm. These ideas were spread through regional economic seminars and by the conference secretariat, gaining widespread endorsement by developed and developing countries. The compromise formula articulated in the Founex Report provided a basis for addressing the development–environment issue, serving as a powerful point of focus in Stockholm institution building.

These empirical findings broadly support the conclusions of scholars who argue that forces other than persuasive scientific evidence govern the entry of ideas into policy.[84] The process is highly complex – especially at the international level – and highly political. The interventionist development economists' activities and influence were linked closely to an entrepreneurial secretariat led by Strong that utilized their conclusions to solve a pressing political and intellectual problem.

Despite this tight relationship, there is good reason to conclude that the economic interventionists had a significant independent influence on conference outcomes. First, the fact that the epistemic community integrated environmental questions within its wider, pre-existing intellectual project suggests that community members seized and shaped the opportunity provided by North–South discord in 1971 to further their own intellectual agenda. They used the Founex Report to co-opt the Stockholm Conference secretariat and major governments at least as much as they were co-opted by them. Secondly, despite a similar level of secretariat support for the involvement and contributions of the International Council of Scientific Unions – discussed in Enora Javaudin's chapter – and the International Union for Conservation of Nature, the recommendations of these expert bodies were of less significance to the conference than those of the interventionist economists.[85] The principal difference between the development experts and these organizations appears to have been the political acumen of community members. Thirdly, though the power of epistemic community ideas rested on the willingness of governments to adopt them, the Founex Report constituted an irresistible focal point during subsequent North–South bargaining. It had a powerful pull in negotiations that, according to observers, was even able to overcome the desire of influential actors to stray from its propositions. Evidence of close epistemic community involvement with policy makers and of the fact that ideas were affected greatly by the political setting does not mean that epistemic communities' policy interventions were simply reactive or inconsequential. The reality is more complex. Research targeting epistemic community influence frequently requires empirical evaluations that can disentangle the precise character of the sometimes knotty relationships of epistemic communities with other major actors like international organizations and governments.

Michael W. Manulak is a postdoctoral fellow at the Balsillie School of International Affairs, University of Waterloo, Canada.

Notes

1. Lars-Göran Engfeldt, *From Stockholm to Johannesburg and Beyond: The Evolution of the International Systems for Sustainable Development Governance and its Implications* (Stockholm: Government Offices of Sweden, 2009), 60.
2. Christian A. Herter Jr. and Jill E. Binder, *The Role of the Secretariat in Multilateral Negotiation: The Case of Maurice Strong and the 1972 UN Conference on the Human Environment* (Washington DC: Johns Hopkins Foreign Policy Institute: 1993), 8.
3. Peter B. Stone, *Did We Save the Earth at Stockholm? The People and Politics in the Conference on the Human Environment* (London: Earth Island, 1973), 102.
4. Peter M. Haas, 'Introduction: Epistemic Communities and International Policy Coordination'. *International Organization* 46(1) (1992): 3.
5. Emanuel Adler and Peter M. Haas, 'Conclusion: Epistemic Communities, World Order, and the Creation of a Reflective Program', *International Organization* 46(1) (1992): 375.
6. Donella H. Meadows et al., *The Limits to Growth* (New York: Universe Books, 1972); Paul R. Ehrlich, *The Population Bomb* (New York: Ballantine Books, 1968); Edward Goldsmith et al., *A Blueprint for Survival* (Boston, MA: Houghton Mifflin, 1972); Rachel Carson, *Silent Spring* (Boston, MA: Houghton Mifflin, 1962). The *Torrey Canyon* foundered, dumping about 120,000 tonnes of crude oil off of the coast of Cornwall, England.
7. P.R. Ehrlich and A.H. Ehrlich, *Population, Resources, Environment* (San Francisco: W.H. Freeman, 1970), 295–319.
8. George F. Kennan, 'To Prevent a World Wasteland', *Foreign Affairs* 48(3) (1970): 401–13; Kingman Brewster Jr., 'Reflections on our National Purpose',

Foreign Affairs 50(3) (1972): 399–415; Leroy S. Wehrle, 'Affluence and the World Tomorrow', *Foreign Affairs* 49(3) (1971): 419–28.

9. Herter and Binder, *The Role of the Secretariat*, 8.
10. Lee M. Talbot, 'A Remarkable Melding of Contrasts and Conflicts', *Uniterra* 1 (1982): 9.
11. Mahbub ul Haq, *The Poverty Curtain: Choices for the Third World* (New York: Columbia University Press, 1976), 84–86.
12. Ibid., 125.
13. Miguel Ozorio de Almeida, 'The Confrontation between Problems of Developed and Environment', in *Environment and Development: The Founex Report* (New York: Carnegie Endowment for Peace, 1972), 45.
14. Herter and Binder, *The Role of the Secretariat*, 9.
15. United Nations General Assembly (hereinafter UNGA), *Resolution 2657 United Nations Conference on the Human Environment*, 7 December 1970.
16. Engfeldt, *From Stockholm to Johannesburg*, 56.
17. Wade Rowland, *The Plot to Save the World: The Life and Times of the Stockholm Conference on the Human Environment* (Toronto: Clark, Irwin & Company Ltd, 1973), 47.
18. Herter and Binder, *The Role of the Secretariat*, 11.
19. Maurice Strong interview, 20 August 2010.
20. United Nations General Assembly, *Preparatory Committee for the United Nations Conference on the Human Environment, Second Session, Geneva, 8–19 February 1971*, A/CONF.48/PC/9, 26 February 1971.
21. Maurice Strong, *Where on Earth are We Going?* (Toronto: Vintage Canada, 2001), 122.
22. Strong, *Where on Earth*, 123–24.
23. Ibid., 123.
24. Gamani Corea, *My Memoirs* (Ratmalana: Gamani Corea Foundation, 2008), 339.
25. 'Memorandum: Panel of Experts on Development and Environment', 31 March 1971, Strong Papers, Harvard University Environmental Science and Public Policy Archives (hereinafter ESPP), Box 37, Folder 371.
26. Clyde Sanger, 'Environment and Development', *International Journal* 28(1) (1972/1973), 106. Corea, *My Memoirs*, 336.
27. UNGA, *Development and Environment: Subject Area V*, Annex I, 26.
28. David Satterthwaite, *Barbara Ward and the Origins of Sustainable Development* (London: International Institute for Sustainable Development, 2006), 15.
29. Strong, *Where on Earth*, 123–24; Corea, *My Memoirs*, 339.
30. Salvatore Schiavo-Campo and Hans W. Singer, *Perspectives of Economic Development* (Boston, MA: Houghton Mifflin, 1970), 70.
31. Haq, *The Poverty Curtain*, 28–47.
32. Gamani Corea, 'Development Strategy and Environment Issue', in *Environment and Development: The Founex Report* (New York, Carnegie Endowment for Peace, 1972), 79–80.

33. Ignacy Sachs, 'Environment and Development', in *Development and Environment: Report and Working Papers of a Panel of Experts Convened by the Secretary-General of the United Nations Conference on the Human Environment (Founex, Switzerland, 4–12 June 1971)* (Paris: Mouton, 1972), 125.
34. World Bank Group, 'Robert Strange McNamara', *World Bank Archives Website*, http://www.worldbank.org/en/about/archives/history/past-presidents/robert-strange-mcnamara, accessed 5 July 2016.
35. Haq, *The Poverty Curtain*, 27–28.
36. David Runnalls interview, 13 August 2010.
37. Stone, *Did We Save the Earth*, 112.
38. Haq, *The Poverty Curtain*, 79–98.
39. Almeida, 'Economic Development', 112–14.
40. Strong, 5 June 1971, Strong Papers, ESPP, Box 37, Folder 371.
41. Wilfred Beckerman interview, 11 June 2010.
42. Strong, 5 June 1971, Strong Papers, ESPP, Box 37, Folder 371.
43. Wilfred Beckerman, 'Economic Development and the Environment: A False Dilemma', in *Environment and Development: The Founex Report* (New York: Carnegie Endowment for Peace, 1972), 61–62.
44. General Agreement on Tariffs and Trade (hereinafter GATT), 'Industrial Pollution Control and International Trade', in *Development and Environment: Report and Working Papers of a Panel of Experts Convened by the Secretary-General of the United Nations Conference on the Human Environment (Founex, Switzerland, 4–12 June 1971)* (Paris: Mouton, 1972), 204.
45. Strong, 5 June 1971, Strong Papers, ESPP, Box 37, Folder 371.
46. Ibid.
47. Enrique Iglesias, 'Development and the Human Environment', Strong Papers, ESPP, Box 40, Folder 395.
48. Corea, 'Development Strategy and Environment Issue', 79–80.
49. Ignacy Sachs, 'Environment and Development', Strong Papers, ESPP, Box 40, Folder 395.
50. Enrique Iglesias, 'Development and the Human Environment', Strong Papers, ESPP, Box 40, Folder 395.
51. Haq, *The Poverty Curtain*, 108.
52. Ignacy Sachs, 'Environmental Concern and Development Planning', in *Environment and Development: The Founex Report* (New York: Carnegie Endowment for Peace, 1972), 74.
53. Haq, *The Poverty Curtain*, 108.
54. GATT, 'Industrial Pollution Control', 215.
55. United Nations Conference on Trade and Development (hereinafter UNCTAD), *Environment and Development, Development and Environment: Report and Working Papers of a Panel of Experts Convened by the Secretary-General of the United Nations Conference on the Human Environment (Founex, Switzerland, 4–12 June 1971)* (Paris: Mouton, 1972), 193; GATT, 'Industrial Pollution Control', 209.
56. UNCTAD, 'Environment and Development', 195–98.

57. UNGA, *Development and Environment: Subject Area V*, Annex I.
58. Ibid., 9.
59. Ibid., 3–4, 8–9, 17, 27–34.
60. Haq, *The Poverty Curtain*, 28–47.
61. Compare, for example, Sachs, 'Environment and Development'. 127, 130, and 134 with UNGA, 'Development and Environment', 6, 13, and 19.
62. Strong, 5 June 1971, Strong Papers, ESPP, Box 37, Folder 371.
63. David Runnalls interview, 13 August 2010.
64. Wilfred Beckerman interview, 11 June 2011.
65. David Runnalls interview, 13 August 2010; Haq, 'The Poverty Curtain', 81.
66. 'Memorandum: Panel of Experts on Development and Environment', 31 March 1971, Strong Papers, ESPP, Box 37, Folder 371.
67. United Nations General Assembly, *Preparatory Committee for the United Nations Conference on the Human Environment, Second Session, Geneva, 8–19 February 1971*, A/CONF.48/PC/9, 26 February 1971.
68. David Runnalls interview, 13 August 2010.
69. Geoffrey Garrett and Barry R. Weingast, 'Ideas, Interests, and Institutions: Constructing the EC's Internal Market', in. *Ideas and Foreign Policy: Beliefs, Institutions, and Political Change*, Judith Goldstein and Robert O. Keohane (eds) (Ithaca: Cornell University Press, 1993).
70. Thomas C. Schelling, *The Strategy of Conflict* (Cambridge, MA: Harvard University Press, 1960), 57.
71. PERMIS-NY to USSEA, 17 August 1971, *Library and Archives Canada*, 68-4-UN-1972, Volume 10.
72. Kitching to Williams, 27 September 1971, The National Archives. FCO55/837/SME2/522/1.
73. CANEMB-Mexico to DEA-ECS, 7 September 1971, *Library and Archives Canada*, 68-4-UN-1972, Volume 11.
74. United Nations Information Service, *Economic Commission for Africa: Addis Ababa Press Release*, ECA/545, 2 September 1971.
75. David Runnalls interview, 13 August 2010.
76. Kitching to Mathieson and Williams, 24 September 1971, The National Archives. FCO55/837/SME2/522/1.
77. Stone, *Did We Save the Earth*, 102.
78. Louis B. Sohn, 'The Stockholm Declaration on the Human Environment', *Harvard International Law Journal* 14 (1973): 423–515.
79. This point corresponds to the Founex Report, paragraphs 3, 4, and 8.
80. See Founex Report, paragraph 54. For analysis, see Sohn, 'The Stockholm Declaration,' 469.
81. See Founex Report, paragraphs 28–33.
82. This was especially so with principle 13. For an analysis, see Engfeldt, *From Stockholm to Johannesburg*, 64.
83. See, for instance, Founex Report, paragraphs 20 and 33.
84. Karen Liftin, *Ozone Discourses: Science and Politics in Global Environmental Cooperation* (New York: Columbia University Press, 2004); Steven Bernstein,

The Compromise of Liberal Environmentalism (New York: Columbia University Press, 2001).

85. Lynton Keith Caldwell, *International Environmental Policy: Emergence and Dimensions* (Durham, NC: Duke University Press, 1990), 44–45.

Bibliography

Adler, Emanuel and Peter M. Haas, 'Conclusion: Epistemic Communities, World Order, and the Creation of a Reflective Research Program', *International Organization* 46(1) (1992): 367–90.

Beckerman, Wilfred, 'Economic Development and the Environment: A False Dilemma', in *Environment and Development: The Founex Report* (New York: Carnegie Endowment for Peace, 1972), 61–62.

——. 'Interview with Michael Manulak', 11 June 2010.

Bernstein, Steven, *The Compromise of Liberal Environmentalism* (New York: Columbia University Press, 2001).

Brewster, Kingman Jr., 'Reflections on Our National Purpose', *Foreign Affairs* 50(3) (1972): 399–415.

Caldwell, Lynton Keith, *International Environmental Policy: Emergence and Dimensions* (Durham, NC: Duke University Press, 1990).

CANEMB-Mexico to DEA-ECS. 7 September 1971. *Library and Archives Canada*, 68-4-UN-1972, Volume 11.

Carson, Rachel, *Silent Spring* (Boston, MA: Houghton Mifflin, 1962).

Corea, Gamani, 'Development Strategy and Environment Issue', in *Environment and Development: The Founex Report* (New York, Carnegie Endowment for Peace, 1972), 79–80.

Corea, Gamani, *My Memoirs* (Ratmalana: Gamani Corea Foundation, 2008).

Ehrlich, Paul R. and A.H. Ehrlich, *Population, Resources, Environment* (San Francisco: W.H. Freeman, 1970).

Ehrlich, Paul R., *The Population Bomb* (New York: Ballantine Books, 1968).

Engfeldt, Lars-Göran, *From Stockholm to Johannesburg and Beyond: The Evolution of the International Systems for Sustainable Development Governance and its Implications* (Stockholm: The Government Offices of Sweden, 2009).

Garrett, Geoffrey and Barry R. Weingast, 'Ideas, Interests, and Institutions: Constructing the EC's Internal Market', in *Ideas and Foreign Policy: Beliefs, Institutions, and Political Change*, Judith Goldstein and Robert O. Keohane (eds) (Ithaca: Cornell University Press, 1993), 173–206.

General Agreement on Tariffs and Trade (GATT), 'Industrial Pollution Control and International Trade', in *Development and Environment: Report and Working Papers of a Panel of Experts Convened by the Secretary-General of the United Nations Conference on the Human Environment (Founex, Switzerland, 4–12 June 1971)* (Paris: Mouton, 1972), 204.

Goldsmith, Edward et al., *A Blueprint for Survival* (Boston, MA: Houghton Mifflin, 1972).

Haas, Peter M. 'Introduction: Epistemic Communities and International Policy Coordination', *International Organization* 46(1) (1992): 1–35.

Haq, Mahbub ul, *The Poverty Curtain: Choices for the Third World* (New York: Columbia University Press, 1976), 84–86.

Herter, Christian A. Jr. and Jill E. Binder, *The Role of the Secretariat in Multilateral Negotiation: The Case of Maurice Strong and the 1972 UN Conference on the Human Environment* (Washington DC: Johns Hopkins Foreign Policy Institute, 1993).

Iglesias, Enrique, 'Development and the Human Environment', Strong Papers, ESPP, Box 40, Folder 395.

Kennan, George F., 'To Prevent a World Wasteland', *Foreign Affairs* 48(3) (1970): 401–13.

Kitching to Mathieson and Williams, 24 September 1971. The National Archives, UK, FCO55/837/SME2/522/1.

Kitching to Williams, 27 September 1971. The National Archives. FCO55/837/SME2/522/1.

Litfin, Karen, *Ozone Discourses: Science and Politics in Global Environmental Cooperation* (New York: Columbia University Press, 2004).

Meadows, Donella H. et al., *The Limits to Growth* (New York: Universe Books, 1972).

Memorandum: Panel of Experts on Development and Environment, 31 March 1971, Strong Papers, Harvard University Environmental Science and Public Policy Archives (hereinafter ESPP), Box 37, Folder 371.

Ozorio de Almeida, Miguel, 'The Confrontation between Problems of Developed and Environment', in *Environment and Development: The Founex Report* (New York: Carnegie Endowment for Peace, 1972), 45.

PERMIS-NY to USSEA. 17 August 1971. *Library and Archives Canada*, 68-4-UN-1972, Volume 10.

Rowland, Wade, *The Plot to Save the World: The Life and Times of the Stockholm Conference on the Human Environment* (Toronto: Clark, Irwin & Company Ltd, 1973).

Runnalls, David, 'Interview with Michael Manulak', 13 August 2010.

Sachs, Ignacy, 'Environment and Development', in *Development and Environment: Report and Working Papers of a Panel of Experts Convened by the Secretary-General of the United Nations Conference on the Human Environment (Founex, Switzerland, 4–12 June 1971)* (Paris: Mouton, 1972), 125.

———. 'Environment and Development', Strong Papers, ESPP, Box 40, Folder 395.

———. 'Environmental Concern and Development Planning', in *Environment and Development: The Founex Report* (New York: Carnegie Endowment for Peace, 1972), 74.

Sanger, Clyde, 'Environment and Development', *International Journal* 28(1) (1972/1973): 103–20.

Satterthwaite, David, *Barbara Ward and the Origins of Sustainable Development* (London: International Institute for Sustainable Development, 2006).

Schelling, Thomas C., *The Strategy of Conflict* (Cambridge, MA: Harvard University Press, 1960).

Schiavo-Campo, Salvatore and Hans W. Singer, *Perspectives of Economic Development* (Boston, MA: Houghton Mifflin, 1970).

Sohn, Louis B., 'The Stockholm Declaration on the Human Environment', *Harvard International Law Journal* 14 (1973): 423–515.

Stone, Peter B., *Did We Save the Earth at Stockholm? The People and Politics in the Conference on the Human Environment* (London: Earth Island, 1973).

Strong, Maurice, *Where on Earth are We Going?* (Toronto: Vintage Canada, 2001).

———. 'Interview with Michael Manulak', 20 August 2010.

Talbot, Lee M. 'A Remarkable Melding of Contrasts and Conflicts', *Uniterra* 1 (1982): 9.

UNGA, *Development and Environment: Subject Area V*, Annex I, 26.

United Nations Conference on Trade and Development (UNCTAD), *Environment and Development, Development and Environment: Report and Working Papers of a Panel of Experts Convened by the Secretary-General of the United Nations Conference on the Human Environment (Founex, Switzerland, 4–12 June 1971)* (Paris: Mouton, 1972).

United Nations General Assembly (UNGA), *Resolution 2657 United Nations Conference on the Human Environment*, 7 December 1970.

———. *Preparatory Committee for the United Nations Conference on the Human Environment, Second Session, Geneva, 8–19 February 1971*, A/CONF.48/PC/9, 26 February1971.

United Nations Information Service, Economic Commission for Africa: *Addis Ababa Press Release*, ECA/545, 2 September 1971.

Wehrle, Leroy S., 'Affluence and the World Tomorrow', *Foreign Affairs* 49(3) (1971): 419–428.

World Bank Group, 'Robert Strange McNamara', *World Bank Archives Website*, http://www.worldbank.org/en/about/archives/history/past-presidents/robert-strange-mcnamara, accessed 5 July 2016.

CHAPTER 4

Only One Earth

The Holy See and Ecology

Luigi Piccioni

On the closing day of the 1972 Stockholm Conference, the final declaration, the official resolutions and recommendations, and the joint declaration by the international non-governmental organizations (INGOs) and their forums, were read from the podium. The most important informal INGO convention meeting outside the Folkets Hus, but recognized by the United Nations (UN), was the Environment Forum. According to one witness, it 'had a certain official standing, but was cut off from the official conference attended by the national delegations. It has been argued that the purpose of the Environment Forum was to provide a real forum for discussion of environmental issues, but also to keep ecologists occupied and not disturbing the official Conference'.[1]

In spite of this, the INGOs' declaration[2] was read in plenary by Barbara Ward, one of the pivotal figures in the preparation and proceedings of the conference.[3] Ward's surprising attitude – a leader behaving, at the same time, as an insider and an outsider – was similar to that of the delegation to which she felt closest: that of the Holy See. Like Ward, in fact, the Vatican was very active at the conference. It was close both to the developed countries that had promoted the summit and to the countries from the Global South and INGOs that were critical towards the summit's philosophy and goals. This apparent ambivalence was not due to the lack of a clear position. Rather, it stemmed from a complex vision of the environmental issue – a vision intimately connected with other issues such as the economic development of the so-called Third World, human rights, peace and disarmament, and education.

The interest of the Holy See and its different structures in addressing environmental concerns is usually dated to Pope John Paul II's intervention, especially in the second half of the 1980s,[4] but the origins of this concern really lie in the early 1970s.[5] While the historical literature on the Catholic Church and ecology is generally limited, to date none has discussed this first stage of the 1960s and early 1970s. The few existing general references to this period in books focused on the more recent past are broadly correct, but superficial and based exclusively on quotations from official documents issued by the Catholic Church such as encyclicals and public messages.

In other words, the way in which the Holy See dealt with the environmental issue before John Paul II remains unexplored territory. Precious but rare and indirect evidence can be traced in some biographical studies.[6] The fact that the Vatican archives so far limit access to pre-1939 documents makes it challenging to carry out research in this field. In an attempt to compensate for this, however, this chapter draws on the rich private archive of one of the members of the Holy See delegation, Giorgio Nebbia, who was nominated consultant of the Justice and Peace Pontifical Commission in June 1972, immediately after the Stockholm Conference.[7] Alongside these private archival sources, oral history interviews with key actors at the time of the Stockholm Conference can also shed light on the Holy See's attitudes to and policies on environmental degradation and linked development issues.

The Catholic Church was then, and still is, a very peculiar institution – at the same time a small sovereign state and a worldwide organization. This chapter will trace why and how it became an important and active player in the Stockholm Conference; how, conversely, the summit was a crucial incentive for the Catholic Church to take on the issue of environment from both a theological and political viewpoint; how the Holy See's official position at the conference was elaborated and who contributed most to this process; and, finally, what the long-term consequences of the Stockholm Conference have been on the Church and on its internal organization.

From the Vatican Council to the Justice and Peace Pontifical Commission

Until the last weeks of 1970, the environment was the great absentee among the important social issues that the Catholic Church had been considering and actively working on over the previous decade. In fact, the Church's statements up to 1970 contained no reference to ecology.[8] The Pope, Church hierarchies and Catholic scholars had shown no interest at all in this new crucial issue. At the beginning of 1970, the number of articles published all over the world by representatives of the Catholic Church about the theology–environment relationship could be counted on one hand. In contrast, Protestant scholars had already published several successful and influential books.[9] The reasons for this Catholic indifference are not easily determined. However, three elements seem to have played a role, two of them mainly philosophical and the third of a more political character.

First, the Catholic view of the universe, which had been reaffirmed in the documents emanating from the Second Vatican Council from 1962 to 1965[10] and in the papal encyclicals, was firmly anthropocentric. Man – as God's image on earth – was considered the peak and ultimate goal of creation; all things on earth had been created to support his existence.[11] The Second Vatican Council articulated this traditional position once more.[12] Such a conceptualization of life on earth made it difficult, however, to connect with a world vision often based on a reflection about the equal dignity of all things: humans, animals and plants.[13]

Secondly, the Vatican Council's attitude towards science and technology shared a widespread amazement about and hope concerning their spectacular progress,[14] an attitude that could also make addressing environmental issues more difficult. Finally, most of the technical and political proposals of the 1960s regarding the resolution of the environmental crisis on a global scale implied some form of birth control, since continued unchecked growth of the world population was considered one of the main factors of growing ecological imbalances.[15] This point was very important for the Catholic hierarchy. The Vatican Council had just reaffirmed the Church's absolute opposition to every form of birth control. Moreover, Pope Paul VI believed that it was necessary to clarify the Church's position on the subject further by issuing a specific encyclical.[16]

Thus, the Catholic Church was not among the minority of institutions and of public opinion fully aware of the growing environmental problems in the 1960s. At the same time, however, the Vatican Council's progressive theological and social preferences made it a potentially important issue to tackle as part of Catholic engagement in the contemporary world. The Vatican Council provoked a profound change in the Church's social vision. This great ecclesiastic global summit, the first in ninety years, had been unexpectedly convened in January 1959 by the newly elected Pope, John XXIII, to renew the Church. The Pope regarded such a renewal as indispensable for a Church that had remained immobile for too long in the face of rapid economic, social and political transformations. He believed that the Church had to thoroughly change its own structure, its way of working and its attitude towards what it considered the 'outer world' if it wanted to preserve the strength and appeal of what it regarded as its eternal truths.

At the time, Pope John XXIII's vision was not entirely novel, and he was not alone in advocating it. Several theologians and ecclesiastical movements – mainly from Northern Europe – had been pushing throughout the twentieth century for changes in the direction of a more collegial, pastoral, ecumenical and social-minded Church. In spite of this, Pius XII's very conservative papacy from 1939 to 1958 had shaped a Church that, generally speaking, was both indifferent to and unprepared for change. Not surprisingly, therefore, the Roman Curia was largely hostile to John XXIII's programme. Charged with the Council's preparation, it tried in different ways to undermine its proceedings and impact. Nonetheless, the Council's long preparation generated great expectations of change. When the roughly three thousand cardinals, bishops and experts gathered in October 1962, a very strong and stable ten-to-one majority for the full realization of John XXIII's programme had formed. This majority was able to overcome the Roman Curia's obstacles to renewal, prevent a too precocious closing of the Council, elect a new Pope not significantly different in outlook from John XXIII when he died in 1963 and approve sixteen important documents.[17] Alongside some papal encyclicals,[18] these would implement an effective (although not irreversible) renewal of the Catholic Church, its internal operations, culture and attitude towards the modern world.

A turning point in the Council's proceedings was the proposal advanced in plenary by the Belgian cardinal Léon-Joseph Suenens on 4 December 1962, after consultation with the Pope and a small group of influential conciliar Fathers. According to Suenens, it was indispensable to give the Council agenda a more organic orientation by adopting a unifying theme, thus going beyond the preparatory documents drafted by the Curia that lacked organization and cohesion. As John XXIII had stated just before the Council opening,[19] this unifying theme should be 'the church of Christ, light to the world'. Suenens went further, however, by proposing the drafting of two distinct documents: the first about the internal reality of the Church and the second about the relationship between the Church and the outer world. Thus, less than two months after the opening of the Council, the assembly enthusiastically welcomed a proposal that was clearly aimed to orient its works towards reforming the Church quite fundamentally. The main consequence of this reorientation was the decision to draft a specific constitution not envisaged in the Council's original programme. This change allowed the issues of peace, social justice and human dignity to become central to the Council agenda.

This event was particularly welcomed by the many conciliar Fathers who had taken part in the formation of a single-issue group focused on poverty.[20] This group, created at the opening of the Council,[21] addressed poverty as an issue for both Catholics and the Church, and it demanded the Church's engagement with the poor on a global scale. The group was in close contact with Suenens and soon became known as the 'Church of the Poor'. Some fifty cardinals, bishops and theologians participated in its work, most of them from France and South America.[22] Many became highly charismatic crucial actors in the Council proceedings: the Italian Cardinal Giacomo Lercaro, the Patriarch of Antioch Maximos IV Sayegh, the bishops Helder Camara and Georges Mercier and the theologians Yves Congar, Bernard Häring and Marie-Dominique Chenu.

Alongside the Church of the Poor, another group soon emerged that was interested in the connections between economy and peace. It was made up mainly of personalities of the international Pax Christi organization.[23] By the end of 1963, finally, a third group established itself. This was a small team of five Anglo-Americans, three clerics, a layman and a laywoman, all working in the fields of poverty relief and economic development and deeply interested in the laity's role in the Church. They too maintained very close relations with some important leaders of the Council's majority such as cardinals Suenens and Giovan Battista Montini. They wanted to reach a simple but very important goal: the creation of 'a specialized unit of the Roman Curia that would focus the Church's efforts in bringing the message of the Gospel to situations of poverty, hunger and exclusion'.[24] Unlike the Church of the Poor, however, they aimed to 'influence the Council in a significant way according to non-direct means of intervention', so much so that they adopted the name of 'conspirators'. Among the founders of the group, two members of the American Catholic Relief Service, James J. Norris and Louis Gremillion, stood out, as well as the British economist, journalist and writer Barbara Ward. They were actively supported, amongst others, by the auxiliary Bishop of New York and Catholic Relief Service Director, Edward Swanstrom, the renowned German theologian Häring, who was also a member of the Church of the Poor, and the

French priest-economist Louis-Joseph Lebret. Many of these individuals would play a crucial role in the writing of the Council's constitution *Gaudium et spes: On the Church in the Modern World* and – in the case of Lebret – of Paul VI's encyclical *Populorum progressio*.

These two documents were among the most visible and innovative results of the conciliar turning point. The Council, in fact, made profound innovations in several fields – liturgy, the relationship between Holy Scripture and theology, ecumenism, Church structure and organization. However, in the political and cultural context of the 1960s, the Church's opening to modernity and its claim of an active engagement for peace, social and human rights, liberty and human dignity had even greater appeal.

In the timespan of only seven years, therefore, John XXIII's encyclicals *Mater et magistra* (1961) and *Pacem in terris* (1963), the conciliar constitution *Gaudium et spes* (1965) and Paul VI's encyclical *Populorum progressio* (1967) redefined both the Church's position towards modernity and its social doctrine. The Church had transformed its attitude of isolation and suspicion towards modernity – dating back to the French Revolution[25] – to resolutely become an active player in a dramatically evolving world. From this perspective, all Catholics should be actively engaged in the resolution of the great problems affecting the modern world (poverty, wars, denial of human rights and dignity, cultural deprivation) in a spirit of openness and dialogue with all men of good will, whatever religion or political creed they professed.

Both the action of the Church of the Poor and of the so-called conspirators contributed significantly to the drafting of *Gaudium et spes*, passed on 8 December 1965, the last day of the Council, with an overwhelming majority of 2,309 to seventy-five. In particular, following the conspirators' recommendations, the conciliar Fathers clearly stated that the Church could not limit itself to proclaiming its engagement with the modern world, but had to actively practise it. This is the reason why paragraph 90 of the encyclical envisaged the creation of an 'organism of the Universal Church ... in order that both justice and the love of Christ towards the poor might be developed everywhere'. This new organism should 'stimulate the Catholic community to promote progress in needy regions and international social justice'.[26]

In January 1967, one year after the closing of the Vatican Council, Paul VI finally inaugurated the institution invoked by *Gaudium et spes*, endowing it with three key features. The new institution was in fact created in the form of a pontifical commission; it was committed mainly to a worldwide application of the *Populorum progressio* encyclical guidelines,[27] and it included in its name the goal of peace. The Justice and Peace Commission started its activities on a global scale in April 1967 with a board of fifteen members, both clergymen and laymen from all continents, aided by a group of twelve so-called consultors (consultants) and three or four clerks working in its headquarters in Rome.[28] In spite of this lean structure, the Commission had national branches everywhere in the world and often at diocesan level too. Justice and Peace thus became the international organization through which not only the Holy See as a state, but also the Catholic Church as a global community addressed a vast array of social issues.

Considering the many international initiatives of the Commission from 1967 to 1971,[29] it is possible to distinguish at least eleven main areas of intervention – development, peace, education, human promotion, ecumenism, social justice, cooperation, human rights, demography, racism and technological change – as well as many other secondary issues. Some of these areas were identified as strategically important from the start, but many of them entered the Commission's agenda only successively, thanks both to internal debate and external stimuli. Indeed, the mission of Justice and Peace was not defined by a well-delineated list of issues. Rather, it was shaped by an open agenda, intended to include any global problem whose resolution could improve the wellbeing and the dignity of human beings.

Moreover, Justice and Peace maintained contacts and collaboration with a large number of other Catholic institutions, ecumenical bodies, INGOs and international organizations such as the UN, the United Nations Educational, Scientific and Cultural Organization (UNESCO), the World Health Organization and the World Bank.[30] Barbara Ward played a pivotal role in the cooperation with international organizations. She was not only a member of the Pontifical Commission, but also advisor of the American President Lyndon B. Johnson, the UN Secretary-General Sithu U Thant and the board of the World Bank.

In the first stages of the preparation of the Stockholm Conference, U Thant faced several difficulties and came to rely on Ward. It was Ward, for example, who helped U Thant in appointing the Canadian Maurice Strong as secretary of the conference.[31] Her report, *Only One Earth*, was published on the eve of the conference.[32] Moreover, as Michael W. Manulak shows in his chapter in this book, in the spring of 1971, Ward, together with Mahbub ul Haq, Gamani Corea, Arthur Lewis and Enrique Iglesias, set the agenda for the crucial Founex seminar of development economists conceived by Strong for bridging environmental needs and Third World development concerns.

The Holy See's Quest for the Environment

In spite of the intensifying global debate, however, the Catholic Church's internal resistance to considering the environmental aspect of the world crisis persisted into the second half of 1970. Moreover, the Church's sudden change of attitude towards the end of 1970 manifested itself mainly through two events that were largely if not entirely independent of Ward's action.

The first event was the publication of two unusual and innovative articles in the Jesuits' fortnightly journal *La Civiltà Cattolica*, the most influential voice in the Catholic press, which is checked and revised by the State Secretariat of the Holy See before publication. Both were written by Father Bartolomeo Sorge, a close collaborator of the Pope on social issues, who contributed to drafting the Apostolic letter *Octogesima adveniens* (1971) and became the journal's editor in 1973. The first article was a reflection on the meaning and implications of technological society considered by Sorge from the perspective of a more radical 'Christian criticism'.[33] The second article, published at the beginning of December 1970, was a wide-ranging, well-informed and reasoned essay on the ecological crisis, considered as 'a problem of conscience and culture'.[34] In fact, the article amounted to a commentary on an even more important novelty: the papal discourse of 16 November 1970 at the twenty-fifth anniversary celebration at the headquarters of the UN's Food and Agriculture Organization (FAO) in Rome, which Sorge had most likely drafted for the Pope. Here, Paul VI had, for the first time, addressed ecology as a global issue.[35]

The second circumstance arose from the necessity for the Holy See as a sovereign state to comply with the diplomatic duties imposed by the preparation of the Stockholm Conference. The most important of these duties was to answer a questionnaire about the possible form and contents of the conference's final declaration.[36] Invited in December 1970 to draw up and send two different documents, a shorter memo and a more complete report, the Holy See finally had to face the environmental issue as a cultural and political priority.

The eight-point questionnaire posed a serious challenge to the Holy See. Following his engagement with the issue in the Jesuit journal, Father

Sorge was charged with forming a small working group made up of both clergy and laymen.[37] Within two months the group defined an official position of the Holy See and the Church at large regarding the global environmental crisis. Their report depended heavily on the drafts prepared by Giorgio Nebbia, a professor of merceology (science of tradable goods) at Bari University who was one of the Italian pioneers of environmentalism. Nebbia's scientific competence combined with his close relationships with several leading figures in world environment, such as Barry Commoner and Aurelio Peccei, also had a strong conciliar inspiration.[38] This is why the group's two documents – the short memo and the long report – were scientifically sound and pervaded by an intense spiritual and social concern.[39]

Eventually, three major documents constituted the Vatican's official contribution to the Stockholm Conference. The first was the report concerning the final declaration, edited by the Sorge group, revised by the State Secretariat and finally signed by the permanent observer of the Holy See to the UN, the Dominican Father Henri De Riedmatten.[40] The second was the Pope's message to the delegates, drafted by Sorge, considerably revised by the Pope and read at the opening of the conference.[41] The third was De Riedmatten's speech in the plenary session of 7 June 1972.[42] Only the official report retained parts of the Sorge group's work, while other sections, such as those about population issues and the request to limit consumption in developed countries, were left out. As a result, the final documents were politically less radical and put stronger emphasis on the theological aspects.

Due to the lack of suitable Church structures specifically dedicated to the environment, the Sorge group's ad hoc work influenced the Vatican's official position in the early stages. In the long term, however, the environmental issue had to be addressed by the Justice and Peace Commission as a part of its mission. During 1971 'a whole host of issues' came 'to the Commission's attention concerning the "high-consumption" societies of the West and of Japan … under the popular name of "pollution of environment" [or] "ecological imbalance"'. But it initially failed to organize a working group to address them.[43] In the meantime, some ecological concerns filtered into two very important Vatican documents: Paul VI's apostolic letter *Octogesima adveniens* and the final document of the bishops' second synod, entitled *Justice in the World*, issued in November 1971.[44]

Acting as interlocutor between the Sorge group and Justice and Peace was Father Philip Land, an American Jesuit who specialized in economic development issues and was a member of both the small team and the Pontifical Commission. However, it took Land until March 1972 to arrange a six-person meeting – including Giorgio Nebbia and the Justice and Peace Secretary Joseph Gremillion – about the issues to be discussed in Stockholm within three months. The participants decided to organize a seminar on the UN conference with Nebbia, the Jesuit Robert Faricy, De Riedmatten and another member of the delegation, Marie Thérèse Grabier-Duvernay, as speakers. This seminar, held in Rome on 4 May 1972, was the first official Justice and Peace initiative in the environmental field. The following month, two members of the Pontifical Commission, Robert Faricy and Marvin Bordelon, were sent to Stockholm – albeit not as members of the official delegation – while Nebbia was appointed as one of twelve Justice and Peace consultors. The Church's growing engagement with the environmental issue finally reached its peak on 15 December 1972, with a lecture by Ward addressed to the entire Roman Curia entitled 'Only One Earth: Its Future and the Church's Responsibilities'.[45]

A Catholic Environmental Vision for Stockholm

The task of the Sorge group and the Justice and Peace Commission was delicate as they had to define the environmental vision of a very particular state. To begin with, the Holy See was a state virtually without territory, so it could not have a proper environmental policy of its own. At the same time, it exerted a huge cultural and political influence all over the world through the multitude of its followers, its local hierarchies and structures, and its diplomacy. Last but not least, the Holy See was a confessional state whose political action had to conform to a very complex religious creed. These three features help to explain both the process of its formation and the peculiarities of the Vatican environmental agenda as it emerged between 1970 and 1972.

Several elements were common to all texts prepared in Rome in the run-up to the Stockholm Conference. They constituted the lowest common denominator of the Holy See's attitude towards the environment and were the main guidelines for its action at Stockholm and in the global arena. One of these elements was a holistic and dialectic vision of the relationship between man and nature. According to this vision, man could not exist without and outside nature, but conversely nature massively bore the mark of man's action. Spirit pervaded matter. Conversely, the environment (that is, matter) was a crucial element for man's cultural and spiritual development.[46] What was at stake here was a cornerstone of Catholic social teaching produced by the Vatican Council and then strongly reaffirmed by Paul VI: the idea of 'integral development', that is, the full realization of all the spiritual and material potential of each human being.[47]

In this context, the Church had to reconsider the role of science and technology. The Church criticized them for growing risks and destructive forces. Science and technology should always consider the cultural and spiritual dimensions of progress and be directed primarily towards the realization of the common good of mankind.[48] These priorities reaffirmed the results of the Vatican Council, but they now assumed a specific implication for environmental conservation. According to the Holy See, the Stockholm Conference should not confine itself to discussing only the geographical, biological or atmospheric aspects of the environmental crisis; rather, it should encompass all of its social and cultural aspects.

From the Church's perspective, integral development meant the right of all men to share all the material and spiritual benefits of Creation. According to the Vatican Council, this right implied the quest for a fairer distribution of global wealth. At Stockholm, then, the need for a healthier and safer environment became part of this quest: the great goal of integral development required that humans lived in a healthy natural environment and could use the natural resources available where they lived. In this way, the Holy See's vision largely overlapped with the philosophy of the Founex report.

Finally, alongside the universal right to integral development featured the universal duty not to threaten earth's resources. As Paul VI wrote in his message to the conference:

> To rule creation means for the human race not to destroy it but to perfect it; to transform the world not into a chaos no longer fit for habitation, but into a beautiful abode where everything is respected. So no one can take possession in an absolute and selfish way of the environment, which is not a 'res nullius' – something not belonging to anyone – but the 'res omnium' – the patrimony of mankind, so that those in possession of it … must use it in a way that redounds to the real advantage of everyone.[49]

Such a complex vision clearly implied a very strong link between environmental protection and another political cornerstone of conciliar thought: the economic development of the poor countries. After all, at the Vatican Council, the Church's engagement against poverty had taken the form of a request for the development of the Global South and for a redesigning of the global economy.[50] The conciliar majority had been able to establish a wholly new social agenda expressed by Paul VI's encyclical *Populorum progressio* and the creation of Justice and Peace.[51] At Stockholm, this engagement translated into a vision that was quite close to that of the delegations from countries from the Global South, stressing in particular two points: the fact that poverty had to be considered the most important form of 'pollution'; and the request for a more balanced, cooperative and responsible development that could incorporate the environmental dimension. In his message to the conference, Paul VI went so far as to imagine a driving role for Third World countries in the common struggle to build a future without ecological dangers:

> Is it utopian to hope that the young nations, who are constructing, at the cost of great efforts, a better future for their peoples … should become the pioneers in the building of a new world, for which the Stockholm Conference is called to give the starting signal? … Thus, instead of seeing in the struggle for a better environment the reaction of fear of the rich, they would see in it, to the benefit of everyone, an affirmation of faith and hope in the destiny of the human family gathered round a common project.[52]

Finally, all the Vatican documents stressed the moral and cultural dimensions of the environmental crisis, demanding not merely technical solutions, but a global effort aimed at changing the human attitude towards nature and its goods.[53] In short, the Holy See took the Stockholm Conference seriously; it used the opportunity to incorporate in its agenda the new environmental issue, now considered very important for integral development; and it connected development and environment by discussing the issues at the centre of the United Nations Conference on Trade and Development (UNCTAD) in Santiago de Chile in March 1972 together with those at the centre of the Stockholm Conference in June of that year.[54]

The only major issue at the Stockholm Conference that the Vatican dismissed was population growth. This issue not only influenced the conduct of the Holy See delegation at Stockholm, but also the Church's attitude towards ecology in subsequent years.[55] The old Malthusian idea that a rapidly growing population could put earth's resources at risk had been widely debated in different ways since the 1930s.[56] The acceleration of population growth during postwar recovery and globalization had fuelled this debate since the second half of the 1940s.[57] Since the second half of the 1960s, the very idea that the environmental crisis could not be solved without effective measures of birth control had become common sense. Many governments and experts from rich countries hoped that some form of family planning through contraception, abortion or sterilization should be adopted on a global scale.[58] Since the 1930s, on the contrary, the Catholic Church had decreed a ban on any form of birth control. In 1968 Paul VI had gone further by issuing a specific encyclical, the *Humanae vitae.*[59]

As it was clear from the outset that the discussion about population growth would play an important role at the Stockholm Conference and that the majority of delegations would be in favour of family planning, the State Secretariat's decision was to play down the issue. The Vatican hoped that the conference would dissociate the environmental crisis and population growth by denying the impact of the latter or even by ignoring it completely.[60] This was not the case, however, resulting in a peculiar position adopted by the Holy See that cut across the alliances that otherwise characterized the conference.

The Holy See shared the vision of the conference organizers about the epochal importance and gravity of the ecological crisis. It also agreed both with the conference's aims and with most of the technical measures proposed by the preparatory committee.[61] Due to the philosophical, cultural and social work of the Vatican Council and the Justice and Peace experience, the Holy See delegation was able to contribute to highlighting problems and dimensions of the ecological crisis that had been neglected before. Moreover, the Holy See delegation refrained from openly accusing the market economies or the capitalist 'system' at large of being mainly responsible for the ecological crisis, as others did both inside and outside the Folkets Hus.[62] This positioning would indicate that the Vatican supported a coherent, enlightened 'Atlantic' position in line with those of countries like Sweden or Canada, for example.

At the same time, however, the strong emphasis on the rights of the poor and on Third World development placed the Holy See in the critics' ranks, which comprised left-wing thinkers like Barry Commoner, most of the INGOs in the alternative forums and most of the delegations from the Global South. Thus, the Holy See could have subscribed almost completely to the speech by the Indian Prime Minister Indira Gandhi, one of the most important and celebrated events of the Stockholm Conference.[63] But this affiliation with the 'progressive' alliance turned out to be only partial when the conference came to discuss the population issue. Most actors from the Global South criticized the idea that birth control should be imposed through coercive methods only in the poor countries. They considered such measures unfair and not very suitable to stop the degradation of earth's resources. The most radical of them even thought that the birth control strategy was part of a plot by the industrialized countries to stop the development of poor countries or to divert attention away from their own political and economic responsibilities. But only very few considered population growth irrelevant for the global environmental crisis. Even Commoner, a kind of 'guru' of the critics, proposed a model in which human population growth (though not as central as in Paul Ehrlich's model) played a remarkable role in the operation of ecological systems.[64]

Thus, in a vote on a Norwegian amendment to the *Recommendations for the Action at the International Level* advocating family planning as a means of limiting the demographic explosion, the Holy See found itself in a small minority opposing it. Forty-five countries voted in favour, twenty abstained and only twelve followed the Argentinian proposal to drop it.[65] Only two opposing countries were democracies (Venezuela and the Republic of Ireland); the others were autocratic states such as Ethiopia or dictatorships. All but one (Romania) were very right-wing, including, alongside Argentina, Brazil, Burundi, Ecuador, Portugal, Spain and Zaire. In addition, China, whose delegate had expressed a radically natalist position in the plenary debate, abstained.

In spite of these and other clashes, the Stockholm Conference succeeded not least thanks to Strong's diplomatic skills. The delegations agreed on the broad and ambitious *Declaration on the Human Environment* and other documents aimed at action. In fact, the Catholic Church expressed its satisfaction with the Conference's success in various ways.[66]

Conclusion

Analysing the Catholic Church's evolving attitudes towards environmental degradation and protection, it turns out that the preparatory work for the Stockholm Conference ran in parallel with the Church's initially moderate interest in the issue. However, the conference promoters' requirements imposed on participants forced the Holy See to define an autonomous and original position, to charge existing groups and commissions with preparing this position and to create long-term structures for debating environmental issues. Thanks to several intermediaries, the Church was able to participate actively in the Stockholm Conference and to create a lasting basis for its engagement with the issue on a global scale.

One pivotal intermediary was Barbara Ward. As mentioned above, she was not only a member of the Pontifical Commission but also advisor of the American President Lyndon B. Johnson, the UN Secretary-General Sithu U Thant and the board of the World Bank. Linking so many institutions and influential individuals, Ward could play a very important role in the Stockholm Conference from an early stage, although she was not yet an environmental expert. While she facilitated Strong's appointment with U Thant as conference secretary, Strong in turn decided to commission Ward with a vast report on the world environmental crisis in preparation of the conference. This report, published on the eve of the conference,[67] made Ward one of the most prominent figures on the global environmental scene.

However, Ward's most important contribution to the success of the conference was chairing, alongside Mahbub ul Haq and Gamani Corea, the Founex seminar of June 1971. There, as Michael W. Manulak shows in much greater detail in his chapter in this book, twenty development economists from all over the world tried for the first time to integrate theoretically environmental and development concerns. The seminar, conceived by Strong, helped overcome the poor countries' resistance to the conference. Ward's sympathy for the development concerns of the Global South were a crucial asset in the seminar.[68] Through Ward's active role in the seminar, all of its debates and conclusions also fed into the activities and positions adopted by the Justice and Peace Commission.

In the years following the Stockholm Conference, however, the Justice and Peace Commission proved unable to bring its work on the environment to the same level as its activities regarding more traditional issues, such as peace and disarmament, economic development, human rights and education.[69] It proved difficult for Catholicism, and for the Catholic Church, to fully grasp the philosophical and anthropological dimensions of the new environmental concern. One crucial reason for this was the Church's suspicion that environmental politics on a global scale would at least potentially entail some form of birth control. Thus, in the two years following the Stockholm Conference, the Holy See made a huge effort to influence the outcome of the first UN conference on world population held in Bucharest in 1974. In this way, the initial impetus from the period from Paul VI's speech at the FAO assembly of November 1970 to the general assembly of Justice and Peace of September 1972, when Barbara Ward and Giorgio Nebbia set out the implications of the Stockholm Conference, was largely lost.

Luigi Piccioni is Lecturer in Economic History, University of Calabria, Italy.

Notes

1. Pontifical Commission Justice and Peace, 'Report on the United Nations Conference on Human Environment (5–16 June 1972, Stockholm) Prepared by Father Robert Faricy, S. J. at the Request of the Secretariat, 1 July 1972', Fondo Giorgio e Gabriella Nebbia, Archivio Fondazione Luigi Micheletti Brescia (hereinafter AFM), Box I&P. For the INGOs' initiatives at the Stockholm Conference, featuring many original documents, see http://www.folkrorelser.org/Stockholm1972/index.html, accessed 20 May 2016. See also Peter Nilsson, 'NGO Involvement in the UN Conference on the Human Environment in Stockholm 1972: Interrelations between Intergovernmental Discourse Framing and Activist Influence' (Political Science D-Essay, University of Linköping, 2003), http://www.ep.liu.se/exjobb/eki/2004/ska/003, accessed 20 May 2016.
2. 'Declaration of the Non-Governmental Organisations', in *Only One Earth: United Nations Conference on the Human Environment. Stockholm, 5–16 June 1972* (Geneva: Centre for Economic and Social Information at the United Nations European Headquarters, 1972), 17.
3. Tord Björk, *The Emergence of Popular Participation in World Politics: United Nations Conference on Human Environment 1972* (Stockholm: University of Stockholm – Department of Political Science, Seminar Paper, 1996), Retrieved from: http://www.folkrorelser.org/inenglish/stockholm72.html, accessed 20 May 2016; Jean Gartlan, *Barbara Ward: Her Life and Letters* (London: Continuum, 2010), 162.
4. See, for example, John Hart, 'Catholicism', in *The Oxford Handbook of Religion and Ecology*, Roger S. Gottlieb (ed.) (New York: Oxford University Press, 2006), 77.
5. Marjorie Keenan, *From Stockholm to Johannesburg: An Historical Overview of the Concern of the Holy See for the Environment* (Vatican City: Libreria Editrice Vaticana, 2002), 15–18; Marjorie Keenan, *Care for Creation: Human Activity and the Environment* (Vatican City: Libreria Editrice Vaticana, 2000), 58.
6. Gartlan, *Barbara Ward*.

7. 'Fondo Giorgio and Gabriella Nebbia', Fondazione Luigi Micheletti, Brescia, http://www.musilbrescia.it/documentazione/dettaglio_fondo.asp?id= 119&sezione=archivio, accessed 20 May 2016.

8. In contrast, see, for example, Keenan, *From Stockholm to Johannesburg*, 14–15, who (unconvincingly) claims that it is possible to find references to the environment in documents issued during the Vatican Council or shortly thereafter.

9. See, inter alia, John B. Cobb Jr., *A Christian Natural Theology* (Philadelphia, PA: Westminster Press, 1965); Conrad Bonifazi, *A Theology of Things: A Study of Man in His Physical Environment* (Philadelphia, PA: Lippincott, 1967); Alfred Stefferud (ed.), *Christians and the Good Earth* (Alexandria, VA: Faith and Nature Papers, 1968). See also Roderick Frazier Nash, *The Rights of Nature: A History of Environmental Ethics* (Madison: University of Wisconsin Press, 1989), 98–106. Among the few Catholic titles, see René Dubos, *A Theology of the Earth* (Washington DC: Smithsonian Institution, 1969). At the beginning of 1972, Giorgio Nebbia, having compiled a vast bibliography of Protestant and Catholic texts, was fully aware of this gap. Giorgio Nebbia to Joseph Gremillion, 22 April 1972, Fondo Giorgio e Gabriella Nebbia, AFM, Box I&P.

10. Cf. Giuseppe Alberigo and Joseph A. Komonchak (eds), *History of Vatican II* (Maryknoll, NY: Orbis-Peeters, 1995–2006). The fiftieth anniversary of the Council has given rise to several new syntheses. See, for example, John O'Malley, *What Happened at Vatican II* (Cambridge, MA: Belknap, 2010).

11. 'Man, created to God's image, received a mandate to subject to himself the earth and all it contains.' Vatican Council II, *Gaudium et spes*, 34. 'This labor, whether it is engaged in independently or hired by someone else, comes immediately from the person, who as it were stamps the things of nature with his seal and subdues them to his will.' Ibid, n. 67; 'The whole of creation is for man, that he has been charged to give it meaning by his intelligent activity, to complete and perfect it by his own efforts and to his own advantage.' Paul VI, *Populorum progressio*, 22.

12. 'It is not enough to increase the general fund of wealth and then distribute it more fairly. It is not enough to develop technology so that the earth may become a more suitable living place for human beings … Economics and technology are meaningless if they do not benefit man, for it is he they are to serve.' Paul VI, *Populorum progressio*, 34.

13. Lynn White's famous article 'The Historical Roots of Our Ecologic Crisis', *Science*, 10 March 1967, 1203–7, is considered a cornerstone in the public debate about the role of anthropocentrism in shaping the attitude of Christian confessions towards nature. It drew upon a well-established critical tradition including Henry David Thoreau, Ralph Waldo Emerson, Charles Darwin, Aldo Leopold and Rachel Carson, but also some Protestant theologians. See also Nash, *The Rights of Nature*, Chapters 2–4.

14. 'The circumstances of the life of modern man have been so profoundly changed in their social and cultural aspects, that we can speak of a new age of human history … New ways are open, therefore, for the perfection and the further

extension of culture. These ways have been prepared by the enormous growth of natural, human and social sciences, by technical progress, and advances in developing and organizing means whereby men can communicate with one another.' Vatican Council II, *Gaudium et spes*, 54.

15. On this debate, despite their differences, see especially Alfred Sauvy, *Croissance zéro?* (Paris: Calmann-Lévy, 1973); Björn-Ola Linnér, *The Return of Malthus: Environmentalism and Post-war Population-Resource Crises* (Isle of Harris: White Horse Press, 2003); Matthew Connelly, *Fatal Misconceptions: The Struggle to Control World Population* (Cambridge, MA: Harvard University Press, 2008).

16. Paul VI's encyclical is the *Humanae vitae*, issued in 1968. See Jean-Louis Flandrin, *L'Église et la contraception* (Paris: Imago, 2006). This study was originally published during the *Humanae vitae* debate. See Jean-Louis Flandrin, *L'Église et le controle des naissances* (Paris: Flammarion, 1970). On this topic, see also Martine Sevegrand, *L'affaire Humanae vitae. L'Église catholique et la contraception* (Paris: Kathala, 2008).

17. The official documentation produced by the conciliar Fathers is gathered in the six volumes of *Acta Synodalia Sacrosancti Concilii Oecumenici Vaticani II* (Vatican City: Typis polyglottis Vaticanis, 1970–99). For the main texts approved by the Council http://www.vatican.va/archive/hist_councils/ii_vatican_council/index.htm, accessed 20 May 2016.

18. Above all, the *Mater et magistra* and the *Pacem in terris*, issued by John XXIII in 1961 and 1963 respectively, and the *Populorum progressio*, issued by Paul VI in 1967.

19. *Radiomessaggio del Santo Padre Giovanni XXIII ai fedeli di tutto il mondo, a un mese dal Concilio Ecumenico Vaticano II*, 11 September 1962, http://www.vatican.va/holy_father/john_xxiii/messages/pont_messages/1962/documents/hf_j-xxiii_mes_19620911_ecumenical-council_it.html, accessed 20 May 2016.

20. Giovanni Turbanti, *Un concilio per il mondo moderno. La redazione della costituzione pastorale 'Gaudium et spes' del Vaticano II* (Bologna: Il Mulino, 2000), 151–56; Andrew Small, 'The Theological Justification for the Establishment of the Pontifical Commission for Justice and Peace (*Iustitia et pax*)' (Ph.D. thesis, Catholic University of America, Washington DC, 2010), 12–16, http://aladinrc.wrlc.org/handle/1961/9228, accessed 5 July 2016.

21. For an account of the process, see the letters of the Bishop of Recife, Helder Camara, in *Vaticano II: corrispondência conciliar. Circulares à familia do São Joaquim 1962-64* (Recife: Editora Universitaria da Universidade Federal de Pernambuco, 2004).

22. For a list of the group's members, see http://www.stefangigacz.com/-church-of-the-poor-group, accessed 20 May 2016.

23. Small, 'The Theological Justification', 14. Except where otherwise indicated, Small's dissertation is the source I have used for information about the 'conspirators'.

24. Ibid., 27.

25. A recent contribution about the evolution of the Church's position towards juridical modernity and human rights is Daniele Menozzi, *Chiesa e diritti umani. Legge naturale e modernità politica dalla Rivoluzione francese ai nostri giorni* (Bologna: Il Mulino, 2012).

26. *Gaudium et spes*, 90.

27. For the text of the encyclical, promulgated by Paul VI on 26 March 1967 and dedicated to the issue of economic development, see: http://www.vatican.va/holy_father/paul_vi/encyclicals/documents/hf_p-vi_enc_26031967_populorum_en.html, accessed 20 May 2016.

28. Among the members and advisors of the Commission were four 'conspirators': Louis Gremillion, Barbara Ward, James J. Norris and Edward Swanstrom. Another 'conspirator', father Arthur McCormak, worked in the Secretariat.

29. 'Pontifical Commission Justice and Peace. Chronicle of Activities 1967–1971', Fondo Giorgio e Gabriella Nebbia, AFM, Box I&P.

30. Ibid.

31. Gartlan, *Barbara Ward*, 161; David Satterthwaite, *Barbara Ward and the Origins of Sustainable Development* (London, Iied, 2006), 11 and 14–17. Strong was actively engaged in ecumenical organizations such as the Young Men's Christian Association (YMCA) and Sodepax. When he was chosen by U Thant as secretary-general of the Stockholm Conference, he had just been appointed to the board of Sodepax, a Joint Committee on Society, Development, and Peace created in 1968 by the World Council of Churches and Justice and Peace. See the sources at http://oasis.lib.harvard.edu/oasis/deliver/~env00004, accessed 20 May 2016.

32. Barbara Ward and René Dubos, *Only One Earth: The Care and Maintenance of a Small Planet* (New York: W.W. Norton, 1972). Forty years later, Maurice Strong gave the same title to an historical assessment of sustainable development as a theory and a practice: Felix Dodds, Michael Strauss and Maurice F. Strong, *Only One Earth: The Long Road via Rio to Sustainable Development* (London: Routledge, 2012).

33. Bartolomeo Sorge, 'Per una critica cristiana della società tecnologica', *La Civiltà Cattolica* 121(3) (1970): 110–20.

34. Bartolomeo Sorge, 'La crisi ecologica. Un problema di scienza e di cultura', *La Civiltà Cattolica* 121(4) (1970): 417–26.

35. *Discours du Pape Paul VI à l'occasion du 25ème anniversaire de la FAO*, 16 November 1970, http://www.vatican.va/holy_father/paul_vi/speeches/1970/documents/hf_p-vi_spe_19701116_xxv-istituzione-fao_it.html, accessed 20 May 2016.

36. The circular letter and the questionnaire were sent by the conference secretariat on 21 December 1970, with the code EC 114/23 (1–3-3).

37. Bartolomeo Sorge to Giorgio Nebbia, 1 January 1971 and 11 January 1971, Fondo Giorgio e Gabriella Nebbia, AFM, Box I&P.

38. See Correspondence Giorgio Nebbia-Barry Commoner, Fondo Giorgio e Gabriella Nebbia, AFM, Box COM; Correspondence Giorgio Nebbia-Aurelio

Peccei, Fondo Giorgio e Gabriella Nebbia, AFM, Box Club of Rome. In 1983 Peccei appointed Nebbia as one of the 100 members of Club of Rome.

39. 'Rapporto della Santa Sede alla Conferenza internazionale di Stoccolma del 1972 su l'environnement', Fondo Giorgio e Gabriella Nebbia, AFM, Box I&P. The document is dated 5 March 1971.

40. 'Rapport du Saint-Siège en vue de la Conférence sur l'environnement', Fondo Giorgio e Gabriella Nebbia, AFM, Box I&P. This final version is not dated, but was distributed in the days immediately before 20 April, as stated in a letter from Father Sorge to Nebbia.

41. For the English version of the message, see Paul VI, 'Message of His Holiness Paul VI to Mr. Maurice F. Strong, Secretary-General of the Conference on the Environment', https://w2.vatican.va/content/paul-vi/en/messages/pont-messages/documents/hf_p-vi_mess_19720605_conferenza-ambiente.html, accessed 5 July 2016.

42. 'Intervention du Chef de la Délégation du Saint-Siège à la Séance Plénière du mercredi 7 juin 1972', Fondo Giorgio e Gabriella Nebbia, AFM, Box I&P.

43. 'Pontifical Commission Justice and Peace. VI General Assembly, 22–28 September 1971, Rome. General Report and Evaluation of the First Experimental Period 1967–71 and Prospectus for the Second Experimental Period 1972–74', Fondo Giorgio e Gabriella Nebbia, AFM, Box I&P.

44. *Octogesima adveniens. Apostolic Letter of Pope Paul VI*, issued on 14 May 1971, http://www.vatican.va/holy_father/paul_vi/apost_letters/documents/hf_p-vi_apl_19710514_octogesima-adveniens_en.html, and *Justice in the World*, at http://www.shc.edu/theolibrary/resources/synodjw.htm, both accessed 20 May 2016. The environmental issue is quoted at paragraph 21 of the first text and paragraph 70 of the second.

45. The text was published in the following year by the Commission under the title *A New Creation? Reflections on the Environmental Issue.* The conference was attended by 150 members of the Roman Curia, among them twelve cardinals and twenty bishops.

46. Rapport du Saint-Siège en vue de la Conférence sur l'environnement, Fondo Giorgio e Gabriella Nebbia, AFM, Box I&P. See also Paul VI, 'A Hospitable Earth for Future Generations'.

47. The concept of 'integral development' was utilized for the first time by Paul VI in the *Populorum progressio* encyclical. See Matthew Clarke, *Development and Religion: Theology and Practice* (Cheltenham: Edward Elgar, 2011), 118.

48. Rapport du Saint-Siège en vue de la Conférence sur l'environnement, Fondo Giorgio e Gabriella Nebbia, AFM, Box I&P.

49. Paul VI, 'A Hospitable Earth for Future Generations'.

50. Turbanti, *Un concilio*, 151–56.

51. Small, 'The Theological Justification', Chapter 2.

52. Paul VI, 'A Hospitable Earth for Future Generations'.

53. See 'Rapport du Saint-Siège en vue de la Conférence sur l'environnement', Fondo Giorgio e Gabriella Nebbia, AFM, Box I&P.

54. Ward highlighted this connection in her speech at the annual assembly of Justice and Peace: 'Pontifical Commission Justice and Peace. VII General Assembly, 20–26 September 1972. Report Pastoral Action for International Justice, Development and Peace Prepared by Lady Jackson (Barbara Ward) at the Request of the Secretariat', Fondo Giorgio e Gabriella Nebbia, AFM, Box I&P.

55. See, in greater detail, Luigi Piccioni, *Forty Years Later: The Reception of the Limits to Growth in Italy, 1971–1974* (Brescia, Fondazione Luigi Micheletti, 2012). Retrieved from: http://www.fondazionemicheletti.it/altronovecento/articolo. aspx?id_articolo=20&tipo_articolo=d_editoriale, accessed 20 May 2016.

56. For an historical overview, see Sauvy, *Croissance zéro?*

57. Yannick Mahrane, Marianna Fenzi, Céline Pessis and Cristophe Bonneuil, 'De la nature à la biosphère. L'invention politique de l'environnement global 1945–1972', *Vingtième Siècle. Revue d'histoire* 39(113) (2012): 129–33.

58. See Connelly, *Fatal Misconceptions*; Paige Whaley Eager, *Global Population Policy: From Population Control to Reproductive Rights* (Aldershot: Ashgate, 2004).

59. Robert McCloy, *Rome et la contraception. Histoire secrète de l'Encyclique Humanae vitae* (Paris: Les Éditions de l'Atelier-Les Éditions Ouvrières, 1998); Sevegrand, *Affaire Humanae vitae*; Connelly, *Fatal Misconceptions*.

60. Henri De Riedmatten to the State Secretariat substitute Giovanni Benelli, 2 February 1971, Bartolomeo Sorge to Giorgio Nebbia, 20 April 1971, and 'Rapport du Saint-Siège en vue de la Conférence sur l'environnement', Fondo Giorgio e Gabriella Nebbia, AFM, Box I&P.

61. Paul VI, 'A Hospitable Earth for Future Generations'.

62. For instance, the *Declaration on the Third World and the Human Environment* issued during the Conference by the International Committee of Young Scientists and Scholars for a Critical and Holistic Approach to Development and the Human Environment, also called 'the Committee'. See the text at http://www.folkrorelser.org/rorelsemapp/dokument/oicommittee.html, accessed 20 May 2016. However, the work of Oi Committee was highly praised in 'Pontifical Commission Justice and Peace. Report on the United Nations Conference on Human Environment (5–16 June 1972, Stockholm) Prepared by Father Robert Faricy, S. J. at the request of the Secretariat', Fondo Giorgio e Gabriella Nebbia, AFM, Box I&P.

63. Indira Gandhi, 'Man and his World', in *On Peoples and Problems*, 2nd edn (London: Hodder & Stoughton, 1983), 91–97.

64. Michael Egan, *Barry Commoner and the Science of Survival: The Remaking of American Environmentalism* (Cambridge, MA: MIT Press, 2007), 136–38.

65. *Report of the United Nations Conference on Human Environment. Stockholm, 5–16 June 1972* (New York: United Nations, 1973), 51–53.

66. See, for example, Joseph Gremillion to Maurice Strong, 17 June 1972, Fondo Giorgio e Gabriella Nebbia, AFM, Box I&P. In this telegram, the Secretary of Justice and Peace invited Strong to 'extend greetings congratulations also to Barbara [Ward] and other associates'.

67. Ward and Dubos, *Only One Earth*.
68. For Ward's role, see also Iris Borowy, *Defining Sustainable Development: A History of the World Commission on Environment and Development (Brundtland Commission)* (London: Routledge, 2014), 33–38.
69. Cf. Keenan, *From Stockholm to Johannesburg*, 23.

Bibliography

Acta Synodalia Sacrosancti Concilii Oecumenici Vaticani II (Vatican City: Typis polyglottis Vaticanis, 1970–99). http://www.vatican.va/archive/hist_councils/ii_vatican_council/index.htm, accessed 20 May 2016.

Alberigo, Giuseppe and Joseph A. Komonchak (eds), *History of Vatican II*, (Maryknoll, NY: Orbis-Peeters, 1995–2006).

Björk, Tord, *The Emergence of Popular Participation in World Politics: United Nations Conference on Human Environment 1972* (Stockholm: University of Stockholm – Department of Political Science, Seminar Paper, 1996). http://www.folkrorelser.org/inenglish/stockholm72.html, accessed 20 May 2016.

Bonifazi, Conrad, *A Theology of Things: A Study of Man in His Physical Environment* (Philadelphia, PA: Lippincott, 1967).

Borowy, Iris, *Defining Sustainable Development: A History of the World Commission on Environment and Development (Brundtland Commission)* (London: Routledge, 2014).

Camara, Helder, 'Letters', in *Vaticano II: corrispondência conciliar. Circulares à familia do São Joaquim 1962–64* (Recife: Editora Universitaria da Universidade Federal de Pernambuco, 2004).

Clarke, Matthew, *Development and Religion: Theology and Practice* (Cheltenham: Edward Elgar, 2011).

Cobb, John B. Jr., *A Christian Natural Theology* (Philadelphia, PA: Westminster Press, 1965).

Connelly, Matthew, *Fatal Misconceptions: The Struggle to Control World Population* (Cambridge, MA: Harvard University Press, 2008).

'Declaration of the Non-Governmental Organisations', in *Only One Earth: United Nations Conference on the Human Environment. Stockholm, 5–16 June 1972* (Geneva: Centre for Economic and Social Information at the United Nations European Headquarters, 1972), 17.

Discours du Pape Paul VI à l'occasion du 25ème anniversaire de la FAO, 16 November 1970, http://www.vatican.va/holy_father/paul_vi/speeches/1970/documents/hf_p-vi_spe_19701116_xxv-istituzione-fao_it.html, accessed 20 May 2016.

Dodds, Felix, Michael Strauss and Maurice F. Strong, *Only One Earth: The Long Road via Rio to Sustainable Development* (London: Routledge, 2012).

Dubos, René, *A Theology of the Earth* (Washington DC: Smithsonian Institution, 1969).

Eager, Paige Whaley, *Global Population Policy: From Population Control to Reproductive Rights* (Aldershot: Ashgate, 2004).

Egan, Michael, *Barry Commoner and the Science of Survival: The Remaking of American Environmentalism* (Cambridge, MA: MIT Press, 2007).

Flandrin, Jean-Louis, *L'Église et le controle des naissances* (Paris: Flammarion, 1970).

———. *L'Église et la contraception* (Paris: Imago, 2006).

'Fondo Giorgio and Gabriella Nebbia', Fondazione Luigi Micheletti, Brescia: http://www.musilbrescia.it/documentazione/dettaglio_fondo.asp?id=119&sezione=archivio, accessed 20 May 2016.

Gandhi, Indira, 'Man and his World', in *On Peoples and Problems*, 2nd edn (London: Hodder & Stoughton, 1983), 91–97.

Gartlan, Jean, *Barbara Ward: Her Life and Letters* (London: Continuum, 2010).

Gremillion, Joseph to Maurice Strong, 17 June 1972, Fondo Giorgio e Gabriella Nebbia, AFM, Box I&P.

Hart, John, 'Catholicism', in *The Oxford Handbook of Religion and Ecology*, Roger S. Gottlieb (ed.) (New York: Oxford University Press, 2006), 77.

Holy See, *Octogesima adveniens. Apostolic letter of Pope Paul VI*, issued on 14 May 1971, http://www.vatican.va/holy_father/paul_vi/apost_letters/documents/hf_p-vi_apl_19710514_octogesima-adveniens_en.html, accessed 20 May 2016.

———. *Justice in the World*, http://www.shc.edu/theolibrary/resources/synodjw.htm, both accessed 20 May 2016.

———. 'Intervention du Chef de la Délégation du Saint-Siège à la Séance Plénière du mercredi 7 juin 1972', Fondo Giorgio e Gabriella Nebbia, AFM, Box I&P.

———. 'Rapport du Saint-Siège en vue de la Conférence sur l'environnement', Fondo Giorgio e Gabriella Nebbia, AFM, Box I&P.

———. 'Rapporto della Santa Sede alla Conferenza internazionale di Stoccolma del 1972 su l'environnement, 5 March 1971', Fondo Giorgio e Gabriella Nebbia, AFM, Box I&P.

International Committee of Young Scientists and Scholars for a Critical and Holistic Approach to Development and the Human Environment, *Declaration on the Third World and the Human Environment*, http://www.folkroreler.org/rorelsemapp/dokument/oicommittee.html, accessed 20 May 2016.

Keenan, Marjorie, *Care for Creation: Human Activity and the Environment* (Vatican City: Libreria Editrice Vaticana, 2000).

———. *From Stockholm to Johannesburg: An Historical Overview of the Concern of the Holy See for the Environment* (Vatican City: Libreria Editrice Vaticana, 2002).

Linnér, Björn-Ola, *The Return of Malthus: Environmentalism and Post-war Population-Resource Crises* (Isle of Harris: White Horse Press, 2003).

Mahrane, Yannick, Marianna Fenzi, Céline Pessis and Cristophe Bonneuil, 'De la nature à la biosphère. L'invention politique de l'environnement global 1945–1972', *Vingtième Siècle. Revue d'histoire* 39(113) (2012): 129–33.

McCloy, Robert, *Rome et la contraception. Histoire secrète de l'Encyclique Humanae vitae* (Paris: Les Éditions de l'Atelier-Les Éditions Ouvrières, 1998).

Menozzi, Daniele, *Chiesa e diritti umani. Legge naturale e modernità politica dalla Rivoluzione francese ai nostri giorni* (Bologna: Il Mulino, 2012).

Nash, Roderick Frazier, *The Rights of Nature: A History of Environmental Ethics* (Madison: University of Wisconsin Press, 1989).

Nebbia, Giorgio to Aurelio Peccei, 'Correspondence', Fondo Giorgio e Gabriella Nebbia, AFM, Box Club of Rome.

Nebbia, Giorgio to Barry Commoner, 'Correspondence', Fondo Giorgio e Gabriella Nebbia, AFM, Box COM.

Nebbia, Giorgio to Joseph Gremillion, 22 April 1972, Fondo Giorgio e Gabriella Nebbia, AFM, Box I&P.

Nilsson, Peter, 'NGO Involvement in the UN Conference on the Human Environment in Stockholm 1972. Interrelations between Intergovernmental Discourse Framing and Activist Influence' (Political Science D-Essay, University of Linköping, 2003), http://www.ep.liu.se/exjobb/eki/2004/ska/003/, accessed 20 May 2016.

O'Malley, John, *What Happened at Vatican II* (Cambridge, MA: Belknap, 2010).

Paul VI, 'Message of His Holiness Paul VI to Mr. Maurice F. Strong, Secretary-General of the Conference on the Environment', https://w2.vatican.va/content/paul-vi/en/messages/pont-messages/documents/hf_p-vi_mess_19720605_conferenza-ambiente.html, accessed 5 July 2016.

Piccioni, Luigi, *Forty Years Later: The Reception of the Limits to Growth in Italy, 1971–1974* (Brescia, Fondazione Luigi Micheletti, 2012). http://www.fondazionemicheletti.it/altronovecento/articolo.aspx?id_articolo=20&tipo_articolo=d_editoriale, accessed 20 May 2016.

Pontifical Commission Justice and Peace, 'Chronicle of Activities 1967–1971', Fondo Giorgio e Gabriella Nebbia, AFM, Box I&P.

———. 'VI General Assembly, 22–28 September 1971, Rome. General Report and Evaluation of the First Experimental Period 1967–71 and Prospectus for the Second Experimental Period 1972–74', Fondo Giorgio e Gabriella Nebbia, AFM, Box I&P.

———. 'VII General Assembly, 20–26 September 1972. Report Pastoral Action for International Justice, Development and Peace Prepared by Lady Jackson (Barbara Ward) at the Request of the Secretariat', Fondo Giorgio e Gabriella Nebbia, AFM, Box I&P.

———. 'Report on the United Nations Conference on Human Environment (5–16 June 1972, Stockholm) Prepared by Father Robert Faricy, S. J. at the request of the Secretariat, 1 July 1972', Fondo Giorgio e Gabriella Nebbia, Archivio Fondazione Luigi Micheletti Brescia (AFM), Box I&P.

Radiomessaggio del Santo Padre Giovanni XXIII ai fedeli di tutto il mondo, a un mese dal Concilio Ecumenico Vaticano II, 11 September 1962, www.vatican.va/holy_father/john_xxiii/messages/pont_messages/1962/documents/hf_j-xxiii_mes_19620911_ecumenical-council_it.html, accessed 20 May 2016.

Riedmatten, Henri De to the State Secretariat substitute Giovanni Benelli, 2 February 1971, Fondo Giorgio e Gabriella Nebbia, AFM, Box I&P.

Satterthwaite, David, *Barbara Ward and the Origins of Sustainable Development* (London: Iied, 2006).

Sauvy, Alfred, *Croissance zéro?* (Paris: Calmann-Lévy, 1973).

Sevegrand, Martine, *L'affaire Humanae vitae. L'Église catholique et la contraception* (Paris: Kathala, 2008).

Small, Andrew, 'The Theological Justification for the Establishment of the Pontifical Commission for Justice and Peace (*Iustitia et pax*)' (Ph.D. thesis, Catholic University of America, Washington DC, 2010) http://aladinrc.wrlc.org/handle/1961/9228, accessed 5 July 2016.

Sorge, Bartolomeo to Giorgio Nebbia, 1 January 1971 and 11 January 1971, Fondo Giorgio e Gabriella Nebbia, AFM, Box I&P.

———. to Giorgio Nebbia, 20 April 1971, Fondo Giorgio e Gabriella Nebbia, AFM, Box I&P.

Sorge, Bartolomeo, 'La crisi ecologica. Un problema di scienza e di cultura', *La Civiltà Cattolica* 121(4) (1970): 417–26.

———. 'Per una critica cristiana della società tecnologica', *La Civiltà Cattolica* 121(3) (1970): 110–20.

Stefferud, Alfred (ed.), *Christians and the Good Earth*, (Alexandria, VA: Faith and Nature Papers, 1968).

Turbanti, Giovanni, *Un concilio per il mondo moderno. La redazione della costituzione pastorale 'Gaudium et spes' del Vaticano II* (Bologna: Il Mulino, 2000).

UN, *Report of the United Nations Conference on Human Environment. Stockholm, 5–16 June 1972* (New York: United Nations, 1973).

'The UN Participatory Rebellion – People's Stockholm Summits '72', http://www.folkrorelser.org/stockholm1972/dokument/Rio20issue1.pdf, accessed 5 April 2016.

Ward, Barbara and René Dubos, *Only One Earth: The Care and Maintenance of a Small Planet* (New York: W.W. Norton, 1972).

White, Lynn, 'The Historical Roots of Our Ecologic Crisis', *Science* 10 March 1967, 1203–7.

CHAPTER 5

Sometimes it's the Economy, Stupid!
International Organizations, Steel and the Environment

Wolfram Kaiser

International organizations (IOs) have played a leading role in identifying and supporting research into environmental hazards. They have propagated transnational collaboration to combat air and water pollution, acid rain and global warming, for example. As several chapters in this book demonstrate, they have also framed environmental issues and developed new policy-making concepts and tools for environmental protection such as the notion of 'sustainable development' in the 1987 Brundtland Report.[1]

Too often, however, research on environmental history focuses only on environmental policy making as a clearly delineated field. Such an approach fails to capture how heavily environmental protection has been contested as a societal goal. Industrialists critical of environmental measures, for instance, routinely highlighted the cost of protection and the loss of competitiveness, especially for energy-intensive industries. In the light of this, this chapter takes a different perspective on the role of IOs in environmental protection, which complements the other chapters in this book. It analyses the significance of environmental issues and concerns in IO policy making for such an energy-intensive industrial sector – steel – to assess their influence and impact across different policy fields and to probe the degree of 'mainstreaming' of environmental concerns. Steel has been a leading sector in industrialization. After the Second World War, it was crucially important for economic reconstruction. Traditionally, steel magnates and companies wielded much political influence in North America and Europe. At the same time, their blue-collar workers were well organized and influential in the trade union movements. Steel, in other words, mattered a great deal politically both to centre-right and left-wing political parties as well as national governments.[2] The sector also saw rapid globalization from the 1970s with a dramatic shift in production to Asia.

Its importance for reconstruction after 1945 induced the pan-European United Nations Economic Commission for Europe (UNECE), the Western Organisation for European Economic Cooperation (OEEC, OECD from 1961–62), as well as the Council for Mutual Economic Assistance (CMEA) in Eastern Europe, to create separate steel committees from the start. The formation of 'core Europe' leading to the present-day European Union (EU) also began in this sector when France, West Germany, Italy and the Benelux countries formed the European Coal and Steel Community (ECSC) in 1952–53. Its High Authority even had independent decision-making and judicial powers,[3] although they were effectively curtailed to a large degree by national governments and transnational business actors, resulting in a pronounced consensus culture and the continued toleration of cartel-type arrangements.[4]

While the IOs were initially interested in increasing production for the postwar reconstruction of Europe, steel mills, alongside the chemical industry, were also major polluters. They emitted particulate matter, or visible soot or dust, and fine particulate matter later identified as having carcinogenic effects. Their emissions also included noxious gases such as sulphur dioxide, carbon monoxide and sulphureted hydrogen. The associated environmental and health hazards initially appeared to remain largely localized. Smoke, dust and the ensuing smog provoked political protests and pressures in industrial areas, which led to the greater distribution (also across borders) of emissions through high-rise chimneys from the 1960s. While the emissions' greater dispersal was initially expected to solve the issue, it in fact aggravated the effects of sulphur dioxide. Sulphuric acid mist, or acid rain, led to the acidification of lakes, notably in Scandinavia, and the destruction of forests far away from the centres of industry.[5] This phenomenon and its effects remained ill-understood until the 1970s, however. Moreover, in the 1980s, it became clear and increasingly part of the scientific consensus how carbon dioxide from the burning of coal for steel production contributed to global warming.[6]

At the same time, the introduction of new process technologies since the early 1950s, especially oxygen steel making and continuous casting, required unprecedented capital expenditure. While these innovations were geared towards productivity gains, they enabled steel companies to claim with some justification – both in the national context and in IOs – that they also significantly increased energy efficiency. In fact, according to a UNECE report from 1989, the average fuel rates fell from more than 800 kg of coking coal per ton of pig iron in the 1950s to approximately 500 kg in the mid 1970s.[7] The industry therefore argued that the increased energy efficiency at least compensated for the rise in output and associated pollution until 1974.

It was in this year that steel production in Western Europe dropped for the first time. The low industrial growth rates after the 1973 oil crisis reduced demand. Steel-consuming industries like shipbuilding moved to Asia in particular, where steel demand and production grew rapidly. Steel increasingly also became substituted by aluminum and plastics, for example in cars. As a result, the sector entered into a prolonged structural crisis in Western Europe and North America as well as in Eastern Europe, although under different auspices, which lasted well into the 1990s.[8]

This chapter's first section analyses how before institutionalizing the environment as a clearly delineated policy field, IOs often discussed environmental issues in various functional contexts and for specific industrial sectors such as steel. For instance, the UNECE and the OECD debated issues of air and water pollution by the steel industry from the mid 1950s onwards. Domestic pressure in Western countries to reduce pollution induced collaboration among experts in IO working groups who enjoyed great freedom to explore issues on the borderline between technology, research, industrial policy and the environment. However, the industry, while interested in transnational learning about technological innovation, sought to use the IOs as platforms for pleading for its treatment as a 'special case'. Industrialists argued that the imposition of regulation requiring expensive technological solutions would be disastrous for an industry with small profit margins and high capital costs. As a result, officials from national ministries and representatives of private and state-owned steel companies succeeded in controlling the committees' larger economic and political conclusions. They limited international cooperation to industry networking, the growth of new expert communities on the borderline between engineering and the natural sciences, and the aggregation and transfer of knowledge about the sources of and strategies for combating environmental degradation.

The second section explores how the institutionalization of environmental policy in the early 1970s changed the game. It zooms in on the OECD, which was the first IO to go down the route of delineating the environment as a new policy field by creating a separate directorate and committee in 1971 – this just after dissolving its Steel Committee in 1970 and bringing the sector into the remit of its general Industry Committee. In light of the increasingly severe steel crisis, OECD member state governments decided to reconstitute the Steel Committee in November 1978.[9] However, as opposed to the earlier Steel Committee, environmental issues were conspicuously absent from the new committee's agenda, from its first meeting through to the late 1980s. Instead, it focused entirely on three interrelated topics: trade issues linked in particular to the negotiation by the European Communities (EC) of 'voluntary' export restraint agreements with countries like Japan; economic issues of the industry's modernization and restructuring; and social issues of unemployment and the retraining of the workforce.

Thus, the steel industry is an excellent example of the limits of IO environmental policy activism in times of economic crisis in a sector that for a long time wielded great political influence across the political spectrum, both nationally and transnationally. The case also shows that the separate institutionalization of environmental policy, as in the OECD in 1971, which reflected the greater politicization of environmental issues, is not necessarily a good indicator of policy change and impact. In fact, the different policy fields became more segregated than before. While IOs, including the OECD, became very proactive in environmental policy, forging transnational expert networks, framing issues and shaping discourses, these networks, frames and discourses actually penetrated IO (as well as national) policy making in other fields, especially industrial policy, only to a very limited extent. In the steel sector, in fact, IOs actually started to focus heavily on socioeconomic issues at the expense of environmental concerns shortly after the 1972 Stockholm Conference. This in turn greatly limited the 'mainstreaming' of environmental policy objectives propagated by the IOs environmental policy makers and environmental activists.

International Organizations and the Steel Industry

Alongside energy and transport, the steel sector was a major concern for IOs and national governments after 1945 due to its centrality for rebuilding infrastructures and for industrial production. Thus, the UNECE, created in 1947, set up a steel committee in the same year. The committee's original terms of reference, dating from November 1947, focused on addressing the steel shortage in Europe.[10] In Western Europe, this shortage was overcome surprisingly quickly. Hence, in 1950, the committee drafted new terms of reference for its work programme. In an act of self-empowerment, it defined them to cover practically every conceivable activity. In future, it could, inter alia, 'collect and transmit all useful information concerning steel production and consumption trends', 'pursue its statistical work', 'undertake … any studies which the Committee may deem of importance' and 'draw any appropriate conclusions arising from its work'.[11]

Just as in the case of its UNECE counterpart, governments never seriously challenged the de facto power of the OEEC Steel Committee to more or less autonomously manage its own agenda and business. Originally formed in 1948 to administer the Marshall Plan, the OEEC was transformed into the OECD in 1961–62 with broader membership including the United States, Canada and Japan.[12] When its Council met in 1965 and decided to prolong the Steel Committee's existence for another five years, there was – in contrast to several other economic sectors – 'unanimous recognition by governments that problems in the iron and steel industry were of importance and did affect government policies and interests'.[13]

The example of the UNECE steel committee illustrates well how its scope changed and grew over time.[14] To overcome the steel shortage in Europe, it first concentrated on reallocating coal and iron ore supplies during 1947–49. Growing Cold War tensions – notably the Soviet Union's unwillingness to disclose the amount of available scrap in its eastern German zone of occupation – led to the shift of discussions about this issue to the Western European OEEC context.[15] Evolving from its concern with the steel shortage, the Steel Committee, in cooperation with the Steel Division of the UNECE Secretariat, also surveyed the development of production and markets in the sector. At the same time, the topic of the steel shortage and then possible overproduction and its likely consequences for the industry also induced statistical work and discussions about investments in the modernization of old, or the construction of new, steel plants. Both themes were closely linked to technological change in the sector that became the topic of several UNECE reports published during the 1950s.

Crucially, as we will see below, the influence of the steel industry on the steel committees of the OEEC and the UNECE was very strong. Representatives in the OECD's steel committee were often ministry officials, but close to heavy industry interests, which in countries such as Germany, Belgium and Luxembourg were well organized and politically influential, or they actually worked for private or state-owned companies, but attended meetings on behalf of the member state concerned. In the UNECE, the role of business and technology experts (as opposed to government officials) in the committee structure was even stronger and many of them were from private or state-owned steel companies.

Just as with the OEEC and the UNECE, the East European CMEA, which began operating in 1950 but was only formally set up under international law in 1959, also had a Standing Committee on Black Metallurgy. Within the very different political context in the Soviet Bloc,[16] it, too, felt free to take up new topics such as air and water pollution in the 1960s. In contrast, the ECSC, despite its strong formal powers, was limited by its founding treaty to health and safety in the workplace alongside its more far-reaching competences for the customs union, competition policy and other aspects of the common market for coal and steel. Its activities initially remained restricted to environmental topics directly affecting workers, not as a larger societal problem. When the three communities were merged administratively in 1967, the European Communities (EC) first passed legislation with environmental implications concerning the classification and labelling of chemical products in 1967. As Jan-Henrik Meyer shows in his chapter in this book, the EC then developed its own broader environmental policy agenda in the 1970s.

International Organizations, Steel, and Air and Water Pollution

Whether in Pittsburgh in the United States, in Liège in Belgium, the Ruhr Area in Germany, Nowa Huta in Poland or Magnitogorsk in the Soviet Union, the inhabitants of urban agglomerations with a high concentration of steel production were well aware of its contribution to air and water pollution. In one of the first IO reports on air and water pollution initiated by its Chemical Products Committee in 1953, the European Productivity Agency (EPA), which operated in the remit of the OEEC, still focused heavily on the chemical industry.[17] Set up to facilitate technology transfer from the United States to Western Europe and to increase productivity levels in the European economy, the EPA highlighted in particular the greater expenditure in American companies on new technical equipment to combat emissions and the greater attention to meteorological conditions in channelling them.[18] While the EPA's two missions in Europe and to North America, as well as the subsequent report, were dominated by scientists and practitioners from the chemical industry, they also included visits to steel plants.

In a report focused entirely on the iron and steel industry published in 1963, the OECD outlined more clearly the industry's particular contribution to air pollution.[19] Clearly, the industry was a 'source of pollution' especially due to 'the large tonnages and the nature of the raw materials that it uses'.[20] The report discussed, in this sequence, the types of pollutants, measurements of pollution, gas cleaning equipment and types of chimneys culminating in some general conclusions. It resolved, falsely (as the carcinogenic effects of fine particulate matter were not well understood at the time), that the industry's emission of particulate matter, or visible dust, 'has mainly a nuisance value'. The emission of gases concerned, in particular, carbon monoxide, sulphur dioxide and sulphureted hydrogen. Sulphur dioxide transformed into sulphuric acid mist, or what later became known as acid rain, 'as a result of phenomena not yet fully understood', as the report admitted.[21]

Without focusing on the steel sector, another OECD report published in 1964 discussed for the first time methods of measuring air pollution.[22] Compiled for the OECD's Committee for Scientific Research by a Working Party of the Sub-Committee for Co-operative Research set up in January 1957, the report listed and discussed scientific methods for measuring the emission from industrial plants especially of smoke, sulphur dioxide, sulphur trioxide and hydrocarbons. This report highlighted the persisting scientific and statistical difficulties in measuring different types of air pollution, which complicated attempts to define more precisely the particular role of industries like steel and to develop targeted national, let alone international, strategies for reducing air pollution.

Initiated at its Steel Committee's thirty-fourth session in 1966, the UNECE published its report on air and water pollution arising from the iron and steel industry in 1970.[23] This report discussed general aspects of the problem, the sources of pollution, the methods to reduce it, and the costs of controlling pollution and future issues with an annex of existing legal regulations. The technical recommendations for suitable methods to reduce different forms of pollution were already far more detailed than in the earlier reports by the OEEC and the UNECE. The report also sought to contextualize air and water pollution by the sector, arguing that 'one of the major difficulties encountered during the preparation of the study was the delimitation of the subject, since problems connected with air and water pollution cover equally social and economic aspects of human activity and enter into many fields of science and technology'.[24] This assessment anticipated, to some extent, the increasing conception of issues such as air and water pollution as transcending different industrial sectors and being part of the larger issue of the 'environment', whilst encompassing the impact on various dimensions of nature and human life.[25]

Uploading: IO Motivations for Deliberating Steel and Pollution

What motivated the specialized steel committees of IOs like the OECD and the UNECE to devote time and resources to the study of air and water pollution? Did the transnational nature of the problem induce the search for common solutions by IOs with interests in the issue of air and water pollution? Were they keen to create new competences at an international level? There is not much evidence for such a functional logic. IOs without decision-making competences were most active up until the early 1970s in discussing the issue. However, they refrained from advocating any particular policy change at any particular level of government or governance. Instead, they concentrated on researching and outlining scientific and technological causes and methods to combat pollution. In their general conclusions they focused on the technology dimension and, increasingly, the economic costs associated with improved environmental protection. Moreover, until the EC's institutional merger in 1967, the ECSC, with its substantial policy-making competences, did not develop a serious interest in the larger environmental agenda.[26]

Instead, two other factors appear to account for the activities of the OECD and UNECE steel committees. First, air and water pollution became a pressing public concern domestically at the subnational level in the 1950s, after the first reconstruction needs had been met. In Western Europe, grassroots groups protested against the environmental effects and health hazards resulting from air and water pollution.[27] The steel industry's own contribution to air pollution in particular was obvious enough. Alongside the chemical industry, steel mills became the main target for such protests. These protests in turn generated national-level legislation. Thus, the 'Great Smog' in London in December 1952, which led to 4,000 premature deaths, resulted in the 1956 British Clean Air Act.[28] An amendment to the West German Industrial Code, or *Gewerbeordnung*, in December 1959, henceforth required new industrial installations to be authorized.[29] Additionally, the federal state government of North-Rhine-Westphalia passed a law in February 1961 that for the first time imposed emissions limits on, for example, brown smoke.[30]

In this changing political and legislative climate, the steel industry primarily feared the imposition of high additional capital costs for measures to reduce pollution. Traditionally, steel companies had multiple links with state institutions and political actors in Western Europe. In the nineteenth century and in interwar Europe, they had closely associated themselves with nationalist and colonialist agendas, and formed the backbone of the war economy during both world wars.[31] After the Second World War, they still had political clout. In West Germany, for example, heavy industry for a long time provided substantial funding for the ruling Christian Democratic Union.[32] In other countries, state institutions and the steel sector became closely linked through the nationalization of the entire sector (as in the United Kingdom in 1967) or individual companies. In this situation, national ministries and steel companies saw their cooperation within the remit of the IOs as a suitable strategy to defend the economic interests of the industry, to plead for exemptions, and to create the public impression that it was concerned about and was actively addressing air and water pollution emanating from steel plants.

Thus, the 1963 OECD report claimed that the steel industry had 'already spent large sums of money on plant that is very costly to install and operate'. If legal rules became 'more exacting', then 'the industry's efforts will add to the cost of iron and steel' – at this stage, a concern more about steel companies' profit margins and their workers' employment security as well as the cost of steel for steel-consuming industries than about comparative advantage. Moreover, the report argued that any future regulation should not be 'inspired by purely abstract concepts based on data that must often be very imprecise' and that it should allow for flexibility, taking into account 'local geography and topography, and climate and weather'. It should 'make allowance for specific cases' and not 'apply everywhere and to all emissions irrespective of their nature, type of discharge and their magnitude'.[33] The industry should be thoroughly consulted over future legislation at the national level. Governments should also consider the costs involved in possible legislation and, if necessary, provide state 'assistance' in the form of tax allowances, long-term loans at low interest or 'even direct contributions to investment expenditure'.[34] In short, the steel experts asked for general legislation that would be flexibly applied to exempt steel plants under varying circumstances, and also for government subsidies to meet any capital costs of investments in environmental protection measures. It was precisely such ad hoc measures that the OECD and the EC problematized and sought to ban in the 1970s when they introduced the polluter pays principle, as Jan-Henrik Meyer demonstrates in his chapter. Expert economists within these IOs argued that subsidies and exemptions, and in particular their flexible application, distorted fair competition.

As a second factor that helps account for the activism of IOs, the experts in the steel committees were keen to use this platform for aggregating information on technological challenges and solutions to facilitate its dissemination and use in new and older steel plants in all member states. Traditionally, technological innovations had become transferred with relative ease in the sector, without any cross-border regulation of technical norms and standards either by private companies in transnational voluntary organizations or by governments in IOs. Unlike some sectors of the Second Industrial Revolution like chemicals and pharmaceuticals, steel production was not 'based on an administered or secret technical knowledge'.[35] New production processes were large scale and impossible to hide. Knowledge about them was widely diffused within the companies. It could easily be transferred to other companies and across borders by businessmen, engineers and specialist workers, as well as through specialized journals like the British *Journal of the Iron and Steel Institute*, published since 1871, and the German journal *Stahl und Eisen*, founded in 1881. The new processes were patented, and the patents were sold and licensed out. Moreover, steel raised no issues of compatibility across borders. What mattered was the type of steel, its purity and hardness. These issues were specified directly between buyers and producers in relationships based on experience and mutual trust as well as formal contracts.

The issue of pollution provided a new challenge, however. Pollution control required far more natural science expertise than the previous engineering-dominated innovations in process technologies. In the 1950s and 1960s, such expertise was still scarce and not at all well integrated with the traditional expert culture in the steel sector, where the roles of engineers and business managers were not clearly delineated.[36] Technical experts also routinely acquired business experience, and managers either came from families with a mining or steel background or had prior technical training or an engineering degree. In this situation, the IOs' steel committees could help to integrate scientific knowledge and expertise with this traditional knowledge base. More specifically, the Soviet Union and other CMEA countries developed a keen interest in using the transnational issue of air and water pollution to reduce technological barriers with the West, not so much to address the even more severe pollution problems there, but as a legitimate strategy to gain access to technological knowhow more generally.[37]

Networks and Transfers

How, then, did the IOs' work in the steel sector impact on the transfer of knowledge or policy solutions? First, the IO committee work enabled networking among companies and experts in the steel industry. This networking facilitated the transfer of knowledge about environmental problems and solutions as well as about more narrowly business-related issues. In addition, collective tours of companies and production sites became popular in postwar Europe. In the West, the Rockefeller Foundation first organized visits by European steel experts to US steel plants and the exchange of trainees and interns.[38] Like several other national organizations, the German Verein Deutscher Eisenhüttenleute (VDEh), the professional organization of steel experts, developed links with CMEA countries like the Soviet Union, Poland and Hungary, for example. Their activities included joint workshops on technology issues that also touched upon air and water pollution.

The UNECE played an important role in facilitating visits and networking across the East–West divide, especially in the second half of the 1950s and during the 1960s. Thus, in 1955, steel experts from several Western European countries went on a two-week trip to the Soviet Union. The group included Pierre van der Rest, the President of the Comité de la sidérurgie belge, who also played a leading political role in the ECSC's Consultative Committee.[39] These visits became a much more regular feature from 1966 onwards.[40] The UNECE organized study tours of steel experts to the Soviet Union in 1966, Italy in 1967, Poland in 1968, the United States and Canada in 1969, Japan in 1970 and Czechoslovakia in 1971.[41] By this time, despite the UNECE's European scope, the groups sometimes travelled to non-European destinations. They also included representatives of countries and companies from outside of Europe – something that reflected the increasing globalization of steel production, trade and networks. Moreover, the groups were also mixed in terms of composition, comprising both officials from ministries and steel experts from state-owned or private steel companies.[42] The study tours were kept informal. The emphasis was on establishing and deepening direct personal contact among steel managers and experts to foster sustainable networks and facilitate knowledge transfer.[43]

Secondly, and more concretely, the IOs' activities at the working group level of experts fostered the emergence of new expert communities in the steel sector on the intersection between technology (engineering) and the natural sciences (physics, chemistry). They reinforced a new trend after the Second World War towards more transnational research collaboration in the sector. In 1951, the first World Metallurgical Congress took place in Detroit.[44] In Europe, national research organizations like the Société française de métallurgie and the VDEh began to organize joint workshops focused on specific technological and scientific issues, with transnational committees for specific research topics including environmental problems such as air and water pollution forming in the 1970s.[45] These professional bodies frequently cooperated with state-funded research institutes that also developed a much greater international orientation. Eventually, these institutes, including, inter alia, the German Max Planck Institute für Eisenforschung, the Benelux Centre National de Recherches Métallurgique, the French Institut de Recherche de la Sidérurgie and the Spanish Centro Naçional de Investigaciones Metalurgicas, formed the Directors of Steel Societies Conference in 1982, which was later reorganized into the European Steel Institutes Confederation in 1990.

Thirdly, however, the steel committees never engaged in a serious debate about environmental policy. What they did discuss and analyse were the causes of air and water pollution or technological solutions for these and other problems. The IOs' reports included comparative information on national legislation, but they made no attempt to identify what later became known as good practice and benchmarking, let alone recommending new commitments under international law. None of the three IOs, the Western OECD, the pan-European UNECE and the Eastern CMEA, promoted concrete emissions targets, technological standards or institutional set-ups for fighting pollution in the steel sector. With no obvious functional economic need for cross-border regulation or standardization, the experts continued to focus on technical solutions. Moreover, as pollution issues were not yet highly politicized, IOs were not under pressure from member state governments or societal actors to play a more proactive role in promoting transnational regulation. However, the main reason for the lack of environmental *policy* activism is the dominant influence of the steel industry – private and state-owned – on the work of the IOs' steel committees – either through their own representatives, as in the UNECE, or via government and ministry officials close to the industry, as in the OECD until the dissolution of its first Steel Committee in 1970. These representatives insisted on avoiding or moderating the regulatory and financial impact of any new laws imposing extra capital costs on the industry.[46]

The combination of two factors resulted in a pronounced ambivalence in the IOs' work on air and water pollution in the steel sector: the industry's interest in international collaboration and its aversion to new environmental legislation that could lead to high capital costs for new anti-pollution measures. At the working group level, experts closely collaborated towards strengthening scientific cooperation and facilitating work on, and the transfer of, new technologies to improve energy efficiency and limit negative environmental impact. At the same time, national or company representatives close to the industry superimposed on these results at the higher steel committee level wider economic and political conclusions reflecting their political and business concerns.

The political influence of steel producers was especially strong in the ECSC, where post-war 'Americanization' was largely limited to the transfer of formal institutions, especially treaty articles on competition policy, modelled to a large extent on the American anti-trust tradition.[47] However, despite these formal institutional changes, informal European institutions and practices persisted, especially Europe's strong cartel tradition. This institutional inertia of European practices was reflected in the lax application by the High Authority of the ECSC's competition law to mergers and acquisitions and in the formation in 1953 of an ECSC *export* cartel, which was not obviously illegal under the treaty. But fifteen years after the formation of the ECSC, the steelmakers in fact created another cartel in 1965–66 in clear breach of the treaty, but tolerated by the High Authority, this time combining a one-year domestic cartel with an export cartel in response to a new crisis of overproduction and dwindling prices that had started in 1961–62. It almost seemed as if 'fifteen years of European [integration] experience had left no noticeable mark' on the industry's behaviour.[48]

In fact, the American anti-trust policy at the time of the Schuman Plan had only strengthened the incipient cooperation among the steel companies in the six ECSC founding member states[49] until they created the Club des sidérurgistes at a meeting in headquarters of the ARBED steel company in Luxembourg in May 1952[50] – an organization later transformed into Eurofer in 1976.[51] With its monthly meetings, it allowed the steel managers and experts to coordinate their policy advice to the High Authority, directly and indirectly, via the Consultative Committee. In this committee the steel industry was regularly represented by influential individuals who combined great technical expertise with excellent political networks. Here, as well as through their close links with the Market Division of the High Authority, which was largely staffed with experts from the industry, the steel companies continuously kept the focus on more narrowly economic issues of the customs union, market integration, price levels and international competition. Environmental concerns remained limited to health and safety issues in the workplace, for which the ECSC had competences.

Crucially, the tightly organized ECSC core of international steel networks, which went back to the formation of the first transnational cartels in the late nineteenth century, had a strong influence on the work of IOs for a long time. These networks' cohesion and political influence largely explains the heavily Eurocentric character of IO policy deliberation in the steel sector.

Sometimes it's the Economy, Stupid!

When he reported on the Stockholm Conference to the OECD Council of Ministers on 30 June 1972, Emiel van Lennep, the OECD secretary-general from 1969 to 1984, observed that 'there was relatively little mention of economic studies ... [regarding the] control or the cost of implementing environmental policies'.[52] Clearly, despite the fact that the OECD had been the first IO to create its own environmental committee in 1970, the IO's core mission remained economic. Discontinuing the Steel Committee in 1970 and bringing the sector within the remit of the Industry Committee reflected the end of postwar reconstruction, the sector's relative health at the time and the desire of the OECD, with its limited financial resources, to expand into new policy fields like the environment.[53] Similarly, the Steel Committee's reconstitution in 1978 was an indicator of shifting priorities: the increasingly severe steel crisis that had set in in 1974 and the sector's great domestic economic, social and political importance for many member states seemed to justify the re-establishment of a separate committee. In fact, IOs like the OECD and the EC with their primarily economic focus rediscovered the growth paradigm after 1973 and, in the light of the economic crisis, put greater emphasis on measures to generate growth than before.[54]

The renewal of the committee was 'due to exceptional circumstances prevailing in the steel sector', emphasized the German Permanent Representative to the OECD, Egon Emmel, on behalf of the EC member states at the Council meeting on 6 November 1978. The committee would submit an annual report that illustrated its 'temporary character'. He insisted that the OECD's 'commitment to expand free world trade and to maintain the free and unimpeded exchange of goods' would not be impaired. As Aldolphe De Baerdemaeker added for the European Commission, 'individual measures were not capable of solving the problem on a world scale'. Discussions within the OECD could 'help to develop a common approach in the search for solutions acceptable to all'.[55]

However, the Steel Committee's Work Programme for 1979 stipulated that the sector's problems were not just of a cyclical, but also of a structural nature and were 'likely to persist for some time'. Clearly, the 'restructuring and modernization of the industry' constituted a more long-term challenge.[56] In fact, it quickly became obvious that the crisis would continue and with it the OECD's Steel Committee. At its first meeting on 19 January 1979, van Lennep outlined the sector's problems, such as rising energy prices since the 1973 oil crisis, the substitution of steel with alternative products including plastics in cars, for example, resulting in overproduction in Western countries and unemployment following the closure of plants. The danger was, the OECD secretary-general emphasized, that each member state would try to 'simply transfer the burden of inevitable adjustments to other countries'. Governments, he predicted, would be 'under pressure from business and labour to take defensive measures', which in the long run would be bad for the global economy.[57]

Against the backdrop of the anticipation of a severe and protracted crisis of the sector, the Steel Committee's Work Programme prioritized economic and social issues, notably studying trade flows, supply and demand, specialized steel subsectors, and questions of the adaptation of production structures and the retraining of labour. Environmental issues were not mentioned at all. The committee's purely economic focus was also reflected in the set-up of its secretariat, which was staffed by the OECD. Alongside a deputy secretary-general and economic and legal advisors, it only comprised officials from the Directorate for Science, Technology and Industry. The Environment Directorate was originally mentioned last in a list of OECD sections that could contribute to the Steel Committee's work. However, no official from the Environment Directorate ever attended meetings of the Steel Committee, nor did the directorate contribute to its work in any other visible way during the 1980s. The preserved Steel Committee documents do not reveal the reasons for their lack of involvement. In any case, the result was that even when the Steel Committee worked on topics with obvious environmental implications, it did not discuss them.

The complete absence of environmental issues from the Steel Committee deliberations can additionally be explained by the continued strong political role of the private and state-owned steel companies. Within the OECD, their influence was institutionally embedded in three different ways. First, with the notable exception of Germany, many Western European companies were state-owned by the early 1980s. In these cases the government representatives on the committee directly represented their interests. Secondly, member state governments were able to co-opt industry experts on their teams at the working group level where much of the analytical work about the sector's challenges was actually done.[58] Finally, the Steel Committee also consulted regularly with the Business and Industry Advisory Committee (BIAC) and the Trade Union Advisory Committee (TUAC) – both independent organizations of business and labour with a privileged advisory status in the OECD since 1962.

In these different ways, the OECD's steel deliberations replicated the tripartite neocorporatist forms of policy and decision making for the sector prevailing at the member state level in different forms, with close consultation among ministries, the (partly state-owned) industry and trade unions.[59] Until the privatization of British Steel in 1988 marked the beginning of more drastic policy measures for the steel sector in Western Europe,[60] this tripartite neocorporatism was effectively geared towards limiting and delaying economic restructuring and minimizing its immediate economic, social and political effects in times of low growth, rising deficits and growing unemployment across the Global North.

Lastly, the exclusion of environmental concerns from the OECD Steel Committee's work also resulted from the continued and very strong influence of the enlarged EC from 1973, when it included the United Kingdom as a major steel producer as well as Denmark and Ireland as new members. Drawing for the first time on the European Commission's strong formal powers in coal and steel policy, the new Belgian Commissioner for Industry, Etienne Davignon, worked towards a Community response to the crisis. From mid 1977 onwards, the Commission, in close cooperation with Eurofer, implemented a policy of mandatory minimum prices, generalized guide prices and controls on investments. When the market situation deteriorated once more in 1980, the system collapsed. It was replaced with more drastic measures from the second Davignon Plan, which included mandatory production quotas for four categories of steel. Even the government in Bonn eventually agreed to this drastic intervention in the market when German private companies also began to run increasing losses. In return, it induced the Commission to begin tackling the politically sensitive issue of state aid for steel companies, which it did from 1981 onwards.[61]

The EC's 'rationalization cartel'[62] was fraught with domestic and international problems. It led to sharp conflicts in Eurofer after 1980 between the state-owned companies and those in private ownership, which received no or limited state aid during the crisis. It also included the threat of anti-dumping measures against third countries to reduce steel imports into the EC, and the negotiation of only apparently voluntary export restraint agreements with countries ranging from Sweden and Austria in Western Europe to Eastern European CMEA countries and Japan and South Korea – agreements that did reduce EC steel imports and mitigated the crisis somewhat. To avoid a direct political clash in the General Agreement on Tariffs and Trade (GATT), the EC sought to upload discussion about its highly interventionist and controversial steel policy and the wider crisis into the OECD as a forum for debating policy.[63] Here, the EC's position was especially strong. Alongside the member states, the European Commission had represented the EC as a whole ever since it had acquired the status of a 'full participant' in the OECD reform of 1961–62. Thus, throughout the late 1970s and early 1980s in particular, the European Commission and the EC governments used the Steel Committee to justify their policies on steel.[64]

All of these factors together help explain the total exclusion of environmental considerations from the work of the Steel Committee throughout the late 1970s and the 1980s. Policy making for the sector at the IO and national levels was completely preoccupied with the domestic economic and social impact of the steel crisis and its implications for international trade (policy). Establishing the environment as a distinctive and institutionalized policy field with its own OECD directorate and committee apparently reflected its greatly increased importance in public opinion and public policy making in the early 1970s. At the same time, it segregated environmental concerns from other policy fields and marginalized it in, or even excluded it from, the deliberation of sector-specific problems, as in the case of the OECD's work on steel.

Against this background, it is not surprising that the Environment Directorate and its director from 1978 to 1984, the Canadian Jim MacNeill, advocated the idea of what later became known as 'environmental mainstreaming'. He called for integrating environmental concerns into all policy making across the OECD, other IOs and at the national level. MacNeill later became secretary-general of the World Commission on Environment and Development, where he played a crucial role in developing the notion of 'sustainable development', as Iris Borowy shows in her chapter in this book. Beginning the struggle for the mainstreaming of environmental concerns, MacNeill submitted a document to the OECD Council in May 1979 that (as many OECD reports) had been prepared by an (unknown) external consultant.[65] It discussed and propagated options for 'integrating environmental concerns into decision-making', precisely at a time when the OECD Steel Committee drifted off into discussing the sector's crisis from a purely economic and social policy perspective without any consideration of environmental matters.

Conclusion

Other chapters in this book provide ample evidence for how IOs have developed their activism in environmental matters, built new institutions, framed new approaches and set agendas that have frequently led to binding international conventions or recommendations that member states are morally obliged to follow. Analysing the case of steel policy, this chapter has inversed the perspective to identify the scope and the limits of IO work on the environment for policy deliberation and decision making in other sectors. It has generated fresh evidence for the early involvement of IOs and their steel committees in discussing air and water pollution from the mid 1950s through to the late 1960s. They mainly contributed to creating networks among officials from ministries and managers and experts from steel companies, both within the Western world and, in the case of the UNECE, across the East–West divide, with increasingly global extensions from the late 1960s onwards.[66] To establish whether, and to what extent, other UN regional economic organizations did the same would require further research on them and any transfers among these organizations within the UN system and beyond. The IOs also facilitated the formation of new transnationally constituted expert communities on the borderline between engineering and the natural sciences, and the aggregation and transfer of knowledge about the causes of air and water pollution, as well as strategies for combating and reducing it.

It has also become clear, however, that through their close links with national ministries and their own direct representation on the IO steel committees, the private and state-owned steel companies chose to upload air and water pollution issues to the international level and not just in order to foster networking and transfer knowledge. Instead, they regarded the IOs as a site for claiming how much they had already done to combat pollution, for arguing how costly any strict new legislation would be and for pleading that they, as an industry, should be treated as a 'special case' in need of government support for any additional anti-pollution measures because of the high costs of future investments. Not surprisingly, therefore, the steel committees refrained from any policy recommendations that could have encouraged stricter national environmental legislation, let alone binding international treaty commitments.

The globalization of the steel sector combined with its protracted structural crisis in Western Europe and North America after 1974 led to a renewed concentration of IOs and their steel committees on socioeconomic issues. This shift was reflected in the OECD unusually co-opting representatives from four industrializing countries onto its reconstituted Steel Committee after 1978, namely Brazil, Mexico, South Korea and India, to defuse political tensions over the trade implications of the EC's slow restructuring and continued state subsidies. The committee, originally conceived as a temporary measure to create a nonconfrontational forum for discussing steel issues, continued its work as the Western steel crisis lasted into the 1990s. The sector's accelerating globalization, with Asian crude steel output eventually outpacing the EU's by roughly eight to one in 2014, guaranteed that the committee became permanently established in the OECD committee structure.[67]

The greater politicization of the environment in the 1960s and early 1970s resulted in its institutionalization as a separate policy field. In fact, in the OECD's new 1970–71 structure for its directorates and committees, the separate steel committee was dissolved and the environment committee was set up at the same time. However, against the background of stagflation in the Western world after the 1973 oil crisis and of the steel crisis after 1974, the environment's politicization and institutionalization as a distinctive policy field had ambivalent effects. IOs and their committees were now freer to study environmental issues horizontally across different sectors. Instead of focusing specifically on new process technologies and their potential for reducing energy consumption and pollution in the steel sector, for example, the new IO committees could now study the potential of technological innovation to reduce environmental degradation across the board. Moreover, they could reduce the influence of business actors on policy-relevant findings and recommendations, and intensify links with the emerging networks of international non-governmental organizations with an interest in the environmental protection agenda.

This new freedom associated with the policy field's separate institutionalization came at a heavy price, however. Unlike its predecessor, the newly created OECD Steel Committee completely ignored environmental issues in its work after 1978. Instead, the sector's crisis induced member states, business actors and trade unions to focus entirely on the economic and social issues of the slow restructuring process. Environmental issues may have been prominent among experts and in international forums. Yet, growing unemployment and social unrest threatened to cost political parties many votes. As a result, the steel crisis had far greater political salience (except for the incipient Green parties) during the 1980s than even the widely reported acid rain problem and the predicted death of forests.[68]

The politicization of the environment combined with the economic crisis after 1973 thus made IO deliberations about environmental issues as low-key 'technical' issues, as it had been practised in the IO steel committees in the 1950s and 1960s, impossible. Paradoxically, the politicization of the environment thus made it easier for governments, business organizations and trade unions to confine its discussion to specialized IOs and committee structures, walling it off in a kind of policy-making ghetto. The steel sector is probably an extreme example. Its crisis was so severe that governments and organized social groups in the Western world began to regard the long-term environmental protection agenda with its demands for costly technological solutions as plainly detrimental to their short- and medium-term socioeconomic priorities of protecting production, profitability and employment.

In the case of the Montreal Protocol – frequently seen as a victory of environmentalists at the international level – the political economy worked to the agreement's advantage. Notably, the U.S. government was willing to sign up to a ban on chlorofluorocarbons (CFCs). Under domestic regulatory pressure, the American chemical industry had already developed and installed production capacity for alternative solutions since the 1970s, so that they could benefit from their technological edge.[69] In the case of steel, however, by the 1970s, the United States lagged behind even Western Europe in the introduction of new process technologies. Moreover, no technological advantage was able to halt the secular processes of material substitution and the shift of production of crude steel in particular to industrializing countries, especially in Asia.

In this situation, political economy concerns for a long time superseded worries about environmental degradation preventing stricter regulation. To the present day, steel is indeed frequently treated as a 'special case', for example, to justify tax exemptions or reductions for this energy-intensive sector with small profit margins. Steel is thus an extreme example of how political economy concerns have often trumped environmental protection in the work of IOs and in national politics and legislation. Sometimes, to paraphrase Bill Clinton's 1992 campaign manager, it is really 'the economy, stupid!'.

Wolfram Kaiser is Professor of European Studies, University of Portsmouth, United Kingdom, and Visiting Professor at the College of Europe, Bruges, Belgium, and at NTNU Trondheim, Norway.

Notes

1. Iris Borowy, *Defining Sustainable Development for Our Common Future: A History of the World Commission on Environment and Development (Brundtland Commission)* (Abingdon: Routledge, 2014).
2. See Yves Mény and Vincent Wright (eds), *The Politics of Steel: Western Europe and the Steel Industry in the Crisis Years (1974–1984)* (Berlin: Walter de Gruyter, 1987).
3. For the history of the ECSC, see Dirk Spierenburg and Raymond Poidevin, *The History of the High Authority of the European Coal and Steel Community: Supranationality in Operation* (London: Weidenfeld & Nicolson, 1994).
4. See Wolfram Kaiser and Johan Schot, *Writing the Rules for Europe: Experts, Cartels, and International Organizations* (Basingstoke: Palgrave Macmillan, 2014), Chapter 7.
5. John McCormick, *The Global Environmental Movement*. (Chichester: John Wiley, 1995), 110, 238–43.
6. See Richard Elliot Benedick, *Ozone Diplomacy: New Directions in Safeguarding the Planet* (Cambridge, MA: Harvard University Press, 1991); Karen Litfin, *Ozone Discourses: Science and Politics in Global Environmental Cooperation* (New York: Columbia University Press, 1994).
7. Cf. OECD, *The Role of Technology in Iron and Steel Developments* (Paris: OECD, 1989).
8. René Leboutte, *Histoire économique et sociale de la construction européenne* (Brussels: PIE Peter Lang, 2008), 464–75; Isabelle Cassiers, 'Le contexte

économique. De l'age d'or à la longue crise', in *Milieux économiques et intégration européenne au XXe siècle. La crise des années 1970. De la conférence de La Haye à la veille de la relance des années 1980*, Éric Bussière, Michel Dumoulin and Sylvain Schirmann (eds) (Brussels: PIE Peter Lang, 2007), 13–32.

9. C/M(78)19(Prov.), 6 November 1978, OECD Archives.
10. UNECE, Industry and Materials Committee, Resolution on the Establishment of a Sub-Committee on Steel, 26 November 1947, E/ECE/IM/9 Rev. I, ARR 14/1360/Box 9, UNOG Archives.
11. Report on the Future Programme of Work of the Steel Committee, Note by the Secretariat, 10 March 1950, E/ECE/Steel/48, UNOG Archives.
12. On the history of the OEEC until 1961, see Richard T. Griffiths (ed.), *Explorations in OEEC History* (Paris: OECD, 1997).
13. 1st Session of the Special Committee for Iron and Steel, 28 September 1965, Sidérurgie, Comité Spécial, 1965–70, OECD Archives.
14. See also UNECE, *Three Decades of the United Nations Economic Commission for Europe* (New York: UNECE, 1978).
15. David Wightman, *Economic Co-operation in Europe: A Study of the United Nations Economic Commission for Europe* (London: Stevens, 1956), 94.
16. Paul R. Josephson et al., *An Environmental History of Russia* (Cambridge: Cambridge University Press, 2013), 178–79, 219–21.
17. OEEC, *European Productivity Agency, Air and Water Pollution: The Position in Europe and in the United States* (Paris: OEEC, 1957).
18. On the multiple purposes of Cold War U.S. meteorological research, see Jacob Darwin Hamblin, *Arming Mother Nature: The Birth of Catastrophic Environmentalism* (Oxford: Oxford University Press, 2013), 108–128.
19. OECD, *Air Pollution in the Iron and Steel Industry* (Paris: OECD, 1963).
20. Ibid., 7.
21. Ibid., 49.
22. OECD, *Methods of Measuring Air Pollution, Report of the Working Party on Methods of Measuring Air Pollution and Survey Techniques* (Paris: OECD, 1964).
23. UNECE, *Problems of Air and Water Pollution Arising in the Iron and Steel Industry* (New York: UNECE, 1970).
24. Ibid., iii.
25. Jens Ivo Engels, 'Modern Environmentalism', in *The Turning Points of Environmental History*, Frank Uekötter (ed.), (Pittsburgh: University of Pittsburgh Press, 2010), 119–31.
26. On the relationship between the UN and the EC over environmental issues and policy, see also Laura Scichilone, 'A New Challenge for Global Governance: The UN and the EEC/EU in the Face of the Contemporary Ecological Crisis', in *Networks of Global Governance: International Organisations and European Integration in Historical Perspective*, Lorenzo Mechi, Guia Migani and Francesco Petrini (eds) (Cambridge: Cambridge Scholars, 2014), 229–48.

27. For a long-term perspective, see David Stradling and Peter Thorsheim, 'The Smoke of Great Cities: British and American Efforts to Control Air Pollution, 1860–1914', *Environmental History* 4(1) (1999): 6–31.
28. For the British context, see Peter Thorsheim, *Inventing Pollution: Coal, Smoke and Culture in Britain since 1800* (Athens, OH: Ohio University Press, 2006).
29. Cf. Frank Uekötter, 'Das organisierte Versagen. Die deutsche Gewerbeaufsicht und die Luftverschmutzung vor dem ökologischen Zeitalter', *Archiv für Sozialgeschichte* 43(1) (2003): 127–50.
30. See also Frank Uekötter, *Naturschutz im Aufbruch. Eine Geschichte des Naturschutzes in Nordrhein-Westfalen 1945–1980* (Frankfurt: Campus, 2004).
31. Although individual steel industrialists like Fritz Thyssen and foreign organizations like the French Comité de Forges provided funding for the National Socialists and most steel magnates were hostile to the Weimar Republic, they did not contribute significantly to Hitler's political rise to power. See (summarizing a long historiographical controversy) Hans-Ulrich Wehler, *Deutsche Gesellschaftsgeschichte*. Vol. 4: *Vom Beginn des Ersten Weltkrieges bis zur Gründung der beiden deutschen Staaten 1914–1949* (Munich: C.H. Beck Verlag, 2003), 293.
32. Frank Bösch, *Die Adenauer-CDU. Gründung, Aufstieg und Krise einer Erfolgspartei 1945–69* (Stuttgart: DVA, 2001).
33. OECD, *Air Pollution*, 91.
34. Ibid., 92.
35. Ervin Hexner, *The International Steel Cartel* (Chapel Hill: University of North Carolina Press, 1943), 15.
36. Cf. Helmut Maier, Andreas Zilt and Manfred Rasch, '150 Jahre Stahlinstitut VDEh. Eine Einführung', in *150 Jahre Stahlinstitut VDEh 1860–2010*, Helmut Maier, Andreas Zilt and Manfred Rasch (eds) (Essen: Klartext, 2010), 1–18, 10.
37. For the wider context of global changes and the crisis of the East-Central and Eastern European socialist systems, see André Steiner, 'The Globalisation Process and the Eastern Bloc Countries in the 1970s and 1980s', *European Review of History* 21(2) (2014): 165–81.
38. Werner Bührer, 'Der Verein Deutscher Eisenhüttenleute, die Internationalisierung und die Montanunion', in Maier, Zilt and Rasch (eds), *150 Jahre Stahlinstitut*, 223–38, 228.
39. See the reports in G.X 18/9/1/102, UNOG Archives.
40. Cf. UNECE, *Three Decades of the United Nations Economic Commission for Europe* (New York: UNECE, 1978).
41. Cf. V.I. Filippov, Director, Industry Division, to J. Stanovnik, Executive Secretary, 17 July 1970, G.X 18/9/1/127, UNOG Archives.
42. List of Participants of the Study Tour in Poland, G.X 18/9/1/127, UNOG Archives.
43. On East–West technology transfer in the Cold War, see also, more generally, Sari Autio-Sarasmo and Katalin Miklóssy (eds), *Reassessing Cold War Europe* (New York: Routledge, 2010); Sari Autio-Sarasmo and Brendan Humphreys (eds), *Winter Kept Us Warm: Cold War Interactions Reconsidered* (Helsinki:

Kikimora Publications, 2010); Sari Autio-Sarasmo, 'Cooperation across the Iron Curtain: Experts, Soviet Transfer of Technology from West Germany in the 1960s', in Martin Kohlrausch, Katrin Steffen and Stefan Wiederkehr (eds), *Expert Cultures in Central Eastern Europe: The Internationalization of Knowledge and the Transformation of National States since World War I* (Osnabrück: Fibre, 2012), 223–39.

44. Cf. William Marsh Baldwin (ed.), *Proceedings of the First World Metallurgical Congress, under the Auspices of the American Society for Metals* (Cleveland, OH: American Society for Metals, 1952).

45. Cf. Bührer, 'Der Verein Deutscher Eisenhüttenleute', 225.

46. On the resulting lack of environmental impact of the wider policy debates, see also, more generally, Paula Schönach, 'Limitations of Environmental Success without Successful Environmental Policy', *Global Environment* 6 (2011): 122–49.

47. Brigitte Leucht, 'Transatlantic Policy Networks in the Creation of the First European Anti-trust Law: Mediating between American Anti-trust and German Ordo-liberalism', in *The History of the European Union. Origins of a Trans- and Supranational Polity, 1950–72*, Wolfram Kaiser, Brigitte Leucht and Morten Rasmussen (eds) (Abingdon: Routledge, 2009), 56–73.

48. Charles Barthel, 'The 1966 European Steel Cartel and the Collapse of the ECSC High Authority', in *Alan S. Milward and a Century of European Change*, Fernando Guirao, Frances M.B. Lynch and Sigfrido M. Ramírez Pérez (eds) (Abingdon: Routledge, 2012), 333–50, 350.

49. Cf. John Gillingham, *Coal, Steel, and the Rebirth of Europe, 1945–1955: The Germans and French from Ruhr Conflict to Economic Community* (Cambridge: Cambridge University Press, 1991), 292.

50. Hans Dichgans, *Montanunion. Menschen und Institutionen* (Düsseldorf: Econ, 1980), 139.

51. See also the brief comments in Charles Barthel, 'Eurofer', in Éric Bussière et al. (eds), *The European Commission 1973–86: History and Memories of an Institution* (Luxembourg: Publications Office of the European Union, 2014), 269–70.

52. C/M(72)17(Prov.), Part I, Paris, 30 June 1972, OECD Archives. This seems to have held, more generally, for the early days of environmental policy, as suggested by contemporary social scientist Anthony Downs in 'Up and Down with Ecology: The "Issue-Attention" Cycle', *Public Interest* 28(1) (1972): 38–50, 39, 46.

53. See also Matthias Schmelzer, 'The Crisis before the Crisis: The "Problems of Modern Society" and the OECD, 1968–74', *European Review of History* 19(6) (2012): 999–1020.

54. Cf. Matthias Schmelzer, 'The Growth Paradigm: History, Hegemony, and the Contested Making of Economic Growthmanship', *Ecological Economics* 118 (2015): 262–71.

55. C/M(78)19(Prov.), Paris, 6 November 1978, OECD Archives.

56. SC(78)1, 24 November 1978, OECD Archives.

57. Opening Statement by the Secretary General, 19 January 1979, SC/M(79)1, Annex 1, OECD Archives.
58. Oral summary by the Chairman of the conclusions and decisions reached at the first meeting of the Steel Committee, SC/M(79)1 Annex 2, OECD Archives; Steel Committee, Summary record of the second meeting held at OECD headquarters, Paris, 30 January 1979, SC/M(79)2, 1 March 1979, OECD Archives.
59. Cf. Yves Mény and Vincent Wright, 'State and Steel in Western Europe', in Mény and Wright (eds), *The Politics of Steel*, 1–110.
60. On British Steel in this period, see Heidrun Abromeit, *An Industry between the State and the Private Sector* (Oxford: Berg, 1986).
61. For the EC's steel policy after 1974, see Lukas Tsoukalis and Robert Strauss, 'Community Policies on Steel 1974–1982', in Mény and Wright (eds), *The Politics of Steel*, 186–221; Leboutte, *Histoire économique et sociale de la construction européenne*, 477–506.
62. Tsoukalis and Strauss, 'Community Policies', 219.
63. See also Statement by Mr A.W. Wolff on taking the Chair of the Steel Committee, 30 January 1979, SC/M(79)2 Annex 1, OECD Archives.
64. See e.g. Steel Committee, Summary Record of the Eleventh Meeting on 14 May 1981, SC/M(81)2, OECD Archives.
65. C/M(79)9(Prov.), 10 May 1979; C(79)58, OECD Archives.
66. On the UNECE, see Yves Berthelot (ed.), *Unity and Diversity in Development Ideas: Perspectives from the UN Regional Commissions* (Bloomington: Indiana University Press, 2004).
67. On the steel industry's globalization, see also Anthony P. D'Costa, *The Global Restructuring of the Steel Industry: Innovations, Institutions, and Industrial Change* (London: Routledge, 1999).
68. For the case of Germany. see Birgit Metzger, *'Erst stirbt der Wald, dann du!'. Das Waldsterben als westdeutsches Politikum (1978–1986)* (Frankfurt: Campus, 2015).
69. R. Daniel Kelemen and David Vogel, 'Trading Places. The Role of the United States and the European Union in International Environmental Politics', *Comparative Political Studies* 43(4) (2010): 427–56, 445 f.

Bibliography

Abromeit, Heidrun, *An Industry between the State and the Private Sector* (Oxford: Berg, 1986).
Autio-Sarasmo, Sari, 'Cooperation across the Iron Curtain: Experts, Soviet Transfer of Technology from West Germany in the 1960s', in *Expert Cultures in Central Eastern Europe: The Internationalization of Knowledge and the Transformation of National States since World War I*, Martin Kohlrausch, Katrin Steffen and Stefan Wiederkehr (eds), (Osnabrück: Fibre, 2012), 223–39.
Autio-Sarasmo, Sari and Brendan Humphreys (eds), *Winter Kept Us Warm: Cold War Interactions Reconsidered* (Helsinki: Kikimora Publications, 2010).

Autio-Sarasmo, Sari and Katalin Miklóssy (eds), *Reassessing Cold War Europe* (New York: Routledge, 2010).

Barthel, Charles, 'The 1966 European Steel Cartel and the Collapse of the ECSC High Authority', in *Alan S. Milward and a Century of European Change*, Fernando Guirao, Frances M.B. Lynch and Sigfrido M. Ramírez Pérez (eds) (Abingdon: Routledge, 2012), 333–50.

———. 'Eurofer', in *The European Commission 1973–86: History and Memories of an Institution*, Éric Bussière et al. (eds) (Luxembourg: Publications Office of the European Union, 2014), 269–70.

Benedick, Richard Elliot, *Ozone Diplomacy: New Directions in Safeguarding the Planet* (Cambridge, MA: Harvard University Press, 1991)

Berthelot, Yves (ed.), *Unity and Diversity in Development Ideas: Perspectives from the UN Regional Commissions* (Bloomington: Indiana University Press, 2004).

Borowy, Iris, *Defining Sustainable Development for Our Common Future: A History of the World Commission on Environment and Development (Brundtland Commission)* (Abingdon: Routledge, 2014).

Bösch, Frank, *Die Adenauer-CDU. Gründung, Aufstieg und Krise einer Erfolgspartei 1945–69* (Stuttgart: DVA, 2001).

Bührer, Werner, 'Der Verein Deutscher Eisenhüttenleute, die Internationalisierung und die Montanunion', in *150 Jahre Stahlinstitut 1860–2010*, Helmut Maier, Andreas Zilt and Manfred Rasch (eds) (Essen: Klartext, 2010), 223–38.

Cassiers, Isabelle, 'Le contexte économique. De l'âge d'or à la longue crise', in *Milieux économiques et intégration européenne au XXe siècle. La crise des années 1970. De la conférence de La Haye à la veille de la relance des années 1980*, Éric Bussière, Michel Dumoulin and Sylvain Schirmann (eds) (Brussels: PIE Peter Lang, 2007), 13–32.

D'Costa, Anthony P., *The Global Restructuring of the Steel Industry: Innovations, Institutions, and Industrial Change* (London: Routledge, 1999).

Dichgans, Hans, *Montanunion. Menschen und Institutionen* (Düsseldorf: Econ, 1980).

Downs, Anthony, 'Up and Down with Ecology: The "Issue-Attention" Cycle', *Public Interest* 28(1) (1972): 38–50.

Engels, Jens Ivo, 'Modern Environmentalism', in *The Turning Points of Environmental History*, Frank Uekötter (ed.) (Pittsburgh: University of Pittsburgh Press, 2010), 119–31.

Filippov, V.I., Director, Industry Division, to J. Stanovnik, Executive Secretary, 17 July 1970, G.X 18/9/1/127, UNOG Archives.

Gillingham, John, *Coal, Steel, and the Rebirth of Europe, 1945–1955: The Germans and French from Ruhr Conflict to Economic Community* (Cambridge: Cambridge University Press, 1991).

Griffiths, Richard T. (ed.), *Explorations in OEEC History* (Paris: OECD, 1997).

Hamblin, Jacob Darwin, *Arming Mother Nature: The Birth of Catastrophic Environmentalism* (Oxford: Oxford University Press, 2013).

Hexner, Ervin, *The International Steel Cartel* (Chapel Hill: University of North Carolina Press, 1943).

Josephson, Paul R. et al., *An Environmental History of Russia* (Cambridge: Cambridge University Press, 2013).

Kaiser, Wolfram and Johan Schot, *Writing the Rules for Europe: Experts, Cartels, and International Organizations* (Basingstoke: Palgrave Macmillan 2014).

Kelemen, R. Daniel and David Vogel, 'Trading Places: The Role of the United States and the European Union in International Environmental Politics', *Comparative Political Studies* 43(4) (2010): 427–56.

Leboutte, René, *Histoire économique et sociale de la construction européenne* (Brussels: PIE Peter Lang, 2008).

Leucht, Brigitte, 'Transatlantic Policy Networks in the Creation of the First European Anti-trust Law. Mediating between American Anti-trust and German Ordo-liberalism', in *The History of the European Union: Origins of a Trans- and Supranational Polity, 1950–72,* Wolfram Kaiser, Brigitte Leucht and Morten Rasmussen (eds) (Abingdon: Routledge, 2009), 56–73.

List of Participants of the Study Tour in Poland, G.X 18/9/1/127, UNOG Archives.

Litfin, Karen, *Ozone Discourses: Science and Politics in Global Environmental Cooperation* (New York: Columbia University Press, 1994).

Maier, Helmut, Andreas Zilt and Manfred Rasch, '150 Jahre Stahlinstitut VDEh. Eine Einführung', in *150 Jahre Stahlinstitut VDEh 1860–2010,* Helmut Maier, Andreas Zilt and Manfred Rasch (eds) (Essen: Klartext, 2010), 1–18.

Marsh Baldwin, William (ed.), *Proceedings of the first World Metallurgical Congress, under the Auspices of the American Society for Metals* (Cleveland, OH: American Society for Metals, 1952).

McCormick, John, *The Global Environmental Movement* (Chichester: John Wiley, 1995).

Mény, Yves and Vincent Wright, 'State and Steel in Western Europe', in *The Politics of Steel: Western Europe and the Steel Industry in the Crisis Years (1974–1984),* Yves Mény and Vincent Wright (eds) (Berlin: Walter de Gruyter, 1987), 1–110.

———. (eds) *The Politics of Steel: Western Europe and the Steel Industry in the Crisis Years (1974–1984)* (Berlin: Walter de Gruyter, 1987).

Metzger, Birgit, *'Erst stirbt der Wald, dann du!'. Das Waldsterben als westdeutsches Politikum (1978–1986)* (Frankfurt: Campus, 2015).

OECD Steel Committee, Summary Record of the Eleventh Meeting on 14 May 1981, SC/M(81)2, OECD Archives.

OECD, *Air Pollution in the Iron and Steel Industry* (Paris: OECD, 1963).

———. *Methods of Measuring Air Pollution, Report of the Working Party on Methods of Measuring Air Pollution and Survey Techniques* (Paris: OECD, 1964).

———. 1st Session of the Special Committee for Iron and Steel, 28 September 1965, Sidérurgie, Comité Spécial, 1965–1970, OECD Archives.

———. C/M(72)17(Prov.), Part I, Paris, 30 June 1972, OECD Archives.

———. C/M(78)19(Prov.), 6 November 1978, OECD Archives.

———. SC(78)1, 24 November 1978, OECD Archives.

———. Opening Statement by the Secretary General, 19 January 1979, SC/M(79)1, Annex 1, OECD Archives.

————. Oral summary by the Chairman of the conclusions and decisions reached at the first meeting of the Steel Committee, SC/M(79)1 Annex 2, OECD Archives; Steel Committee, Summary record of the second meeting held at OECD headquarters, Paris, 30 January 1979, SC/M(79)2, 1 March 1979, OECD Archives.

————. Statement by Mr. A.W. Wolff on taking the Chair of the Steel Committee, 30 January 1979, SC/M(79)2 Annex 1, OECD Archives.

————. C/M(79)9(Prov.), 10 May 1979; C(79)58, OECD Archives.

————. *The Role of Technology in Iron and Steel Developments* (Paris: OECD, 1989).

OEEC, *European Productivity Agency, Air and Water Pollution: The Position in Europe and in the United States* (Paris: OEEC, 1957).

Schmelzer, Matthias, 'The Crisis before the Crisis: The "Problems of Modern Society" and the OECD, 1968–74', *European Review of History* 19(6) (2012), 999–1020.

————. 'The Growth Paradigm: History, Hegemony, and the Contested Making of Economic Growthmanship', *Ecological Economics* 118 (2015): 262–71.

Schönach, Paula, 'Limitations of Environmental Success without Successful Environmental Policy', *Global Environment* 6 (2011): 122–49.

Scichilone, Laura, 'A New Challenge for Global Governance: The UN and the EEC/EU in the Face of the Contemporary Ecological Crisis', in *Networks of Global Governance: International Organisations and European Integration in Historical Perspective*, Lorenzo Mechi, Guia Migani and Francesco Petrini (eds) (Cambridge: Cambridge Scholars, 2014), 229–48.

Spierenburg, Dirk and Raymond Poidevin, *The History of the High Authority of the European Coal and Steel Community: Supranationality in Operation* (London: Weidenfeld & Nicolson, 1994).

Steiner, André, 'The Globalisation Process and the Eastern Bloc Countries in the 1970s and 1980s', *European Review of History* 21(2) (2014): 165–81.

Stradling, David and Peter Thorsheim, 'The Smoke of Great Cities: British and American Efforts to Control Air Pollution, 1860–1914', *Environmental History* 4(1) (1999): 6–31.

Thorsheim, Peter, *Inventing Pollution. Coal, Smoke and Culture in Britain since 1800* (Athens, OH: Ohio University Press, 2006).

Tsoukalis, Lukas and Robert Strauss, 'Community Policies on Steel 1974–1982', in *The Politics of Steel: Western Europe and the Steel Industry in the Crisis Years (1974–1984)*, Yves Mény and Vincent Wright (eds) (Berlin: Walter de Gruyter, 1987), 186–221.

Uekötter, Frank, 'Das organisierte Versagen. Die deutsche Gewerbeaufsicht und die Luftverschmutzung vor dem ökologischen Zeitalter', *Archiv für Sozialgeschichte* 43(1) (2003): 127–50.

————. *Naturschutz im Aufbruch. Eine Geschichte des Naturschutzes in Nordrhein-Westfalen 1945–1980* (Frankfurt: Campus, 2004).

UNECE, Industry and Materials Committee, Resolution on the Establishment of a Sub-Committee on Steel, 26 November 1947, E/ECE/IM/9 Rev. I, ARR 14/1360/Box 9, UNOG Archives.

———. Report on the Future Programme of Work of the Steel Committee, Note by the Secretariat, 10 March 1950, E/ECE/Steel/48, UNOG Archives.

———. *Problems of Air and Water Pollution Arising in the Iron and Steel Industry* (New York: UNECE, 1970).

———. *Three Decades of the United Nations Economic Commission for Europe* (New York: UNECE, 1978).

Wehler, Hans-Ulrich, *Deutsche Gesellschaftsgeschichte. Vol. 4: Vom Beginn des Ersten Weltkrieges bis zur Gründung der beiden deutschen Staaten 1914–1949* (Munich: C.H. Beck, 2003).

Wightman, David, *Economic Co-operation in Europe: A Study of the United Nations Economic Commission for Europe* (London: Stevens, 1956).

⚜ CHAPTER 6

Making the Polluter Pay
How the European Communities Established Environmental Protection

Jan-Henrik Meyer

When the Environmental Committee of the Organisation for Economic Co-operation and Development (OECD) held a meeting at ministerial level in Paris at the end of 1974, Carlo Scarascia-Mugnozza, the Vice-President of the European Commission and Italian Commissioner responsible for environmental and transport policy, addressed the ministers.[1] Barely one year after the start of the European Communities' (EC) environmental policy and the declaration of the first Environmental Action Programme,[2] the Commissioner also called for a more ambitious environmental policy that went 'beyond the battle against pollution' at the level of the OECD. He highlighted the need for a more comprehensive conception of economic growth towards improving the 'quality of life'. He emphasized that the EC was committed to its new Environmental Action Programme even in a period of economic trouble. The 1973 oil crisis had pushed Western economies into recession.[3] He promised EC member states' support for the OECD's environmental objectives and highlighted that the EC ministers had just agreed on an (albeit non-binding) recommendation on the polluter pays principle, which had also been an OECD priority.[4]

This episode from the early days of the new EC environmental policy seems very much in line with the familiar image of the present-day European Union (EU). At the international level, the EU has been presenting itself as a global environmental leader, providing an example of good practice and pushing others to follow.[5] However, the EC was actually a latecomer to environmental policy. The governments of the EC member states only agreed in principle on establishing a new policy in the autumn of 1972 in the wake the United Nations' (UN) Stockholm Conference on 'The Human Environment' in June 1972.

Rather than being a leader among international organizations (IOs), the EC initially followed the lead of others. This allowed it to draw on the work of the different IOs already engaged in the new policy area.[6] Scarascia-Mugnozza's ideas about 'quality of life', that is, going beyond a purely quantitative conception of growth, had been widely discussed across different IOs, including the OECD.[7] In his address, he also openly acknowledged that the EC owed the definition of the polluter pays principle to the OECD.[8]

This chapter explores how the EC as a latecomer carved out its own place among IOs in the new policy area. It demonstrates that the already-existing environmental policies of other IOs provided important models and points of reference. At the same time, different institutional and individual actors within the EC mattered strongly in shaping an environmental policy. However, they were bound by the specific opportunities and constraints of a regional organization committed to market integration in order to foster prosperity.

The chapter is organized as follows. The first part will provide a brief overview of the creation of EC environmental policy, its contents and focus, as well as the actors and institutions involved in this process. The second part will analyse how actors from different EC institutions put one of the core principles of environmental policy on the EC agenda.[9] Drawing on a variety of sources, including other IOs, these actors laid the basis for one of the first concrete pieces of EC environmental policy: the recommendation on the polluter pays principle. The chapter will conclude by exploring how and why the EC carved out its role despite – or possibly even profiting from – being a latecomer to environmental policy, and how it thus laid the foundations for an increasingly prominent role.

Institutionalizing a Policy: The Environmental Action Programme

That the EC started to become involved in the area of environmental protection in the early 1970s seems at least as surprising as in the case of the OECD discussed by Iris Borowy in her chapter. First and foremost, the EC was an economic community, committed to facilitating economic growth by unleashing the productive forces of competition via the creation of a Common Market across its Western European member states. When the EC was created in a merger in 1967 of the European Economic Community (EEC), the European Coal and Steel Community

(ECSC) and the European Atomic Community (EURATOM), it incorporated two Communities designed to fuel Europe's economy by providing unlimited and equal access to what were considered the necessary energy sources of the past, present and future.

Today, coal combustion and nuclear energy are controversial issues. Already in the 1950s and 1960s, the negative side-effects of the use of coal (not only in steel production) were well known, as Wolfram Kaiser demonstrates in his chapter. However, at the time, the main issue of concern were sulphur dioxide emissions leading to acid rain and the acidification of Scandinavian lakes, as highlighted in the first chapter of this book. The impact of CO_2 emissions on the global climate – though not unknown in the early 1970s – was not yet a major environmental issue. Nuclear energy only became more controversial in the second half of the 1970s when environmentalists started to challenge the EC on the large-scale use of nuclear power for electricity generation.[10]

When the Treaties of Rome were formulated in 1956–57, the environment had not yet emerged as a policy area – as explained in the first chapter in this book. Hence, there was no explicit legal basis for an EC environmental policy until the Single European Act of 1987. The new policy was thus only weakly enshrined in a series of Environmental Action Programmes from 1973 onwards. However, the scope and contents of the new policy were surprisingly wide and comprehensive, reflecting the state of the art on the issue in the aftermath of the Stockholm Conference.[11]

The first Environmental Action Programme defined what EC environmental policy was to be about. It integrated two different conceptions of the environment. On the one hand, the Action Programme's definition was in line with the new ecological notion of the environment as a globally interconnected, border-crossing problem of nature universalized as the 'biosphere'. On the other hand, the description of the environmental problem harks back to a more anthropocentric conception of the environment, subsuming the environment under more traditional socioeconomic concerns. Its declared objective was to 'improve the setting and quality of life' and 'surroundings and living conditions of the peoples of the Community'. The aim was to 'reconcile' economic 'expansion with the increasingly imperative need to preserve the natural environment', foreshadowing to some extent the notion of 'sustainable development' that Iris Borowy discusses in her chapter.[12]

The Action Programme's six more concrete objectives related first and foremost to the most pressing contemporary issue of 'prevent[ing]', 'reduc[ing]' and 'eliminat[ing] pollution'; secondly, and more generally, to 'maintaining a satisfactory ecological balance'; and, thirdly, to the venerable conservationist concern about 'the sound management of … resources'. The fourth aim referred to social concerns of a 'quality'-oriented management of 'development', with a view to 'improving the working conditions and the settings of life'. These issues relate to what environmental historians have more recently problematized as environmental justice concerns.[13] Fifthly and more concretely, the policy aimed at the integration – or what Wolfram Kaiser discusses in his chapter as mainstreaming – of 'environmental aspects in town planning and land use'. The sixth and final aim related to global cooperation to 'seek common solutions to environmental problems with States outside the Community, particularly in international organisations'.[14] International relations and the cooperation with other IOs were thus a declared priority of EC environmental policy right from the start. This indicates how the EC was deeply embedded in an already-existing world of IOs dealing with environmental issues.

These objectives translated into a number of policy principles. Apart from the polluter pays principle, they included the principle of prevention of pollution 'at the source', counting on 'technological progress' and the integration (or mainstreaming) of environmental impacts 'at the earliest possible stage in all the technical planning and decision making process' in line with a concern for precaution.[15]

The concrete contents of the Action Programme reflected these objectives and concerns as well as the European and cross-border scope of the policy. The Action Programme specified three main areas of action, the first of which was pollution control. This included the evaluation of health risks of chemicals, an area in which the EC had already been active since the 1960s. It comprised the setting of common standards and quality objectives, but also more concrete issues of border-crossing pollution, notably the pollution of the Rhine basin. Waste also featured as a pollution issue, including 'the handling and storage of nuclear wastes'. Secondly, 'action to improve the environment' related to nature protection and resource exploitation and conservation, 'urban development' and 'improvement of the working environment'. In line with the sixth policy objective, the third area concerned international cooperation. The EC was to develop a coordinated stance in various IOs, including the OECD, the Council for Europe and the United Nations.[16]

This choice of issues demonstrates that EC policy makers at the time did not intend to replace but to complement, harmonize and coordinate the new national environmental policies of its (from 1973) nine member states. The Action Programme clearly stated what, in EU jargon, is now known as the subsidiarity principle, namely that anti-pollution action is to be taken at the most appropriate level (ranging from the local to the European level).[17] Moreover, the focus on setting common standards and rules and harmonizing policies was well in line with the ultimate rationale of the EC, namely ensuring the functioning of the Common Market.

Legal rules and competences specified in the EC treaties played a role for the areas in which the EC became active and strongly influenced which actors were to take the lead. Most measures had to be justified by reference to the Common Market due to the lack of a separate legal basis for the new policy. Similarly, the strong focus on international environmental policy was not only a result of the activism of IOs in the new policy area and the EC's self-assertion – it was also a consequence of the European Commission's special role at the international level. In a number of IOs, the Commission represented the EC. For instance, from 1961, the Commission already represented the then three communities within the OECD, alongside the member states.[18]

The scope and contents of environmental policy at the EC level not only depended on existing concepts and legal and institutional structures, but also on the activism and cooperation of individuals and institutions. As I will discuss in greater detail below, the two supranational institutions of the EC, the European Commission and the European Parliament, both promoted the new policy as well as its establishment within the EC. The European Commission's role as a central administration, executive and the single initiator of legislative projects predestined it to continuously look out for European problems in search of EC solutions. Moreover, most Commission officials and Members of the European Parliament (MEPs) shared a federalist spirit and sought to expand the scope of EC action.[19] Environmental issues seemed to lend themselves to such action.[20]

At the time, the European Parliament was still an unelected consultative assembly with no legislative or budgetary competences. Its members (MEPs) held dual mandates: they were delegated from their national parliaments. Consequently, usually only those members of national parliaments most committed to the idea of European integration volunteered to go to Strasbourg. At the same time, self-selection and dual mandates also meant that they were closely connected to national political debates and agendas. In many cases, they were well versed in parliamentary procedures and creatively used all instruments available to place what they considered to be their citizens' concerns on the European agenda.[21]

As a constitutionally weak but activist assembly, the European Parliament organized its work in committees. In 1970, the Committee for Public Health and Social Affairs was the first EC body to place the environment on the EC agenda. This Committee declared the pollution of Europe's main river, the Rhine, a scandal and an environmental disaster of cross-border scope, and called for EC action. An environmental committee was only established in the spring of 1973 as the Committee of Public Health and the Environment. A number of individual MEPs, including the German Christian Democrat Hans-Edgar Jahn,[22] who were involved in these early activities, soon became central actors and specialists on environmental issues. They continued to support the policy and demanded legislative action throughout the 1970s. They thus helped to expand EC environmental policy and to actually implement it.

Individual actors also mattered within the European Commission. At the highest political level, two of the nine (from 1973 thirteen) Commissioners played a central role. However, they differed in their analysis of and approach to the environmental problem.[23] The Commissioner for Agriculture, Sicco Mansholt, a Dutch social democrat and a farmer himself, was well known as a reckless technocrat and promoter of economic modernization. In 1968, the so-called Mansholt Plan, developed by the European Commission under his leadership, envisaged changing the structure of European agriculture from the family farm to large-scale industrialized farming, a shocking vision to most European farmers.[24] From 1971 onwards, however, Mansholt grew increasingly sceptical of large-scale technological solutions – including nuclear power – and became one of the key supporters of EC environmental action. His membership in the Club of Rome crucially contributed to this rapid change of attitude. He had advance access to, and was deeply impressed by, the scenarios of the study on the 'Limits to Growth'[25] that charted an environmental cataclysm.[26] He became increasingly critical of development and growth as the primary aims of modern society, and advocated an ecological conception of the environment.[27]

However, the ardent European federalist Altiero Spinelli, the Italian Commissioner responsible for industry and trade, considered economic development indispensable in order to fight poverty and ensure human progress. Rather than curtailing growth, policy makers should count on technology and human ingenuity to provide solutions to environmental problems. This view was prominently reflected in the Environmental Action Programme. Spinelli was probably the individual actor most instrumental in establishing the issue of the environment at the EC level.[28] He initiated and promoted the new policy as an instrument both for further integration and for demonstrating to European citizens that the EC took social and environmental concerns seriously.[29] In his anthropocentric perspective, the environment was just one of those socioeconomic issues necessary to improve the 'quality of life'.

In February 1971, the European Commission formally established a 'Working Group on the Environment' at the highest political level to discuss the options for a new policy. Spinelli presided over the group that included Mansholt as well as the Commissioners responsible for the Internal Market, Energy, Competition, Regional Policy, Social Affairs, Transport and the Budget. As these plans were to have implications for existing policy areas, the relevant Directorates-General were involved too.[30]

The differences in priority between Mansholt and Spinelli were reflected in the coexistence of two different definitions of the environment within the Action Programme. Thus, with a view to the focus of the policy, it had important implications insofar that its bureaucratic organization took place within Spinelli's area of competence. He established a special administrative unit to prepare EC work on environmental issues. Robert Toulemon, the responsible French Director-General of the Commission's Directorate-General for Science, Technology and Industry (DG III), set up a unit that prepared the Commission communications and proposals that formed the basis for the eventual Environmental Action Programme. The French former European Atomic Energy Community (EURATOM) official Michel Carpentier became its first director.[31]

This unit did not act in isolation, but consulted a broad range of experts, policy makers and institutions in different forums in order to design proposals that would garner sufficient political support. Carpentier and his team visited national ministries dealing with environmental issues. They worked closely with the officials from the Council of Minister's Working Group on Environmental Issues established in February 1972.[32] The unit's officials also avidly collected different materials from a variety of sources. They travelled to international meetings and conferences on the new environmental issue on both sides of the Atlantic, discussing issues of international cooperation. The division of roles between IOs, notably the EC, the

General Agreement on Tariffs and Trade (GATT) and the OECD, was an important consideration in this respect.[33] Initially mainly observing the work of other IOs dealing with environmental concerns, Carpentier and his team took an increasingly assertive role on issues they dealt with when representing the EC within other IOs, for instance, in the International Commission for the Protection of the Rhine against Pollution founded in 1950.[34]

From 1973 onwards, Carpentier's unit transformed into the Commission department in charge of environmental action, officially called the Service of the Environment and Consumer Protection (SEPC).[35] Its small number of staff included Commission officials such as Claus Stuffmann, who had already dealt with environmental issues in Spinelli's *cabinet* or private office,[36] a number of officials who had previously worked at EURATOM's Joint Research Centre at Ispra in Italy,[37] and 'parachuted in' national experts, such as the former World Bank official Stanley P. Johnson. From the perspective of the British Conservative government, Johnson qualified for the job of a head of unit on pollution control since he had contributed as environmental expert to the successful Conservative election campaign in 1970. He was part of the international environmental 'scene'. He had attended the Stockholm Conference representing the International Planned Parenthood Federation, a population control non-governmental organization (NGO), and had written one of the first books on environmental politics.[38] Carpentier's group of expert officials continued to shape and advance EC environmental policy in the 1970s and 1980s, for example, in the areas of bird protection and water and air pollution control.[39]

Whatever the European Parliament and the Commission were advocating, the support of the governments of the member states in the Council of Ministers was indispensable to expand EC policy in the area of the environment without a treaty base. However, even if most member state governments were engaged in establishing environmental protection in their own country at the time, they were divided over the scope and shape of environmental policy at the EC level. While, for instance, the German government supported the EC initiative as a way to expand and project its own national policy to the European level, the French government was notably more sceptical. In the Gaullist tradition, environmental minister Robert Poujade initially favoured inter-governmental cooperation for common policy measures to address the problem of new obstacles to trade through national environmental legislation.[40]

The Stockholm Conference acted as a catalyst for common action. In February 1972, the Council of Ministers not only set up its working group on the environment mentioned above, but also an 'ad hoc' group to coordinate a common position for Stockholm.[41] Still under the impression of the Stockholm Conference, the heads of state and government, and subsequently the Council eventually endorsed establishing an Environmental Action Programme in October 1972 and defined a number of core policy principles.[42] Among these principles, the polluter pays principle featured centrally. How did this principle enter the EC debate?

Setting the Agenda for a Policy Principle: The Polluter Pays!

'The polluter pays principle, this I affirm, has been invented by the Commission, followed by the Council', Carpentier highlighted in an oral history interview in 2010.[43] Contemporary documents tell a different story. Indeed, while the EC is widely acknowledged as its main promoter during the past forty years, the principle did not originate in Brussels. When they put it on the European agenda in the early 1970s, the European Parliament and the European Commission explicitly drew on external examples. Both institutions referred to academic debates and national precedents as well as the work of the Council of Europe and the OECD. The Parliament and the Commission's approach differed in terms of emphasis and timing: the Parliament's Economic Committee first flagged the polluter pays principle in 1970, but the Commission only included it in its second Communication of March 1972.

The notion that those causing harm to someone else, for instance, by spoiling water or air, should be held responsible for the damage is an old legal principle. Already in the nineteenth century, industrialists were occasionally taken to court and forced to pay compensation.[44] However, the polluter pays principle is more complex and comprehensive.[45] It is an idea informed by environmental economics. The principle essentially states that any pollution comes at a price because it causes damage to third parties. Hence, the cost of pollution should be allocated to – and paid for by – those who cause it. Effectively, the principle requires that (industrial) polluters include what economists call negative externalities into the costs of production. This is to prevent the scenario of the community affected by the pollution having to bear the cost. From a normative perspective, the principle seeks to ensure fairness and environmental justice. From a behavioural economics perspective, it is central for market economies, in two respects: first, only if pollution comes with a price tag is there an incentive to avoid it and to seek efficient solutions; and secondly, to ensure fair competition in the market, all participants have to include the cost of pollution. Externalizing costs de facto amounts to an undue subsidy to the polluter that distorts competition. The polluter pays principle was thus considered highly relevant to the EC as an economic community.

The discussion of the polluter pays principle dates back to the earliest instances of the environmental debate in an EC context. In the summer of 1969, the chemical Thiodan that the German chemical company Hoechst routinely and illegally released into the river Main, a major tributary to the Rhine, caused the death of huge numbers of fish.[46] This environmental catastrophe raised great concern about the quality of drinking water downstream in the Netherlands. The European Parliament politicized the issue and used the popular concern to call for EC action.[47]

The European Parliament's Committee for Public Health and Social Affairs produced an own initiative report, the Parliament's most powerful instrument for agenda setting. The rapporteur in charge, the Dutch Christian Democrat MEP Jacob Boersma,[48] demanded EC action on the 'protection of inland waterways with special reference to the pollution of the Rhine'.[49] The report stressed not only the dangers of polluted water to the health of citizens, but also the economic costs for companies relying on clean water as a resource in both agriculture and industry. The report criticized, for instance, the routine practice of dumping large amounts of salt, a byproduct of mining in the Alsace, into the river. Boersma equally warned against overheating the river once the ambitious Swiss, French and German programmes for building nuclear power plants were completed. Images of a steaming river had triggered heated debates in the media in 1970, preceding the conflict about nuclear power and the dangers of radiation by half a decade.[50]

The Boersma report referred to the different activities of IOs – such as the International Commission for the Protection of the Rhine against Pollution, discussed in the first chapter of this book – but considered them insufficient. Instead of providing practical solutions, they were limited to research and nonbinding recommendations. The report thus called for the harmonization of anti-pollution legislation. Divergent rules in different countries and government subsidies for environmental measures distorted competition among polluting industries in the Common Market. Consequently, the Commission, as the guardian of the treaties, was obliged to act.[51] The report strategically stressed the economic implications. From earlier responses to parliamentary questions[52] the MEPs knew that issues relating to the Common Market would provide the Commission with a legal basis for Community action. Eventually this argument proved decisive. Even the most hesitant governments, including the French government, preferred Community rules to contradictory national environmental legislation leading to new obstacles to trade in the Common Market.[53]

The Dutch socialist and trained mining engineer Adriaan Oele[54] was the first to explicitly refer to the polluter pays principle. Oele was the responsible rapporteur of the Economic Committee, which added its opinion to Boersma's report.[55] Oele agreed that the pollution of rivers – and specifically of the Rhine – constituted an important cross-border economic problem. Different rules and government subsidies distorted fair competition.

Remarkably, in what we tend to consider the heyday of technocratic command and control, a Social Democrat advocated market solutions to environmental problems. Oele argued that restrictive rules and fines were less effective than economic incentives. He reluctantly advocated using bans and fines in the short run, given they were the only measures readily available.[56] However, drawing on the debate in economics, he was by no means a precocious advocate of neoliberal market radicalism. His source of inspiration was the book *The Cost of Economic Growth* by Ezra J. Mishan, an economist from the London School of Economics and at the time well-known critic of the conventional notion of economic growth.[57] In line with the definition of the polluter pays principle, Oele advocated including the cost of pollution into the calculation of production costs. If universally applied, he argued, this would help to solve the pollution problem in a way that was compatible with market principles.[58]

The polluter pays principle also featured very prominently in the second parliamentary report on environmental issues. In 1971, the Committee for Social Affairs and Public Health produced its own initiative report on the second pressing cross-border pollution issue, namely air pollution.[59] While Boersma started writing the report, subsequently Jahn took over and completed it when Boersma left the European Parliament to join the Dutch government as a minister in 1971. The accompanying parliamentary resolution called for basing new Community rules on the polluter pays principle and warned that subsidies should be limited to the solution of specific problems.[60]

Jahn's report itself traced the principle back to national legislation and IOs. When reviewing existing legislation, the report went back to the 1962 Clean Air Act of the German federal state of North Rhine Westphalia. West Germany's most populous state included the Ruhr area, its heavily polluted industrial heartland of coal pits and steel mills, but also the Federal Republic's capital city Bonn, where Jahn resided. Almost a decade before environmental protection became a distinct area of policy, this pioneering act had already stipulated that – as a matter of principle – the polluter should bear the cost of pollution.[61] Additionally, the report linked the idea to the world of IOs. It highlighted that the principle had been part of the Council of Europe's Declaration of Principles on Air Pollution Control.[62]

While stressing the relevance of the principle, Jahn's report equally emphasized the potential problems of international competition with third countries. If polluting industries in other countries received subsidies for buying filters, for example, this would put European producers at a disadvantage.[63] Considering these problems, the statement by the Economic Committee accompanying this report, again written by Oele, made detailed suggestions as to how to flexibly implement the principle.[64] In the Parliament's subsequent report on the Commission's First Communication of 1971, Jahn reiterated this demand for flexibility and criticized the European Commission for not even mentioning the principle.[65]

Indeed, the European Commission only included the polluter pays principle in its second Communication on an Environmental Action Programme of March 1972.[66] The Commission did not claim to have invented the principle, nor did it refer to any of the sources the European Parliament had mentioned. Instead, it solely pointed to the work of the OECD. The Communication quoted extensively from the international agreement on the matter that the OECD's Environmental Committee had reached only a few weeks earlier.[67]

The OECD had started to get involved in international environmental policy early on. Building on the already-existing Committees on Research, Water and Air Pollution, established in 1967 and 1968, it set up an Environmental Committee in 1970. This was staffed by senior officials from its member states in Western Europe (including Turkey and Yugoslavia), Canada, the United States, Japan and the European Commission representing the EC as a whole. The new committee's purpose was to discuss the economic implications of environmental policy as well as encouraging cooperation in environmental research and management.[68]

Just like the EC, the OECD was devoted to economic growth through free trade and economic development, at the global level of (Western) industrialized countries. The OECD Environmental Committee addressed environmental issues and defined the polluter pays principle with a view to impact of national environmental legislation on international competition. In May 1972, the OECD issued a *Recommendation of the Council on Guiding Principles Concerning International Economic Aspects of Environmental Policy* based on the consensus reached in its Environmental Committee.[69] The OECD sought to commit its members to this principle, because the allocation of environmental costs and state subsidies for anti-pollution measures had serious consequences for international trade. In the course of the following years, the OECD continued to work on this issue. In 1974, in a follow-up recommendation, it specified guidelines for the implementation of the polluter pays principle.[70]

From the spring of 1972 onwards, the Commission wholeheartedly embraced the polluter pays principle. Reiterating the Council of Ministers' declaration of principles of the EC policy in October 1972, the Commission included this principle as a core tenet in its proposal for the Environmental Action Programme. However, EC policy makers did not swallow the OECD principle hook, line and sinker. In the Council of Ministers' discussion on the Environmental Action Programme in July 1973, Commissioner Scarascia-Mugnozza as well as a number of the national ministers, such as the responsible Dutch and the Irish ministers, explicitly highlighted that the principle was to be defined 'flexibly', allowing, for instance, for the problems of certain industries and regional disparities.[71] This preference for a softer stance on the principle was also reflected in the lengthy definition in the eventual Action Programme, which indeed highlighted 'regional imbalances' as a justification for exemptions from the overall tenet that the 'cost of preventing and eliminating nuisances must in principle be borne by the polluter.'[72]

Quite swiftly, in the spring of 1974, the Commission also proposed a Council recommendation that again drew on the OECD's work on this matter, but adapted it to the EC conditions. EC policy making not only involved the EC institutions, but also experts and interest groups. In early 1973, when preparing the recommendation, the Commission had obtained two expert studies prepared by economists from Belgium and Germany, respectively.[73]

As the environment turned into an issue of ordinary policy making, the Commission also had to take into account the views of affected industries, such as the European textile industry association COMITEXTIL.[74] European textile lobbyists argued that the OECD members were implementing OECD rules differently and that competition from outside the OECD was important in their sector. Thus, they supported those exemptions for 'industrial, social and technical reasons' included in the proposals. They even demanded more extensive exemptions and warned against an overly rigid application of the principle in Europe, which would lead to distortion of competition. When Commissioner Scarascia-Mugnozza gave his speech at the OECD in November 1974, the EC ministers had just reached agreement on this recommendation, which was formally adopted in March 1975.[75]

Why, then, did the Commission borrow so heavily from the OECD? First, actors and their role in networks that linked the EC with other IOs mattered. Overlapping committee memberships in both IOs facilitated the transfer of the polluter pays concept from the OECD. The same high-ranking Commission officials – such as Toulemon and Carpentier, who were prominently involved in shaping and lobbying for an EC environmental policy – were also members in the OECD Environmental Committee. Scarascia-Mugnozza, who became the first EC Commissioner responsible for environmental issues in 1973, had represented Italy on the OECD Environmental Committee in 1970. Leading national officials, who were subsequently responsible both for international and EC environmental policy, were equally part of this Committee.[76] These officials acted as mediators of concepts and solutions.

Secondly, drawing on the OECD made it possible to import a consensual idea. The officials who participated in the drafting of the polluter pays principle at the level of the OECD subsequently transferred it to the EC. Indeed, in the OECD level discussions in April 1970, it was the Commission that emphasized the impact of environmental rules as obstacles to trade.[77] The transfers of goals and ideas clearly went both ways, with network ties and structures spanning both institutions. This not only facilitated the mediation of the idea to the EC, but also offered very practical political advantages to the EC policy makers. Concerning an increasingly contested issue, they were able to present the precedent of an emerging international consensus. Already within the OECD, where the EC members were prominently represented, they had only reached agreement 'after long negotiations'.[78]

Thirdly, the 'fit' of ideas mattered. The EC and the OECD only differed in terms of geographical scope. The two IOs shared the same priorities and faced the same problems: how to encourage an environmental policy while not harming their overall goal of opening markets and freeing up competition. Hence, the OECD definition resonated well with the drafters of the EC environmental policy. In quite practical terms, it made sense to replicate within the EC the rules set by the OECD internationally. Thus, EC members could avoid the cost of adapting to them for trade with third countries. In addition, the prestige of the OECD as a major player in international economic policy would rub off on the fledgling EC policy. At the same time, the definition of the polluter pays principle pioneered by the OECD was sufficiently flexible to allow for exemptions, which the EC extended even further in its own legislation.[79]

However, why did the European Parliament draw on the Council for Europe as the relevant IO, while the Commission took its cues from the OECD? And why was the latter apparently more consequential? Time and place seem to be relevant here. In terms of timing, the European Parliament reports date from 1970 and 1971, when the OECD had not yet started working on the polluter pays principle, but was only gathering economic expertise on the issue.[80] Conversely, the declarations of the Council for Europe were still recent and its role and prestige in environmental policy probably reached its peak with the European Conservation Year of 1970. Sharing the same buildings in Strasbourg, overlapping memberships in the Consultative Assembly of the Council of Europe and the European Parliament facilitated greater exchange among its members too.

The role of the OECD as a more prominent practical and rhetorical point of reference is the result not only of strong overlaps in priorities and objectives, in officials and staff, but also reflects a certain coevolution of the OECD and the EC involvement on the polluter pays principle. The EC allowed subsidies more liberally. At the same time, the EC member states and the Commission were able to coordinate their stance internally and to represent it collectively in the OECD. In any case, a more flexible stance internally than externally seemed favourable from the EC's perspective, both politically and economically.

Conclusions

When the EC as a regional IO in Cold War Western Europe started to discuss environmental protection and subsequently included it among its policies, it was entering an increasingly crowded political space that was already inhabited by a number of other IOs. This chapter has explored how the EC defined its own policy under these specific conditions. Three conclusions can be drawn.

First, being a latecomer to environmental policy actually had certain advantages. EC actors and institutions were able to learn from the growing body of expertise on the issue. The contents of the EC Environmental Action Programme clearly reflected the contemporary state of the art of the environmental debate. EC policy makers were also able to learn from existing rules and agreements that other IOs had established. They were able to assess not only what was scientifically desirable and technically feasible, but also what was politically attainable. This was particularly relevant as environmental objectives increasingly competed with traditional and well-entrenched socioeconomic goals in the wake of the oil crisis. The EC's adoption of the polluter pays principle is a case in point here. Drawing on the consensus reached in the OECD on the polluter pays principle helped mobilize political support among the member states. Giving in to demands for a more flexible interpretation, EC policy makers enshrined a core tenet of environmental policy in an early piece of environmental legislation.

Secondly, the case of the polluter pays principle also demonstrates the importance of institutional features and legal competences. The European Commission's role as representative of the EC within the OECD, for instance, facilitated the transfer of ideas between these IOs. The overlap in objectives among both institutions was conducive to including the polluter pays principle. When the institutionally powerless European Parliament responded to current environmental catastrophes of cross-border scope and was the first institution to call for EC action on the environment, its rapporteurs quickly reframed the issue as it related to the Common Market. This was the area where the EC was competent to act. Similarly, the European Commission convinced the member states that EC rule making was preferable to national rules obstructing the Common Market. This also had implications for the content of the Environmental Action Programme, with its focus on pollution and international cooperation. Until the formal establishment of an EC environmental policy in the 1980s, most environmental legislation had to be justified by reference to the Common Market. However, once set in motion, EC environmental policy quickly expanded its scope beyond immediate market concerns to include issues such as bird protection, for example.[81]

This, thirdly, points to the importance of individual and institutional actors. In many cases they creatively interpreted existing rules and legal bases in order to expand the scope of 'their' policy area. At the highest political level, Commissioners like Spinelli and Mansholt sought to carve out a role for the EC in environmental policy making and sought to shape the definition of the environment in the EC context according to their political preferences. A small group of officials and MEPs specialized in the new policy. They gathered expertise and kept promoting the policy for the next decade or more. These actors also helped facilitate the transfer of concepts and solutions from other IOs that they were involved with.

Arguably, it was the interplay of these three factors that helped the EC to establish an increasingly prominent role among IOs in the area of environmental protection, despite its late arrival. The capacity of the EC to make binding laws, which both the Commission and the European Parliament kept highlighting,[82] was less crucial at this early stage. Clearly, the member states preferred a nonbinding recommendation on the polluter pays principle. However, binding legislation subsequently became more important.

Jan-Henrik Meyer is a senior researcher at the Max-Planck-Institute for the History of European Law, Frankfurt, Germany and an associate researcher at the Centre for Contemporary History, Potsdam.

Notes

1. This chapter is based on multi-archival research. It is part of a larger project on the origins of the EC environmental policy, supported by the German Science Foundation-funded KFG 'The Transformative Power of Europe' at FU Berlin, a Marie Curie Intra-European Fellowship and a Marie Curie Reintegration Grant within the 7th European Programme, by the Danish Research Council for Culture and Communication (FKK) within the project 'Transnational History' at Aarhus University, and by a Rachel Carson Fellowship of the Rachel Carson Center, Munich.

2. Council of the European Communities, 'Declaration of the Council of the European Communities and of the Representatives of the Governments of the Member States Meeting in the Council of 22 November 1973 on the Programme of Action of the European Communities on the environment', *Official Journal*

of the European Communities (hereinafter OJEC) 16, C 112, 20 December 1973: C 112, 1 ff.

3. On the oil crises of the 1970s, see, most recently, Frank Bösch and Rüdiger Graf, 'Reacting to Anticipations: Energy Crises and Energy Policy in the 1970s. An Introduction', *Historical Social Research* 39(4) (2014): 7–21 and the contributions to this special issue.

4. Carlo Scarascia-Mugnozza, 'Intervention de M. Scarascia-Mugnozza, Vice-Président de la Commission des Communautés européennes, session du Comité de l'Environnement réuni au niveau ministériel, OECD Paris, 13 novembre 1974', Historical Archives of the European Commission, Brussels, (hereinafter HAEC), Speeches Collection: Box S, Scarascia-Mugnozza, 1–4, 2 (quote), 4.

5. Katharina Holzinger and Thomas Sommerer, 'EU Environmental Policy – Greening the World?', in *EU Policies in a Global Perspective. Shaping or Taking International Regimes?*, Gerda Falkner (ed.) (Abingdon: Routledge, 2014), 111–29.

6. European Commission, First Communication of the Commission about the Community's Policy on the Environment. SEC (71) 2616 final, 22 July 1971, Archive of European Integration, http://aei.pitt.edu/3126/1/3126.pdf, accessed 25 May 2016, Annex C: Principal International Organizations Concerned with Environmental Problems, C1–C6.

7. 'The Need for Intergovernmental Co-operation and Co-ordination in Environmental Policy. Summary of a Paper Read by Mr. Gérard ELDIN, Deputy Secretary General, OECD, 6 January 1971', HAEC, BAC 35/1980: 199: USA. Eldin served as Deputy Secretary General from January 1970 to March 1980; OECD, 'List of OECD Secretaries-General and Deputies since 1961', 2014, http://www.oecd.org/about/secretary-general/listofoecdsecretaries-generalanddeputiessince1961.htm, accessed 25 May 2016.

8. Scarascia-Mugnozza, 'Intervention', 1. The relevant OECD Recommendations are: OECD, Recommendation of the Council on Guiding Principles Concerning International Economic Aspects of Environmental Policies, 26 May 1972 – C(72)128, 1972, http://acts.oecd.org/Instruments/ShowInstrumentView.aspx?InstrumentID=4&Lang=en&Book=False, accessed 25 May 2016; OECD, Recommendation of the Council on the Implementation of the Polluter Pays Principle, 14 November 1974 – C(74)223 1974, http://acts.oecd.org/Instruments/ShowInstrumentView.aspx?InstrumentID=11&InstrumentPID=9&Lang=en, accessed 25 May 2016.

9. Jan-Henrik Meyer, 'Getting Started: Agenda-Setting in European Environmental Policy in the 1970s', in *The Institutions and Dynamics of the European Community, 1973–83*, Johnny Laursen (ed.) (Baden-Baden: Nomos, 2014), 221–42.

10. Jan-Henrik Meyer, 'Challenging the Atomic Community: The European Environmental Bureau and the Europeanization of Anti-nuclear Protest', in *Societal Actors in European Integration. Polity-Building and Policy-Making 1958–1992*, Wolfram Kaiser and Jan-Henrik Meyer (eds) (Basingstoke:

Palgrave Macmillan, 2013), 197–220; Jan-Henrik Meyer, "'Where Do We Go from Wyhl?" Transnational Anti-nuclear Protest Targeting European and International Organisations in the 1970s', *Historical Social Research* 39(1) (2014): 212–35.

11. Jan-Henrik Meyer 'Appropriating the Environment: How the European Institutions Received the Novel Idea of the Environment and Made it Their Own', *KFG 'The Transformative Power of Europe' Working Paper* 31 (2011): 1–33, http://edocs.fu-berlin.de/docs/receive/FUDOCS_document_000000012522, accessed 25 May 2016.

12. Communities, 'Declaration of the Council of 22 November 1973, Part I, Title I.

13. See, e.g. Robert Gottlieb, *Forcing the Spring: The Transformation of the American Environmental Movement. Revised and Updated Edition* (Washington DC: Island Press, 2005); Julie Sze, *Noxious New York: The Racial Politics of Urban Health and Environmental Justice* (Cambridge, MA: MIT Press, 2007).

14. Communities, 'Declaration of the Council of 22 November 1973, Part I, Title I.

15. Ibid.

16. Ibid., Part II.

17. Ibid., Part I, Title II, no. 10.

18. OECD, 'Supplementary Protocol No. 1 to the Convention on the Organisation for Economic Co-Operation and Development, [Bundesanzeiger] No. 44, Bonn, 24 August 1961', Archive of the Council of Ministers (hereinafter ACM), CM2 1972.515 (1961): 18.

19. Henry H. Kerr, Jr., 'Changing Attitudes through International Participation: European Parliamentarians and Integration', *International Organization* 27(1) (1973): 45–83.

20. Jan-Henrik Meyer, Interview with Claus Stuffmann, former Head of Unit at the European Commission's Service for the Environment and Consumer Protection, Brussels, 10 June 2009; Eric Bussière and Arthe van van Laer, Entretien avec Michel Carpentier, Paris, 5 janvier 2004, HAEU, Florence, 14.

21. Ann-Christina L. Knudsen, 'Modes de recrutement et de circulation des premiers membres britanniques et danois du Parlement européen', *Cultures & Conflits* 85–86(1–2) (2012): 61–79; Ann-Christina L. Knudsen, 'The European Parliament and Political Careers at the Nexus of European Integration and Transnational History', in *The Institutions and Dynamics of the European Community, 1973–83*, Johnny Laursen (ed.) (Baden-Baden: Nomos, 2014), 76–96.

22. On Jahn, see Jan-Henrik Meyer, 'A Good European. Hans Edgar Jahn – Anti-Bolshevist, Cold-Warrior, Environmentalist', in *Living Political Biography: Narrating 20th Century European Lives*, Ann-Christina L. Knudsen and Karen Gram-Skjoldager (eds) (Aarhus: Aarhus University Press, 2012), 137–59.

23. Laura Scichilone, 'The Origins of the Common Environmental Policy: The Contributions of Spinelli and Mansholt in the *Ad Hoc* Group of the European Commission', in *The Road to a United Europe: Interpretations of the Process of European Integration*, Morten Rasmussen and Ann-Christina Lauring Knudsen (eds) (Brussels: PIE, 2009), 335–48, 341–43.

24. Carine Germond, 'An Emerging Anti-reform Green Front? Farm Interest Groups Fighting the "Agriculture 1980" Project, 1968–72', *European Review of History* 22(3) (2015): 433–50.
25. Dennis Meadows et al., *The Limits to Growth* (New York: Universe Books, 1972).
26. Thorsten Schulz-Walden, *Anfänge globaler Umweltpolitik. Umweltsicherheit in der internationalen Politik (1969–1975)* (Munich: Oldenbourg, 2013), 165 f.
27. Sicco Mansholt, 'Wachstum für wen? Brief an den Kommissionspräsidenten Franco Maria Malfatti, 9.2.1972', *Forum E Bulletin der Jungen Europäischen Föderalisten* 2(4) (1972): 14–24.
28. Scichilone, 'The Origins of the Common Environmental Policy', 342.
29. Schulz-Walden, *Anfänge globaler Umweltpolitik*, 170; Arthe van van Laer, Entretien avec Michel Carpentier, Bordeaux, 22 octobre 2010, HAEU, Florence, http://archives.eui.eu/en/files/transcript/16415.pdf, accessed 25 May 2016, 3.
30. Laura Scichilone, *L'Europa e la sfida ecologica. Storia della politica ambientale europea 1969–1998* (Milan: Il Mulino, 2008), 49 f.
31. Ibid., 51.
32. Entretien avec Carpentier, 2004, 14 f.
33. European Commission, 'Political and Institutional Aspects of Environmental Management (Summary)', Washington, 15 January 1971 (probably presented at Atlantic Council and Battelle Institute Joint Conference on Goals and Strategy for Environmental Quality in the Seventies, 15–17 January 1971, JHM), HAEC, BAC 35/1980: 199: USA.
34. Schulz-Walden, *Anfänge globaler Umweltpolitik*, 155–165; Entretien avec Carpentier, 2004, 15. On Protection of the Rhine: Mark Cioc, 'Europe's River: The Rhine as a Prelude to Transnational Cooperation and the Common Market', in *Nation-States and the Global Environment. New Approaches to International Environmental History*, Erika Marie Bsumek, David Kinkela and Mark Atwood Lawrence (eds) (Oxford: Oxford University Press, 2013), 25–42, 27.
35. Only when the Single European Act established a legal basis for environmental policy in 1987 was this service transformed into a fully fledged Directorate-General.
36. Interview with Stuffmann.
37. Entretien avec Carpentier, 2010, 6.
38. Stanley P. Johnson, *The Politics of the Environment. The British Experience* (London: Tom Stacey, 1973); Stanley P. Johnson, *Stanley, I Presume* (London: Fourth Estate, 2009), 234–37, 242–49, 258–63; Christian van de Velde, Interview with Stanley Johnson, London, 18 October 2011, 2011, HAEU, Florence, http://archives.eui.eu/en/files/transcript/16515.pdf, accessed 25 May 2016.
39. Stanley P. Johnson, *The Pollution Control Policy of the European Communities* (London: Graham and Trotman, 1983); Jan-Henrik Meyer, 'Saving Migrants: A Transnational Network Supporting Supranational Bird Protection Policy in the 1970s', in *Transnational Networks in Regional Integration: Governing Europe*

1945–83, Wolfram Kaiser, Brigitte Leucht and Michael Gehler (eds) (Basingstoke: Palgrave Macmillan, 2010), 176–98.

40. Jan-Henrik Meyer, 'Un faux départ? Les acteurs français dans la politique environnementale européenne des années 1970', in *Une protection de l'environnement à la française, XIXe–XXe siècles*, Jean-François Mouhot and Charles-François Mathis (eds) (Seyssel: Editions Champ Vallon, 2013), 120–30, 124–26; Représentation Permanente de la France auprès des Communautés européennes, 'Mémorandum du Gouvernement francais rélatif au développement d'une Coopération Européenne pour la protection de l'environnement, Bruxelles, 20 janvier 1972', ACM, CM2 1972.513: 1–19.

41. Coreper, 'Compte-rendu du Coreper 634, 29 Feb–3 March 1972', ACM, CM2 1972.513.

42. Norman Pohl, 'Grün ist die Hoffnung – Umweltpolitik und die Erwartungen hinsichtlich einer Reform der Institutionen der Europäischen Gemeinschaft um 1970', in *Natur- und Umweltschutz nach 1945. Konzepte, Konflikte, Kompetenzen*, Franz-Josef Brüggemeier and Jens Ivo Engels (eds) (Frankfurt: Campus, 2005), 162–82, 179 f.

43. Van Laer, Entretien avec Carpentier, 2010, 4. My translation from the French original, JHM.

44. Joachim Radkau, *Nature and Power: A Global History of the Environment* (Cambridge: Cambridge University Press, 2008), 240, 244; Jean-Baptiste Fressoz, 'Payer pour polluer. L'industrie chimique et compensation des dommages environnementaux, 1800–1850', *Histoire & Mesure* 28(1) (2013): 145–86.

45. Priscilla Schwartz, 'The Polluter Pays Principle', in *Research Handbook on International Environmental Law*, Malgosia Fitzmaurice, P. Merkouris and David M. Ong (eds) (Cheltenham: Edward Elgar, 2010), 243–61; Sharon Beder, 'The Polluter Pays Principle', in *Environmental Principles and Policies: An Interdisciplinary Introduction*, Sharon Beder (ed.) (London: Earthscan, 2006), 32–46.

46. Kai F. Hünemörder, *Die Frühgeschichte der globalen Umweltkrise und die Formierung der deutschen Umweltpolitik (1950–1973)* (Stuttgart: Franz Steiner, 2004), 84–87; Mark Cioc, *The Rhine. An Eco-Biography, 1815–2000* (Seattle: University of Washington Press, 2002), 141.

47. Jan-Henrik Meyer, 'Green Activism: The European Parliament's Environmental Committee Promoting a European Environmental Policy in the 1970s', *Journal of European Integration History* 17(1) (2011): 73–85.

48. Boersma, 'Oud-minister Jaap Boersma overleden', *De Telegraaf*, 6 March 2012, http://www.telegraaf.nl/binnenland/article20123594.ece, accessed 25 May 2016.

49. Jacob Boersma, 'Bericht im Namen des Ausschusses für Sozial- und Gesundheitsfragen über die Reinhaltung der Binnengewässer unter besonderer Berücksichtigung der Verunreinigung des Rheins, 11 November 1970', CARDOC, PEO-AP RP/ASOC.1967 AO-0161/70 (hereinafter Boersma-Report).

50. E.g. Spiegel, 'Tod im Strom. Industrie Kernkraftwerke', *Der Spiegel*, 23 February 1970, 46; Theo Löbsack, 'Wenn der Rhein dampft. Zu den geplanten Atommeilern darf nicht geschwiegen werden', *Die Zeit*, 24 April 1970, 67.
51. Boersma-Report, 13 (§38–40).
52. Ibid., 11 (§32).
53. Meyer, 'Un faux départ', 125–27.
54. Model European Parliament, 'Dr. A.P. Oele', 2012, Montesquieu Instituut, http://www.mepnederland.nl/9353000/1/j9vvincioiml3zp/vg09ll3pxwyl, accessed 25 May 2016; 'Adriaan Oele', in *Who's Who in Europe. Dictionnaire biographique des personnalités européennes contemporaines*, Edward A. de Maeijer (ed.) (Brussels: Europ-élite, 1972), 2261; Agnes Koerts, 'Ad Oele, spagaat tussen ingenieurs en politici', *Binnenlands Bestuur* 20: 42 (1999): 20–23.
55. Incidentally, Boersma was a member of both committees.
56. Market instruments became central elements of EU environmental policy from the 1990s; c.f. Andrew Jordan et al., 'European Governance and the Transfer of "New" Environmental Policy Instruments (NEPIs) in the European Union', *Public Administration* 81(3) (2003): 555–74.
57. Ezra J. Mishan, *The Costs of Economic Growth* (London: Staples Press, 1967).
58. Boersma-Report, 19 (§15).
59. Hans Edgar Jahn, 'Bericht im Auftrag des Ausschusses für Sozial- und Gesundheitsfragen über die Notwendigkeit einer Gemeinschaftsaktion zur Reinhaltung der Luft, 15.12.1971', *CARDOC* PE0 AP RP ASOC.1967 0181/71 (hereinafter Jahn-Report).
60. Ibid., 4 (§8).
61. Ibid., 12 (§32).
62. Ibid., 15 (§44); Council of Europe, *Resolution (68)4, Adopted by the Ministers' Deputies on 8 March 1968, Approving the 'Declaration of Principles' on Air Pollution Control* (Strasbourg, 1968).
63. Jahn-Report, 22 (§80).
64. Ibid., 29f. (§13–19).
65. Jahn, Hans Edgar, 'Bericht im Namen des Ausschusses für Sozial- und Gesundheitsfragen über die Erste Mitteilung der Kommission der Europäischen Gemeinschaften über die Politik der Gemeinschaft auf dem Gebiet des Umweltschutzes', 14.04.1972, doc 9/72', *CARDOC* PE0 AP RP ASOC.1967 0009/72, 7 (§16), 59 (§80).
66. European Commission, 'Communication from the Commission to the Council on a European Communities' Programme concerning the Environment (submitted on 24 March 1972)', *Bulletin of the European Communities. Supplement* 5(5) (1972): 1–69. In the final version available from the Commission archive, this 'second' communication is actually called the 'deuxième' communication, as was contemporary parlance.
67. Ibid., 35–37.

68. OECD, 'Comité de l'Environnement. Grandes lignes du programe proposé pour 1971 [note du Secrétariat], ENV (70) 15, 4.11.1970', ACM, CM 2 1970.429: 1–12, 2.
69. OECD, Recommendation, 26 May 1972 – C(72)128.
70. OECD, Recommendation, 14 November 1974 – C(74)223
71. Council of the European Communities, 'Entwurf eines Protokolls über die 251. Tagung des Rates am 19. und 20. Juli 1973 in Brüssel, doc 1586/73, 30. Juli 1973', ACM, CM 2 1973.15 71: 1–50, 10, 12, 15.
72. Council of the European Communities, 'Declaration of the Council of 22 November 1973, Part I, Title II, No. 5.
73. Achille Hannequart, 'A Study of the Economic Tools for an Environmental Policy, Brussels, 24 May 1973, ENV/49/73 A', HAEC, BAC 58/1992: 319 (1972–74): 1–53; Harald Jürgensen and Kai-Peter Jaeschke, 'Study to Determine the Social Cost of Pollution, Hamburg, April 1973, ENV/63/73 d', HAEC, BAC 58/1992: 319 (1972–74): 1–55.
74. COMITEXTIL, 'Prise de Position de COMITEXTIL (Comité de Coordination des Industries Textiles de la Communauté Economique Européenne) relative au projet de recommandation du Conseil en ce que concerne l'allocation des coûts et de l'intervention des pouvoirs publiques en matière de l'environnment (Principe pollueur-payeur), 5 November 1974', HAEC, BAC 68/1984: 201 (Groupe des experts économiques) (1974): 1–2.
75. European Communities, '75/436/Euratom, ECSC, EEC: Council Recommendation of 3 March 1975 Regarding Cost Allocation and Action by Public Authorities on Environmental Matters, *OJEC*: L 194, 25/07/1975 (1975): 1–4.
76. OECD, 'Comité préparatoire ad hoc sur les activités de l'organisation concernant les problèmes de l'enivironnement liés à la croissance économique. Liste des participants (note de Secrétariat), ENV (70) 12, 11.05.1970', ACM CM 2 1970.430: 1–8.
77. OECD, 'Comité préparatoire ad hoc sur les activités de l'organisation concernant les problèmes de l'enivironnement liés à la croissance économique. Conclusions générales (note de Secrétariat), ENV (70) 11, 30.04.1970', ACM, CM 2 1970.430: 1–8, 3.
78. Commission, 'Communication from the Commission', 24 March 1972, 35–37.
79. Communities, '75/436/Euratom, ECSC, EEC: Council Recommendation of 3 March 1975.
80. E.g. OECD, 'Environmental Committee. Sub-committee of Economic Experts. Problems of Environmental Economics. Record of the Seminar Held at the OECD (Summer 1971), AEU/ENV 71.19', OECD Archives, Paris, AEU-ENV 1971.
81. Meyer, 'Saving Migrants'.
82. Commission, 'First Communication', 22 July 1971, 27.

Bibliography

'Adriaan Oele', in *Who's Who in Europe. Dictionnaire biographique des personnalités européennes contemporaines*, Edward A. de Maeijer (ed.) (Brussels: Europélite, 1972), 2261.

Beder, Sharon, 'The Polluter Pays Principle', in *Environmental Principles and Policies. An Interdisciplinary Introduction*, Sharon Beder (ed.) (London: Earthscan, 2006), 32–46.

Boersma, Jacob, 'Bericht im Namen des Ausschusses für Sozial- und Gesundheitsfragen über die Reinhaltung der Binnengewässer unter besonderer Berücksichtigung der Verunreinigung des Rheins, 11 November 1970', CARDOC, PEO-AP RP/ASOC.1967 AO-0161/70.

Boersma, 'Oud-minister Jaap Boersma overleden', *De Telegraaf*, 6 March 2012, http://www.telegraaf.nl/binnenland/article20123594.ece, accessed 25 May 2016.

Bösch, Frank and Rüdiger Graf, 'Reacting to Anticipations: Energy Crises and Energy Policy in the 1970s: An Introduction', *Historical Social Research* 39(4) (2014): 7–21.

Bussière, Eric and Arthe van Laer, Entretien avec Michel Carpentier, Paris, 5 janvier 2004, HAEU, Florence, 14.

Cioc, Mark, *The Rhine. An Eco-Biography, 1815–2000* (Seattle: University of Washington Press, 2002).

Cioc, Mark, 'Europe's River: The Rhine as a Prelude to Transnational Cooperation and the Common Market', in *Nation-States and the Global Environment: New Approaches to International Environmental History*, Erika Marie Bsumek, David Kinkela and Mark Atwood Lawrence (eds) (Oxford: Oxford University Press, 2013), 25–42.

COMITEXTIL, 'Prise de Position de COMITEXTIL (Comité de Coordination des Industries Textiles de la Communauté Economique Européenne) relative au projet de recommandation du Conseil en ce que concerne l'allocation des coûts et de l'intervention des pouvoirs publiques en matière de l'environnment (Principe pollueur-payeur), 5 November 1974', HAEC, BAC 68/1984: 201 (Groupe des experts économiques) (1974): 1–2.

Coreper, 'Compte-rendu du Coreper 634, 29 Feb–3 March 1972', ACM, CM2 1972.513.

Council of Europe, *Resolution (68)4, Adopted by the Ministers' Deputies on 8 March 1968, Approving the 'Declaration of Principles' on Air Pollution Control* (Strasbourg, 1968).

Council of the European Communities, 'Declaration of the Council of the European Communities and of the Representatives of the Governments of the Member States Meeting in the Council of 22 November 1973 on the Programme of Action of the European Communities on the Environment', *Official Journal of the European Communities* (hereinafter OJEC) 16, C 112, 20 December 1973: C 112, 1 ff.

Council of the European Communities, 'Entwurf eines Protokolls über die 251. Tagung des Rates am 19. und 20. Juli 1973 in Brüssel, doc 1586/73, 30. Juli 1973', ACM, CM 2 1973.15 71: 1–50.

European Commission, 'Communication from the Commission to the Council on a European Communities' Programme concerning the Environment (submitted on 24 March 1972)', *Bulletin of the European Communities. Supplement* 5(5) (1972): 1–69.

European Commission, 'Political and Institutional Aspects of Environmental Management (Summary)', Washington DC, 15 January 1971 (presented at Atlantic Council and Battelle Institute Joint Conference on Goals and Strategy for Environmental Quality in the Seventies, 15–17 January 1971), HAEC, BAC 35/1980: 199: USA.

European Commission, First Communication of the Commission about the Community's Policy on the Environment. SEC (71) 2616 final, 22 July 1971, Archive of European Integration, http://aei.pitt.edu/3126/1/3126.pdf, accessed 2 July 2016, Annex C: Principal International Organizations Concerned with Environmental Problems, C1–C6.

European Communities, '75/436/Euratom, ECSC, EEC: Council Recommendation of 3 March 1975 Regarding Cost Allocation and Action by Public Authorities on Environmental Matters', *OJEC*: L 194, 25/07/1975 (1975): 1–4.

Fressoz, Jean-Baptiste, 'Payer pour polluer. L'industrie chimique et compensation des dommages environmentaux, 1800–1850', *Histoire & Mesure* 28(1) (2013): 145–86.

Germond, Carine, 'An Emerging Anti-reform Green Front? Farm Interest Groups Fighting the "Agriculture 1980" Project, 1968–72', *European Review of History* 22(3) (2015): 433–50.

Gottlieb, Robert, *Forcing the Spring: The Transformation of the American Environmental Movement. Revised and Updated Edition* (Washington DC: Island Press, 2005).

Hannequart, Achille, 'A Study of the Economic Tools for an Environmental Policy, Brussels, 24 May 1973, ENV/49/73 A', HAEC, BAC 58/1992: 319 (1972–74): 1–53.

Holzinger Katharina and Thomas Sommerer, 'EU Environmental Policy – Greening the World?', in *EU Policies in a Global Perspective: Shaping or Taking International Regimes?*, Gerda Falkner (ed.) (Abingdon: Routledge, 2014), 111–29.

Hünemörder, Kai F., *Die Frühgeschichte der globalen Umweltkrise und die Formierung der deutschen Umweltpolitik (1950–1973)* (Stuttgart: Franz Steiner, 2004).

Jahn, Hans Edgar, 'Bericht im Namen des Ausschusses für Sozial- und Gesundheitsfragen über die Erste Mitteilung der Kommission der Europäischen Gemeinschaften über die Politik der Gemeinschaft auf dem Gebiet des Umweltschutzes', 14.04.1972, doc 9/72', *CARDOC* PE0 AP RP ASOC.1967 0009/72.

Jahn, Hans Edgar, 'Bericht im Auftrag des Ausschusses für Sozial- und Gesundheitsfragen über die Notwendigkeit einer Gemeinschaftsaktion zur Reinhaltung der Luft, 15.12.1971', *CARDOC* PE0 AP RP ASOC.1967 0181/71.

Johnson, Stanley P., *Stanley, I Presume* (London: Fourth Estate, 2009).

Johnson, Stanley P., *The Politics of the Environment: The British Experience* (London: Tom Stacey, 1973).

Johnson, Stanley P., *The Pollution Control Policy of the European Communities* (London: Graham and Trotman, 1983).

Jordan, Andrew et al., 'European Governance and the Transfer of "New" Environmental Policy Instruments (NEPIs) in the European Union', *Public Administration* 81(3) (2003): 555–74.

Jürgensen, Harald and Kai-Peter Jaeschke, 'Study to Determine the Social Cost of Pollution, Hamburg, April 1973, ENV/63/73 d', HAEC, BAC 58/1992: 319 (1972–74): 1–55.

Kerr, Henry H. Jr., 'Changing Attitudes through International Participation: European Parliamentarians and Integration', *International Organization* 27(1) (1973): 45–83.

Knudsen, Ann-Christina L., 'Modes de recrutement et de circulation des premiers membres britanniques et danois du Parlement européen', *Cultures & Conflits* 85–86(1–2) (2012): 61–79.

Knudsen, Ann-Christina L., 'The European Parliament and Political Careers at the Nexus of European Integration and Transnational History', in *The Institutions and Dynamics of the European Community, 1973–83*, Johnny Laursen (ed.) (Baden-Baden: Nomos, 2014), 76–96.

Koerts, Agnes, 'Ad Oele, spagaat tussen ingenieurs en politici', *Binnenlands Bestuur* 20(42) (1999): 20–23.

Löbsack, Theo, 'Wenn der Rhein dampft. Zu den geplanten Atommeilern darf nicht geschwiegen werden', *Die Zeit*, 24 April 1970, 67.

Mansholt, Sicco, 'Wachstum für wen? Brief an den Kommissionspräsidenten Franco Maria Malfatti, 9.2.1972', *Forum E Bulletin der Jungen Europäischen Föderalisten* 2(4) (1972): 14–24.

Meadows, Dennis et al., *The Limits to Growth* (New York: Universe Books, 1972).

Meyer, Jan-Henrik, 'A Good European: Hans Edgar Jahn – Anti-Bolshevist, Cold-Warrior, Environmentalist', in *Living Political Biography. Narrating 20th Century European Lives*, Ann-Christina L. Knudsen and Karen Gram-Skjoldager (eds) (Aarhus: Aarhus University Press, 2012), 137–59.

Meyer, Jan-Henrik, 'Appropriating the Environment: How the European Institutions Received the Novel Idea of the Environment and Made it Their Own', *KFG 'The Transformative Power of Europe' Working Paper*, 31 (2011): 1–33, http://edocs.fu-berlin.de/docs/receive/FUDOCS_document_000000012522, accessed 25 May 2016.

Meyer, Jan-Henrik, 'Challenging the Atomic Community: The European Environmental Bureau and the Europeanization of Anti-nuclear Protest', in *Societal Actors in European Integration. Polity-Building and Policy-Making*

1958–1992, Wolfram Kaiser and Jan-Henrik Meyer (eds) (Basingstoke: Palgrave Macmillan, 2013), 197–220.

Meyer, Jan-Henrik, 'Getting Started: Agenda-Setting in European Environmental Policy in the 1970s', in *The Institutions and Dynamics of the European Community, 1973–83*, Johnny Laursen (ed.) (Baden-Baden: Nomos, 2014), 221–42.

Meyer, Jan-Henrik, 'Saving Migrants: A Transnational Network Supporting Supranational Bird Protection Policy in the 1970s', in *Transnational Networks in Regional Integration. Governing Europe 1945–83*, Wolfram Kaiser, Brigitte Leucht and Michael Gehler (eds) (Basingstoke: Palgrave Macmillan, 2010), 176–98.

Meyer, Jan-Henrik, 'Un faux départ? Les acteurs français dans la politique environnementale européenne des années 1970', in *Une protection de l'environnement à la française, XIXe–XXe siècles*, Jean-François Mouhot and Charles-François Mathis (eds) (Seyssel: Editions Champ Vallon, 2013), 120–30.

Meyer, Jan-Henrik, '"Where Do We Go from Wyhl?" Transnational Anti-nuclear Protest Targeting European and International Organisations in the 1970s', *Historical Social Research* 39(1) (2014): 212–35.

Meyer, Jan-Henrik, Interview with Claus Stuffmann, former Head of Unit at the European Commission's Service for the Environment and Consumer Protection, Brussels, 10 June 2009.

Meyer, Jan-Henrik, 'Green Activism: The European Parliament's Environmental Committee promoting a European Environmental Policy in the 1970s', *Journal of European Integration History* 17(1) (2011): 73–85.

Mishan, Ezra J. *The Costs of Economic Growth* (London: Staples Press, 1967).

Model European Parliament, 'Dr. A.P. Oele', 2012, Montesquieu Instituut, http://www.mepnederland.nl/9353000/1/j9vvincioiml3zp/vg09ll3pxwyl, accessed 25 May 2016.

OECD, 'Supplementary Protocol No. 1 to the Convention on the Organisation for Economic Co-Operation and Development, [Bundesanzeiger] No. 44, Bonn, 24 August 1961', Archive of the Council of Ministers (ACM), CM2 1972.515 (1961): 18.

———. 'Comité préparatoire ad hoc sur les activités de l'organisation concernant les problèmes de l'environnement liés à la croissance économique. Conclusions générales (note de Secrétariat), ENV (70) 11, 30.04.1970', ACM, CM 2 1970.430: 1–8, 3.

———. 'Comité préparatoire ad hoc sur les activités de l'organisation concernant les problèmes de l'enivironnement liés à la croissance économique. Liste des participants (note de Secrétariat), ENV (70) 12, 11.05.1970', ACM CM 2 1970.430: 1–8.

———. 'Comité de l'Environnement. Grandes lignes du programe proposé pour 1971 (note de Secrétariat), ENV (70) 15, 4.11.1970', ACM, CM 2 1970.429: 1–12.

———. 'Environmental Committee. Sub-committee of Economic Experts. Problems of Environmental Economics. Record of the Seminar Held at the OECD (Summer 1971), AEU/ENV 71.19', OECD Archives, Paris, AEU-ENV 1971.

———. 'The Need for Intergovernmental Co-operation and Co-ordination in Environmental Policy. Summary of a Paper Read by Mr. Gérard ELDIN, Deputy Secretary General, OECD, 6 January 1971', HAEC, BAC 35/1980: 199: USA.

———. Recommendation of the Council on Guiding Principles Concerning International Economic Aspects of Environmental Policies, 26 May 1972 – C(72)128, 1972, http://acts.oecd.org/Instruments/ShowInstrumentView.aspx?InstrumentID=4&Lang=en&Book=False, accessed 25 May 2016.

———. Recommendation of the Council on the Implementation of the Polluter Pays Principle, 14 November 1974 – C(74)223 1974, http://acts.oecd.org/Instruments/ShowInstrumentView.aspx?InstrumentID=11&InstrumentPID=9&Lang=en accessed 25 May 2016.

Pohl, Norman, 'Grün ist die Hoffnung – Umweltpolitik und die Erwartungen hinsichtlich einer Reform der Institutionen der Europäischen Gemeinschaft um 1970', in *Natur- und Umweltschutz nach 1945. Konzepte, Konflikte, Kompetenzen*, Franz-Josef Brüggemeier and Jens Ivo Engels (eds) (Frankfurt: Campus, 2005), 162–82.

Radkau, Joachim, *Nature and Power: A Global History of the Environment* (Cambridge: Cambridge University Press, 2008).

Représentation Permanente de la France auprès des Communautés européennes, 'Mémorandum du Gouvernement francais rélatif au développement d'une Coopération Européenne pour la protection de l'environnement, Bruxelles, 20 janvier 1972', ACM, CM2 1972.513: 1–19.

Scarascia-Mugnozza, Carlo, 'Intervention de M. Scarascia-Mugnozza, Vice-Président de la Commission des Communautés européennes, session du Comité de l'Environnement réuni au niveau ministériel, OECD Paris, 13 novembre 1974', Historical Archives of the European Commission, Brussels, (HAEC), Speeches Collection: Box S, Scarascia-Mugnozza, 1–4.

Schulz-Walden, Thorsten, *Anfänge globaler Umweltpolitik. Umweltsicherheit in der internationalen Politik (1969–1975)* (Munich: Oldenbourg, 2013).

Schwartz, Priscilla, 'The Polluter Pays Principle', in *Research Handbook on International Environmental Law*, Malgosia Fitzmaurice, P. Merkouris and David M. Ong (eds) (Cheltenham: Edward Elgar, 2010), 243–61.

Scichilone, Laura, 'The Origins of the Common Environmental Policy: The Contributions of Spinelli and Mansholt in the *Ad Hoc* Group of the European Commission', in *The Road to a United Europe: Interpretations of the Process of European Integration*, Morten Rasmussen and Ann-Christina Lauring Knudsen (eds) (Brussels: PIE, 2009), 335–48.

Scichilone, Laura, *L'Europa e la sfida ecologica. Storia della politica ambientale europea 1969–1998* (Milan: Il Mulino, 2008).

Spiegel, 'Tod im Strom. Industrie Kernkraftwerke', *Der Spiegel*, 23 February 1970, 46.

Sze, Julie, *Noxious New York: The Racial Politics of Urban Health and Environmental Justice* (Cambridge, MA: MIT Press, 2007).

Van Laer, Arthe, Entretien avec Michel Carpentier, Bordeaux, 22 octobre 2010, HAEU, Florence, http://archives.eui.eu/en/files/transcript/16415.pdf, accessed 25 May 2016.

Velde, Christian van de, Interview with Stanley Johnson, London, 18 October 2011, HAEU, Florence, http://archives.eui.eu/en/files/transcript/16515.pdf, accessed 25 May 2016.

CHAPTER 7

(Re-)Thinking Environment and Economy

The Organisation for Economic Co-operation and Development and Sustainable Development

Iris Borowy

The Organisation for Economic Co-operation and Development (OECD) is a paradox.[1] It has existed for more than fifty years, its current members produce approximately 60 per cent of global goods and services, and it is one of the most important sources of international data and analyses on a broad range of topics. Nonetheless, as an organization, it has attracted surprisingly little research interest. As recently as 2010, it was characterized as 'a rarely researched and analysed organization'.[2] Historical scholarship has been particularly scarce. Only recently have a few historians turned their attention to the OECD.[3] Otherwise, the organization has mostly aroused the curiosity of political scientists who have tried to come to terms with its unusual structure and working strategies.[4]

Consequently, comparatively little is known about how the OECD came to be the first international organization to have a division dedicated to the environment, and how it had a formative role in the creation of the concept of sustainable development. Since sustainable development addresses challenges of development at the intersection of economic and environmental concerns, it is perhaps not surprising that it should have links to the environmental division of an economic agency (or vice versa). But the dynamics of how these links played out and how the OECD came to establish an environmental division in the first place is far less clear. This chapter explores the connections between the OECD and the conceptualizations of sustainable development, in particular its ambivalent role in both promoting and then containing its criticism of mainstream economics.

From the OEEC to the OECD

The OECD has its origins in the Organisation for European Economic Co-operation (OEEC), which was established in April 1948 in order to organize the implementation of the Marshall Plan and to bolster Western European cooperation against the perceived threat of communist expansion. Its members at the time were Austria, Belgium, Denmark, France, Greece, Iceland, Ireland, Italy, Luxembourg, the Netherlands, Norway, Portugal, Sweden, Switzerland, Turkey, the United Kingdom and (as a sovereign state from 1949) West Germany. Canada and the United States were associated members. With European recovery, the creation of the North Atlantic Treaty Organization (NATO) in 1949 and of the European Economic Community (EEC) in the Treaty of Rome in 1957/58, the OEEC lost its original purpose. It was transformed into the OECD in 1961, designed to coordinate economic planning and policy among its member states, and between them and the outside world. The change of name articulated the shift of scope towards longer-term economic development and beyond Europe. Japan (1964), Australia (1971), Finland (1969) and New Zealand (1973) joined during the following years, and additional countries followed in the 1990s.[5] By its Convention, the OECD committed members to adopt policies designed 'to achieve the highest sustainable economic growth and employment and a rising standard of living', a phrase that, at the time, did not indicate an integration of environmental considerations, but denoted stable long-term growth without inflation.[6]

Decisions were made at the OECD Council, which consisted of one delegate per member state, usually permanent representatives at the OECD, and one representative of the EEC, later the European Communities (EC after 1967) or the European Union (EU after 1993), respectively. The Council met at the ministerial level once a year to discuss the broad outlines of its work programme. Its decisions directed the work of thematic committees, supported by their respective directorates in the Secretariat. The OECD's budget of between $124 million (1964) and $200 million (1980) (in 2010 US dollars) came almost entirely from the member states, mostly as regular contributions calculated according to the size of their economies, and they were supplemented by voluntary payments to support specific programmes. The United States consistently made the highest contributions by far, which considerably strengthened its position during decision-making processes. However, its financial potency was counterbalanced by strong

European influence: Western European countries held a majority in the Council, which took decisions unanimously, and the EU and its institutional predecessors held weekly meetings of delegates to the OECD in order to coordinate common positions.[7] This situation ensured that the Anglo-American turn towards economic neoliberalism during the early 1980s was mitigated by Keynesian viewpoints prevalent in mainland Europe with its strong social democratic tradition. Besides, OECD decision making was more subtle than a simple confrontation of politicians coming from different countries. Committees were expected to take decisions by consensus, thus encouraging compromise positions. At least until the early 1990s, such decisions were facilitated by a substantial homogeneity of views among members, since all national ministries and academic institutions in the member states embraced a broadly similar programme of liberalism and a market economy.[8] In addition, OECD activities relied heavily on the Secretariat, whose staff consisted of hundreds of highly qualified experts, mostly economists, but also lawyers, biologists, sociologists or chemists. OECD staff at the Paris headquarters sought additional expertise from outsiders from academia, business, trade unions and international non-governmental organizations (INGOs). These specialists contributed papers and took part in meetings and discussions, and their impact through informal contact on these occasions was probably at least as important as their formalized input in the preparation of documents.[9] In this expert culture, perceived scientific knowledge formed an essential prerequisite for any policy recommendation.

These characteristics have remained remarkably stable throughout the OECD's existence and to this day. Consequently, the function of the OECD in environmental work (and elsewhere) has been described as that of a 'think tank for government officials, producing analytical work, generating data for decision making, and disseminating information about lessons learned about good practices based on experience with environmental policy'.[10] Thereby, the OECD affects existing and produces new discourses, influencing which topics are considered relevant by policy makers and beyond, how they should be perceived and how potential solutions should be evaluated according to which norms.[11]

Data and analytical work are crucial because, in material terms, the OECD has remarkably little to offer. It does not conduct practical projects or provide grants, and it has only limited legal instruments, no funds and no executive power to enforce its decisions. International agreements, conventions or decisions become legally binding when voted on by the OECD Council of Ministers, while recommendations, declarations, arrangements or understandings do not.[12] Of all the subjects addressed at the OECD, environmental work has yielded by far the most legally binding decisions, resulting largely from its studies on chemicals and waste management, while on the whole, the Council makes only sparing use of this tool.[13] For most of its influence, the OECD relies on 'its reputation, good arguments, and persuasion'.[14] A lot of OECD work has been dedicated to discussing appropriate future policy directions, where this claim to superior expertise could have an exhilarating effect on staff and visitors, as pointed out by Jim MacNeill, head of the Environment Directorate from 1978 to 1984:

> The working climate was stimulating. It was the West's official center of economic thought which meant that it could command the presence of the West's best minds. So it was an intellectual crossroads, not only for the advanced industrialized countries but also, on many occasions, people from the so-called developing countries.[15]

Clearly, OECD members, employees as well as politicians, have frequently considered themselves a model for the rest of the world.[16] This was particularly true for the environment, where: 'Arguments are the only resource the secretariat and the environment directorate have at their disposal.'[17] Four main policy mechanisms for the Environment Directorate can be identified: (1) defining principles in environmental policy; (2) framing discourses, diffusing ideas and changing perceptions of problems; (3) setting agendas; and (4) distributing knowledge and arguments through popular and scientific publications.[18]

In many ways, this constellation was similar to the World Bank, whose focal point was also economic growth and that also produced large quantities of data and high-quality analyses. But while the World Bank was dominated by the United States and interacted closely with low-income countries around the world, the OECD had a far more European character and remained more limited to a small group of high-income countries, sometimes derided as a 'rich man's club'. Surprisingly, in the late 1960s, this club came close to questioning the very system on which its wealth depended.

From Growth to Environment

The environment did not form part of the OEEC mandate, which had been written years before the environment entered the political agenda. However, problems subsequently categorized as environmental issues, mainly regarding pollution control, slipped into its work programme almost inadvertently when an OEEC commission of government representatives established a Committee for Applied Research in 1957. Designed to address economic challenges common to all member states, the Committee initiated several small studies on industrial water and air pollution. After 1961, its successor body, the Committee for Scientific Research (soon renamed Committee for Research Cooperation), continued these activities, adding a few studies on noise and pesticides, and cooperating with other OECD directorates.[19]

It was a period when environmental concerns first arrived on the international agenda. A series of accidents and scandals such as the smog episodes in Donora, Pennsylvania (1948), London (1952) and Los Angeles (1954), the contamination of Minamata Bay in Japan with mercury in the 1950s and 1960s, and the explosion of a chemical plant in Seveso in Italy in 1976 alerted the public to the health risks connected with industrial production.[20] These and other events gave rise to a strong environmental movement and to environmental legislation in high-income countries. Similarly, by the early 1970s, numerous international organizations, such as the United Nations (UN), the World Health Organization (WHO) and NATO had begun to include the new environmental issue in their work.[21] Most efforts dealt with air and water pollution, although other issues such as nuclear energy were also included.[22] At the same time, studies critical of the prevailing forms of economic development began to appear. Kenneth Galbraith's *Affluent Society* (1958), Rachel Carson's *Silent Spring* (1962) and Ezra Mishan's *The Costs of Economic Growth* (1967) were only a few of a growing number of bestselling books that warned of the environmental and social price of the existing economic system.[23]

Plausibly, it was the combination of a focus on economic performance and a strong scientific programme that made people working at the OECD more sensitive to such studies and made them realize earlier than other agencies that economic policies necessarily had an environmental impact and required environmental measures. A 1966 meeting on 'Research on the Unintended Occurrence of Chemicals in the Environment', attended by senior officials from eighteen countries and several international organizations, including the Food and Agriculture Organization (FAO) and the WHO, turned out to be a seminal moment, which triggered studies into the use of and trade with hazardous chemicals, to become one of the largest and longest-running programmes of the future Environment Directorate.[24]

Two years later, two prominent members of the OECD Secretariat took their concern a step further to a more fundamental critique of the development that modern industrial nations were taking. Discussions about the possible need for a new direction in economic development began in a small group of people around Danish Secretary-General Thorkil Kristensen and the Scottish head of the Directorate for Scientific Affairs, Alexander King, who had been with the OEEC/OECD since 1957. Frustrated by what they perceived as the inadequate response of member governments to problems resulting from the unprecedented economic growth since 1945, they found common ground with the Italian industrialist Aurelio Peccei. In 1968, they and other like-minded intellectuals from industry and academia as well as from within the OECD founded the Club of Rome, designed to stimulate serious study into these questions. Initially, its Executive Committee consisted mainly of members of the OECD Secretariat and the Committee on Science and Technology.[25] In February 1969, Kristensen took these activities into the core of the OECD when he presented a note to the OECD Ministerial Council, entitled 'Problems of a Modern Society'. Citing social upheaval and protests by students, women and minorities, the paper described a widespread feeling in OECD countries that rapid economic development created new problems that required new solutions.[26]

The topic was taken up by Kristensen's successor, the new OECD Secretary-General Emiel van Lennep. Van Lennep was a former treasurer general of the Dutch Ministry of Finance with years of experience at the OECD's Development Assistance Committee (DAC) and the Economic Policy Committee, where, at the request of the United States, he had become chair of the Working Party in charge of monetary coordination. He combined a keen appreciation of the broader social and environmental implications of economic growth with a trust in markets and supply-side economics.[27] In September 1969, he presented a thirty-page position paper in which he summed up the present situation and its challenges: economic growth in the 1960s had surpassed expectations and a similar performance was expected in the 1970s. Rather than purely positive, this exceptional growth had ambivalent implications, because it caused profound transformations, including a rising living standard as well as extensive social problems, which alienated young people especially from ways of life that they blamed for environmental destruction and for the fragmentation of society. In the list of challenges, environmental issues appeared as one problem among others, albeit an important one since anti-pollution efforts threatened to incur substantial costs.[28]

By December 1969, van Lennep had moved from naming the problematic repercussions of economic growth to proposing suitable counterstrategies. Intelligent policies should move beyond a purely quantitative view and 'put *more emphasis on welfare, and less on growth for its own sake*'.[29] Current economic calculations were not only insufficient but also possibly plainly wrong. Market prices of goods ignored the damage they inflicted on people and the environment, and were therefore likely to overrate their values. Rethinking value assessments required a fundamental reconceptualization of growth, including long-term policies and visions of the future and, in practical terms, necessitated expensive research and investments. Van Lennep therefore suggested setting up working groups in charge of assessing the costs and benefits of various strategies in different fields and advocated a thoroughly holistic approach.[30]

These suggestions were generally well received by the OECD Council in January 1970, eliciting only limited and mild criticism. Some representatives, including the one from the United States, cautioned that economic growth remained the central policy goal and its benefits should not be forgotten. Van Lennep reassured the delegates that economic growth indeed remained the OECD's fundamental aim, but that it was time to rethink its form. Most delegates actively welcomed the initiative.[31] At van Lennep's suggestion, the Council agreed to set up a preparatory committee, comprised of senior representatives of member governments, which was to decide on the number, mandates, compositions and terms of reference of the specific working groups.[32] These working groups should assess the dimensions and economic significance of existing problems, identify major options open to governments and, most importantly, different ways in which costs could be shared among enterprises, consumers, taxpayers and users. His list of relevant issues demonstrated how the environment had moved to become the centre of attention. The majority of topics addressed issues of chemical pollution caused by the industrialization of modern societies: increases in atmospheric pollution through CO_2 and SO_2 due to fossil fuels (already a prominent issue at the time) and through aircraft vapour; eutrophication of international waters due to detergents, fertilizers and thermal pollution; and the effects of industrial chemicals, notably persistent pesticides, in the environment. Some other topics, such as recycling and the disposal of non-degradable solid waste, the problems of urban transportation, noise pollution and the preservation of wilderness areas and natural beauty were approached more systematically, questioning the ways in which life in modern societies was being organized.[33]

Thus, between September 1969 and February 1970, the aim of reflections regarding future OECD work had shifted from a basic reconsideration of the character of economic development, taking into account its various weaknesses including its effect on the environment, to the consideration of a list of tangible environmental concerns with a focus on the distribution of necessary costs among stakeholders. In the process, the function of the environment shifted from being a relevant factor in the overriding aim of social wellbeing on a par with economic growth to being a factor whose conflict with the overriding aim of economic growth needed to be studied to identify possible solutions.

The tension between these two approaches showed during the meetings of the preparatory committee in March and April 1970. Most members proposed a comparatively conservative list of topics based on a continuation of those that were already being studied within the OECD (water pollution, air pollution, noise, transport and urban development). But some delegations suggested new topics such as the accumulation of solid waste, land use, town planning and the use of natural resources, which could be linked the studies on economic growth.[34] As a new item that somewhat reconciled the two different approaches, the committee proposed the establishment of a central environment commission designed to ensure a connection between the specific environmental issues and the overall OECD work on economic development. It should be complemented by specific working groups on old and new topics and a central unit within the Secretariat. Apparently, disagreement over this proposal ran deep. Two – unnamed – delegations went on record stating that they were unable to support it at that stage.[35] Nevertheless, in May 1970, the Ministerial Council endorsed this plan, thereby creating the OECD Environment Committee. Its first head was Christian Herter (United States), who had just been appointed deputy assistant secretary of state for environmental and population affairs after having worked as a senior executive for the Mobil Oil Corporation since 1961.[36] Choosing him as head of the Environment Committee presumably helped placate

U.S. concerns about an overly critical stand on economic growth. The same was true of the committee's mandate, whose principal task, according to van Lennep, was to determine who was to pay for pollution reduction and environmental improvement 'while retaining the values of the market economy'.[37] The original plan for a holistic approach to the environment and related issues was safeguarded to some extent through the establishment of a corresponding Environmental Directorate in 1971, which, upon van Lennep's dogged prodding, received nineteen staff from other directorates and fifteen new positions.[38]

Thus, the OECD became the first international organization to establish a department specifically in charge of environmental questions, a year before the creation of the United Nations Environment Programme (UNEP). The move clearly upgraded the environment within the OECD's work profile and beyond, as it provided a precedent to which other international organizations would have to react. Immediately, the Environment Committee provided a venue for its members to discuss issues on the agenda of the upcoming UN Conference at Stockholm.[39] In the long run, it also served as an instrument to leave the concept of a growth-based market economic system intact and to avoid discussions on fundamental systemic changes.

The change of perspective is especially clear when compared to another outcome of the initial OECD discussions in 1968. In 1972, the spectacular study initiated by the Club of Rome, *Limits to Growth*, was published and unleashed a lively discussion on the benefits and risks of continued economic growth. It was widely read and it profoundly affected the economic-environmental discourse. But its impact was weakened when vehement and often unfair criticism, mainly by economists, left the study discredited. For years, belief in a possible systemic collapse was widely considered politically radical, naïve or both, and despite its early ties to the Club of Rome, the OECD never took up the rhetoric and data of *Limits to Growth*.[40]

From Environment to Sustainable Development

During the following years, the OECD Environment Directorate initiated an impressive range of studies, relying on the ready support of its member countries, most of which were developing their own national policies and programmes at the time. The impact of energy production and consumption, transportation, biodiversity and especially hazardous waste and chemicals, which got its own committee, received priority attention.[41] Thus, the relationship between environmental and economic concerns formed only one of its areas of work, albeit an important one.

Early on, van Lennep insisted that the economic aspects of environmental issues would have to be the main focus of attention of an economic agency like the OECD.[42] This assignment immediately exposed the new Environment Committee to the tension between environmentalism and free market principles. In an unregulated market, measures taken by some economic actors to reduce the environmental burden of production and/or consumption might place them at a disadvantage in relation to their less conscientious competitors. International standards were therefore required. However, environmental standards, unless controlled for fairness and evenness, could also be used to exclude some actors from the market, threatening to distort them. These debates reflected the uneasy relationship between the Environment Committee and the established economists at the OECD, who were sceptical of these new environmental studies.[43] By contrast, the OECD Trade Committee took a relaxed view of environmental standards as possible trade barriers. Members considered the issue as no different in principle from any other potential trade discrimination and one that could be addressed in similar ways.[44] These differences in approach affected the position of the new Environment Committee. The much-coveted cooperation with the Economic Policy Committee never materialized, but officials from other directorates began regularly attending its sessions. Deprived of more intensive cooperation, the Environment Directorate founded its own Group of Economic Experts, which studied the relations between environmental and economic policies and goals (in addition to other groups addressing energy, water and air management, waste, noise or chemicals).[45]

Initial studies unfolded under the influence of the upcoming UN Conference on the Human Environment in Stockholm. An extended seminar on the Problems of Environment Economics, held between April and June 1971, tested early ideas, which fed into the 'Guiding Principles Concerning International Economic Aspects of Environmental Policies'.[46] They suggested several mechanisms to integrate environmental concerns into market activities such as agreed environmental standards, the reconciliation of environmental policies with the rules of the General Agreement on Tariffs and Trade (GATT) and compensatory trade fees. These guidelines, it was hoped, would serve as inspiration and as a model for the negotiations in Stockholm, given that the OECD countries accounted for two-thirds of international trade as well as for most of the environmental burden.[47]

However, events surrounding the Stockholm conference showed that these considerations met the concerns of the high-income OECD countries, but neglected those of the majority of countries and people of the world. The former were interested in international regulations, which would affect everybody equally, thus allowing environmental protection within a functioning market. As Michael W. Manulak shows in his chapter in this book, representatives of low-income countries tended to view these objectives as efforts to safeguard existing market privileges for industrialized countries, thereby limiting the economic development of the South, and to gain neocolonial control over Southern economies. They demanded that Northern governments honour earlier commitments for development assistance and pay for the majority of environmental degradations, notably those that the policies of high-income countries had caused. At this crossroads of contradictory interests, policy makers began looking for a new developmental concept, which would reconcile economic, environmental and social/distributional goals.[48]

Within this triangular challenge, the OECD was at the forefront of efforts to reconcile the first two of these three goals. Probably its most important achievement in this regard was to discuss and eventually establish consensus on the principle that whoever was responsible for causing environmental damage would also be compelled to pay for the damages caused or for its rehabilitation, the polluter pays principle (PPP). This concept had already been advocated by the Council of Europe in its Charter on Air Pollution Control in 1968.[49] As Meyer discusses in his chapter on the polluter pays principle in this volume, it became a central element of EC policy in the mid 1970s. The concept was meant to internalize external costs in domestic production so that the price of a product would fully reflect the real costs incurred during production, operation and disposal, including those of environmental protection or clean-up. The main aim was to prevent governments from resorting to subsidies to protect their export. The effective implementation of the polluter pays principle required a similar use in all countries, and while the OECD had no executive powers, a mixture of the obvious need for coordination and of group pressure led to broad acceptance of the principle among and, to some extent, beyond OECD countries.[50]

Shortly after, attention turned to more preventive forms of economic-environmental reconciliation. The first meeting of the OECD Environment Committee at Ministerial Level took place in November

1974. It was chaired by Gro Harlem Brundtland, who had become the Norwegian environment minister just a few weeks earlier.[51] Here, the interest in an integration of economic and environmental demands gave rise to the concept that environmental considerations needed to be incorporated into decision making in all economic sectors at an early stage, an idea that became the mainstay of further OECD studies and would form the core of 'sustainable development' some years later.[52] This approach was spurred when Jim MacNeill became director of the Environment Directorate in 1978. As engineer-economist responsible for coordinating the South Saskatchewan River Development Project, who later held several senior positions in the Canadian public service including environmental advisor to Prime Minister Pierre Trudeau, MacNeill had firsthand experience of the connection between environmental and economic concerns. In 1971, he argued that it made little sense to accept economic development in its current form as a given and to address the environment only when the damage was done. Instead, the two should be perceived and treated as being intrinsically interlinked.[53] Under his leadership, the Environment Committee returned to questions of principle regarding the relationship between the environment and the economy, questioning the conceived wisdom of the time that environmental protection was detrimental to economic performance.

In 1979, a lengthy study on *Environment and Current Economic Issues* discussed in detail how economic growth during recent years had related to environmental policies. After analyses of a broad range of sectors, it concluded that 'environmental aims and economic growth objectives are or should be made complementary' and that measures designed to decrease the burden on the environment had 'made significant contributions to improvements in welfare in OECD communities'.[54] OECD environment ministers largely followed this line of argument. In a nonbinding *Declaration of Anticipatory Environmental Policies,* the OECD member states agreed in 1979 to integrate at an early stage environmental considerations into discussions of all economic and social policy, which were likely to have a significant environmental effect, and to conduct environment assessment requirements in development assistance projects.[55] As a follow-up, the OECD Environment Committee launched a range of related studies on this question. With the cooperation of several governments, industries and other OECD directorates, OECD staff analysed the environmental aspects of multinational investment, the economic value of genetic materials, the implementation of the Recommendation on Guiding

Principles on the International Economic Aspects of Environmental Policies, and the effectiveness and efficiency of environmental regulations and of tax and fiscal policies.[56]

These studies were timely because they used the data that were becoming available after approximately a decade of experience with environmental policies in many countries. Thus, they offered a chance to discuss the effects of early timid efforts to reconcile environmental and economic concerns in industrial countries through end-of-pipe measures, while similar issues were being discussed on a more theoretical level with regard to low-income countries in the global South, as analysed by Stephen Macekura in this volume. But they were also increasingly at odds with the political climate of the time as the period of exceptional economic prosperity was coming to an end. Instead, the years from the mid 1970s onwards were marked by oil crises, declining growth rates, the breakdown of the Bretton Woods system and, in 1979/80, the elections of Margaret Thatcher in Britain and Ronald Reagan in the US, who personified the international turn to neoliberalism. The change in attitude of the U.S. government was noticeable at the OECD Environment Directorate, though the U.S. ambassador and his staff remained personally supportive.[57] As neoliberalism gained international traction, the OECD at large also shifted its economic focus away from Keynesian orthodoxy towards neoclassical supply-side approaches and its opposition to governmental regulations.[58] In contrast to the years of high growth rates, which had given rise to the OECD Environment Committee, environmental work was now perceived in the context of efforts to get economies back on to a growth track. These circumstances may have increased the acceptability of an approach that tested environmental policies in terms of cost-benefit considerations. These considerations together with the renewed emphasis on the environment-economy nexus resulted in a restructured Environment Directorate in 1980, when an Environment and Economy Division formed one of four basic divisions.[59] A year later, the OECD Environmental Directorate and member countries declared its work regarding economic-environmental management one of its priorities.[60]

The results of the studies, begun in 1979, were presented at the Conference on the Economics of Environment in June 1984. It confirmed earlier research that argued that the environment and the economy were 'mutually reinforcing' and 'supportive of and supported by technological innovation' if 'properly managed'.[61] Such management required the integration of environmental and economic policies, widespread 'anticipate-and-prevent' strategies as well as adequate institutional arrangements and information policies. This was a crucial proviso: the environment and the economy were not automatically mutually reinforcing. Making them so required a change of attitude and of policy.[62] At that time, MacNeill was already actively preparing his departure from the OECD to pursue these ideas in a different framework.

From Sustainable Development back to Growth

Ten years after the 1972 Stockholm Conference and the creation of the UNEP, disillusionment about the lack of tangible successes in international environmental work provided the ground for the idea of a new initiative that would stimulate international discussions once more. Mostafa Tolba, head of the UNEP, in particular, lobbied for this idea, and in December 1983, the UN General Assembly called for the institution of an independent Special Commission 'to propose long-term environmental strategies for achieving sustainable development to the year 2000 and beyond', which would be financed by donations from governments, but would report to the UN.[63] As chairperson, the UN chose Brundtland, former Norwegian Prime Minister and Minister of the Environment. In early 1984, Brundtland, whose introduction into international environmental policies had been at the OECD, asked MacNeill to act as Secretary-General of the World Commission on Environment and Development (WCED). Better known as the Brundtland Commission, it was to make *sustainable development* a well-known concept.

The scope of the WCED mandate differed markedly from that of the OECD Environment Directorate in that it was to consider economic-environmental relations on a global scale instead of only within rich countries. Consequently, questions of the global distribution of income and opportunities, which were of negligible importance at the OECD, took a central role at the WCED. Nevertheless, crucial considerations of how economic wellbeing could be reconciled with environmental protection remained, and MacNeill relied heavily on his OECD experience when drafting the agenda of the Commission's work programme. His proposal, soon endorsed by the rest of the WCED, was to address development and environment as interlinked and to study this relationship in a series of issues including population, energy, industry, agriculture, human settlements, international economic relations, global environmental monitoring and international cooperation.[64] This plan, supplemented by a few topics that were added later (biodiversity, peace and security and the commons), formed the basis for the discussions of the WCED during the following three years.[65]

In 1987, the Commission presented its report called *Our Common Future* (OCF) to the UN General Assembly. It became most famous for its definition of sustainable development as 'development that meets the needs of the present without compromising the ability of future generations to meet their own needs'.[66] However, a more important conclusion of the text, stated repeatedly, was its insistence that:

> Sustainable development objectives should be incorporated in the terms of reference of those cabinet and legislative committees dealing with national economic policy and planning as well as those dealing with key sectoral and international policies.[67]

This principle, which echoed earlier OECD findings, was expanded by the demand that major governmental agencies should be made 'fully accountable for ensuring that their policies, programmes, and budgets support development that is ecologically as well as economically sustainable'.[68] The final recommendations of the Commission report included, among other points, an improvement of international planning and coordination through the institution of a UN board for sustainable development, a global risk assessment programme and a 'steering group' of eminent individuals; reformed national and international legal structures by giving the state the responsibility to 'observe the principle of optimum sustainable yield in the exploitation of living natural resources and ecosystems' and to assess all major new policies regarding their effect on sustainable development, while giving the International Court of Justice a mandate for environmental and resource management problems; and enhanced funding possibilities by establishing a special international banking facility for the development and protection of critical habitats and ecosystems, and a procedure for automatic financial transfers through the taxation of various seemingly free global goods and services. The points also included implementing OECD recommendations regarding environmental assessment policy and their application for bilateral aid programmes.[69]

Though its roots in OECD activities were unmistakable, these recommendations went further, in terms of both their geographical and political scope, although in a different way from the World Conservation Strategy (WCS) published seven years earlier. Instead of placing *sustainable development* within a North–South context (although this element certainly loomed large in their background discussions), the Brundtland Commission conceptualized it as a principle relevant everywhere at all levels of decision making, ranging from the individual to the international, highlighting the shared responsibility and experience of the 'common future'. In contrast to the WCS, the Brundtland Report was widely circulated around the world and came to be accepted as the defining reference for *sustainable development.*

The report also had a major impact on international organizations, at least for a while. In December 1987, the UN General Assembly called upon all governments and other governing institutions and all UN bodies to place its policies in all sectors on a sustainable footing.[70] As a result, virtually all UN agencies and several governments reviewed their policies with regard to their compatibility with sustainable development.[71] Thus, the OECD was only one of many organizations that undertook a process of reconsideration of its work in light of WCED recommendations in 1988, but it did so on a comparatively large scale. At the request of the Norwegian government, it organized a high-level meeting of delegates of OECD countries in November 1988, and the event was preceded by several intra-agency assessments prepared for the Secretary-General.[72] A first appraisal found that almost all departments were already working on issues addressed by the WCED, including ways to include natural resources into national accounting systems, coastal zones, climate change, chemical products, financial North–South transfers, debt issues, nuclear energy, air pollution, etc. However, the list ended with the sobering remark that it remained difficult to make use of those projects in any coordinated way.[73] A more detailed assessment in October found that numerous governments, international organizations, industrial corporations and INGOs were actively considering the WCED's findings, but that their responses often focused on a narrow environmental reading of sustainable development while ignoring other aspects such as social equality between and within countries, the redirection of military expenditures into the development sector, the need for UN restructuring and for new mechanisms to finance the transition to sustainable practices in low-income countries.[74] Supposedly, these issues were considered too politically sensitive in most UN agencies – as they were at the OECD, where they were similarly overlooked. Thus, a position paper of the Environment Directorate proposed studies on better systems of natural resource accounting and intersectoral studies on several horizontal topics such as transport, technology, coastal zone management and rural development, none of which addressed social equity or any of the other neglected themes, let alone the possibility of raising funds for a transition to sustainable development by taxing international trade or financial operations.[75]

These discrepancies also showed during a meeting in November 1988. MacNeill, who had been invited, singled out as the principal theme of the WCED report that major national and international economic institutions should become fully responsible and accountable for ensuring that their policies and budgets supported a development that was sustainable.[76] In more tangible terms, he listed resource accounting, debt management, guidelines for transnational corporations and technology transfer, development assistance, addressing the transfer of environmental damage costs from North to South, and forms of automatic financing as important issues of sustainable development.[77] Only two of these topics (resource accounting and debt management) were being considered for further OECD studies at the time.[78] Despite such clear discrepancies, all participants agreed that the OECD should embrace the 'sustainable development' concept into its work programme.[79]

A summary of this seminar was distributed to the OECD directors for inspiration, and several committees subsequently proposed specific projects in this field.[80] Some issues, such as energy, environmental economics and technology, emerged as integrated themes that should continue to receive special attention across several directorates.[81]

At first sight, the introduction of 'sustainable development' into the OECD as an advanced version of its earlier studies appeared a remarkable success. Following the 1989 Ministerial Council meeting, the Environment Committee declared the 'Economics of Sustainable Development' one of its five core themes of its medium-term programme.[82] However, its considerations often remained on a technical level. A serious attempt at implementing the entirety of the WCED recommendations, especially those regarding legal reforms, was obviously politically sensitive. After all, they would have entailed that all government and corporate agencies would be held accountable for all their domestic and international policy decisions with regard to whether economic and environmental aspects had been properly reconciled. Arguably, the idea discussed in OECD circles that came closest to this principle was to make the polluter pays principle more relevant to sustainable development and to extend the concept to a user-pays principle – which would have accounted for the fact that high-income countries were responsible for a disproportionate use of resources – for the exploitation of not fully renewable raw materials.

In the form that it was actually used, however, the polluter pays principle was rather toothless. Making it consistent with WCED recommendations and extending it to user payments would have transformed it into a forceful international programme, which would have unleashed a North–South flow of payments and would have revolutionized international relations, both economically and politically. There is no indication that the idea was seriously pursued. It reappeared in OECD discussions regarding domestic policies in 1990 and again in 2006, but without coming close to the wide acceptance of the polluter pays principle.[83]

But OECD bodies continued to pay lip service to the concept of sustainable development. In 1989, at a Ministerial Council meeting, the participating ministers restated the importance of work on sustainable development. A year later, in 1990, the Environment Directorate and the Economic and Statistics Department coorganized expert meetings that produced a brochure on *The Economics of Sustainable Development - A Progress Report*. Participants recommended leaving aside fruitless discussions on how sustainable development should be defined. Rather they should focus pragmatically on goals that, in most people's views, formed part of a transition towards sustainable development. Subsequently, the Environment Committee endorsed nine goals, which would shape the OECD interpretation of sustainable development during the following years: promoting worldwide economic growth; integrating economic and environmental policies; pricing goods and resources to reflect environmental costs; promoting technological change to support 'clean' growth; placing unique resources under sound management; controlling population growth where excessive; upgrading factors supporting sound environmental management; expanding international cooperation; and monitoring progress towards sustainable development objectives.[84] This list is noteworthy as much for what it included as for what it omitted: all recommendations echoed in some form those of the WCED report, but they left out several key items, such as the integration of environmental issues in high-level international jurisdiction or the responsibility of national governments to keep their economic activities within sustainable yields and to be accountable for the outcome.

If adopting work on the environment had been a means to avoid addressing systemic changes in a growth-based economic concept in the early 1970s, adopting a selective view of sustainable development that excluded discussions on economic limits and redistribution and on profound changes in economic structures served the same purpose in 1990.

Conclusion

The OECD environmental programme and conceptualization of sustainable development grew from considerations regarding the relation of the existing growth-based market economy with its physical and social contexts. Given the increasingly evident impact of economic activities on the environment, both local and global, it may have been inevitable that the OECD, an economic organization, would one day approach the question of how economic performance could be reconciled with safeguarding environmental integrity, the core of sustainable development. But how it would approach the issue and how these discussions would evolve over time was no foregone conclusion.

In the course of four decades, OECD involvement in work on the environment and sustainable development has been marked by ambivalence. On the one hand, it has repeatedly adopted an avant-garde role, addressing crucial issues before they were mainstream: it established the Environment Committee and Directorate before most governments even had corresponding ministers, it established central principles of conditions for the compatibility of economy and environment before they were discussed as sustainable development, and the Environment Committee began studies on acid rain, ozone depletion, hazardous waste and climate change before they figured prominently on the international agenda.[85] Typically, these initiatives were taken by high-level OECD staff, who then sought and received the endorsement and active cooperation from government representatives and various outside experts and stakeholders. On the other hand, these initiatives have consistently been integrated into the OECD's overall economic credo and subjugated to an apparently unshakable belief in economic growth as the principal goal of development and in market forces as a strategy to reconcile environmental resources and services with this goal.

This attitude reflected conditions in member countries, where major stakeholders had their reputations and careers tied to increasing output and profits, as well as the professional background of the majority of OECD staff as experts trained in mainstream economics. But it also represents a missed opportunity at the beginning of OECD engagement, when considerations began by questioning not the form of economic growth, but its purpose. This early approach was the result of a serendipitous meeting of concerned individuals and a contingency of environmental concern and robust economies, and it stood little chance of changing the character of the OECD.

Thus, in contrast to Macekura's findings about the connected approach of environmentalism and developmental thought in general, overall OECD arguments aimed at integrating the environment into the logic of the economy rather than vice versa. Accordingly, it recommended mechanisms typically used as 'economic instruments', such as taxes, charges and tradable permits.[86] By calling for and, to some extent, implementing such tools, OECD efforts have served to postulate that, in principle, continued economic growth was compatible with environmental protection, a position that it has consistently held since 1970, but that appears increasingly doubtful.[87]

Implicitly, some of its actions contributed to the common misconception of sustainable development as stipulating that economic growth and environmental protection *were* compatible rather than that they *could be made* compatible through a fundamental change of economic thinking and acting, even though early reports by the OECD and the Brundtland Commission made this amply clear. Ironically, in the process, the OECD helped stave off discussions about the limits to growth, a discussion in whose beginning it had an essential role.

Meanwhile, the growing endorsement of sustainable development shows signs of being replaced by a new change of terminology. In 2009, in the midst of a global financial and economic crisis, the Ministerial Council meeting moved 'sustainable development' from a context of environment to 'green growth', declaring that '"green" and "growth" could go hand-in-hand' and calling on the OECD to 'develop as a horizontal project, a Green Growth Strategy in order to achieve economic recovery and environmentally and socially sustainable economic growth'.[88] At present, its strategy platform addresses numerous important issues, including the environmental dimension of the quality of life and the socioeconomic context of growth, but clearly its main focus is the preservation of exactly this growth.[89] The change of name can therefore be seen as a fitting next step in the one decade-old struggle of the OECD between the recognition of the need for changes in the economy-environment complex and the inability to accept a form of change that would involve a serious change of attitude towards economic growth.

The final evaluation of the role of the OECD, the environment and sustainable development therefore remains ambivalent. Despite its loyalty to a growth and market-based system, the OECD has nevertheless been at the forefront of establishing demanding environmental standards and procedures. Tellingly, in the early twenty-first century, a network of INGOs formed OECD Watch, an initiative designed:

to ensure that business activity contributes to sustainable development and poverty eradication and that corporations are held accountable for their impacts around the globe. In the absence of a globally binding framework for corporate accountability, the OECD Guidelines are one of the few mechanisms available for holding corporations to account for their international operations.[90]

The group places information about the compliance of corporations with the OECD guidelines on a website and in a newsletter.

The OECD's advanced position on questions of regulations and accountability contrasts with its slow or nonexistent advance on other fields. In 2013, Jonathon Porrit and Peter Maddon of the OECD Forum for the Future declared: 'Investment strategies almost never consider external costs to the environment when calculating potential returns. But incorporating environmental risk and sustainability into investor mindsets is possible – and urgent.'[91] This finding repeats almost verbatim similar calls some forty-five years earlier. It would be wrong to see this long history of repetitions of – more or less – the same recommendations of how to reconcile environmental and economic exigencies as a weakness of sustainable development as a concept or as a sign of incompetence of the OECD. Instead, it demonstrates that while the activism of determined persons and institutional networking is important in establishing a new concept, a point highlighted by Macekura in his discussion of sustainable development, they are not enough to translate a concept into behaviour and policy when short-term private, economic and political advantages are overwhelmingly stacked against such change. Eventually, it is a sign of the weakness of our social systems and, ultimately, of the human spirit that prevents the implementation of what at least two generations of experts at the OECD and elsewhere have found to be true.

Iris Borowy is Distinguished Professor at Shanghai University, China.

Notes

1. I would like to thank the editors and Jim MacNeill for helpful comments on an earlier draft of the chapter.
2. Kerstin Martens and Anja P. Jakobi, 'The OECD as an Actor in International Politics', in *Mechanisms of OECD Governance: International Incentives for National Policy-Making?* Kerstin Martens and Anja P. Jakobi (eds) (Oxford: Oxford Scholarship Online, 2010), 1–25, 5 and 2.

3. These researchers are mostly affiliated with the Geneva-based international research project on 'OECDhistoryproject. Warden of the West. The OECD and the global political economy, 1948 to present'. See http://oecdhistoryproject. net/, accessed 5 July 2016.

4. Rianne Mahon and Stephen McBride (eds), *The OECD and Transnational Governance,* (Vancouver: UBC Press, 2008); Peter Carroll and Aynsley Kellow, *The OECD: A Study of Organisational Adaptation* (Cheltenham: Edgar Elgar, 2011).

5. Robert Wolfe, 'From Reconstructing Europe to Constructing Globalization: The OECD in Historical Perspective', in *The OECD and Transnational Governance,* Rianne Mahon and Stephen McBride (eds) (Vancouver: UBC Press, 2008), 25–42, 26–33.

6. Convention on the Organisation for Economic Co-operation and Development, Article 1, 14 December 1960, http://www.oecd.org/general/conventionontheo rganisationforeconomicco-operationanddevelopment.htm, accessed 20 May 2016; cf. Bill Long, *International Environmental Issues and the OECD 1950– 2000* (Paris: OECD, 2000), 2.

7. Carroll and Kellow, *The OECD,* 15–17; Matthias Schmelzer, 'The Hegemony of Growth: The Making and Remaking of the Economic Growth Paradigm and the OECD' (Ph.D. dissertation, European University Viadrina, Frankfurt/ Oder, 2013), 49–58.

8. Matthias Schmelzer, *Freiheit für Wechselkurse und Kapital. Die Ursprünge neoliberaler Währungspolitik und die Mont Pélerin Society* (Marburg: Metropolis, 2010).

9. Martens and Jakobi, 'The OECD as an Actor', 14; Carroll and Kellow, *The OECD,* 21–30; Rianne Mahon and Stephen McBride, 'Introduction', in *The OECD and Transnational Governance,* Rianne Mahon and Stephen McBride (eds) (Vancouver: UBC Press, 2008), 3–24, 14–15; Kenneth G. Ruffing, 'The Role of the Organization for Economic Cooperation and Development in Environmental Policy Making', *Review of Environmental Economic and Policy* 4(2) (2010): 199–220, 202.

10. Ruffing, 'Role of the Organization', 200.

11. Martens and Jakobi, 'The OECD as an Actor', 7–10; Ruffing, 'Role of the Organization', 214.

12. Martens and Jakobi, 'The OECD as an Actor', 6; Mahon and McBride, 'Introduction', 7–8.

13. Richard Woodward, 'Towards Complex Multilateralism? Civil Society and the OECD', in *The OECD and Transnational Governance,* Rianne Mahon and Stephen McBride (eds) (Vancouver: UBC Press, 2008), 77–95, 86; Ruffing, 'Role of the Organization', 213.

14. Martens and Jakobi, 'The OECD as an Actor', 2.

15. MacNeill to the author, 1 February 2012.

16. Tony Porter and Michael Webb, 'Role of the OECD in the Orchestration of Global Knowledge Networks', in *The OECD and Transnational Governance,*

Rianne Mahon and Stephen McBride (eds) (Vancouver: UBC Press, 2008), 43–59, 44–45.

17. Per-Olof Busch, 'The OECD Environment Directorate: The Art of Persuasion and its Limitations', *Global Governance Working Paper* 20 (2006), http://www.glogov.org/images/doc/wp20.pdf, accessed 5 July 2016,, 11.

18. Busch, 'The OECD Directorate', 4–6.

19. Present Status of Work of the Committee for Research Cooperation and Other OECD Bodies Relating to the Environment, ENV (70)4, 4 March 1970, OECD archive (OECDA); Long, *International Environmental Issues*, 28–31.

20. John R. McNeill, *Something New under the Sun* (New York: W.W. Norton, 2000), 66–67, 129–39, 337–39.

21. Jan-Henrik Meyer, 'Appropriating the Environment: How the European Institutions Received the Novel Idea of the Environment and Made it Their Own', *KFG Working Paper* 31 (2011), Research College 'The Transformative Power of Europe', Free University Berlin, http://www.polsoz.fu-berlin.de/en/v/transformeurope/publications/working_paper/WP_31_Meyer_neu.pdf, accessed 20 May 2016; Thorsten Schulz, 'Transatlantic Environmental Security in the 1970s? NATO's "Third Dimension" as an Early Environmental and Human Security Approach', *Historical Social Research* 35(4) (2010): 309–28.

22. List prepared by Ad Hoc Preparatory Committee on the Activities of the Organisation on Environmental Problems Related to Economic Growth, ENV (70)3, 27 February 1970, OECDA.

23. Kenneth Galbraith, *The Affluent Society* (Boston, MA: Houghton Mifflin Harcourt, 1958); Rachel Carson, *Silent Spring* (Boston, MA: Houghton Mifflin, 1962); Ezra Mishan, *The Costs of Economic Growth* (New York: Staples Press, 1967).

24. Long, *International Environmental Issues,* 29.

25. Matthias Schmelzer, 'The Crisis before the Crisis: The "Problems of Modern Society" and the OECD, 1968–1974', *European Review of History* 19(6) (2012): 999–1020.

26. Problems of Modern Society. Note by the Secretary-General, C(69)168, 12 December 1969, OECDA.

27. Carroll and Kellow, *The OECD*, 70.

28. Problems of Modern Society. Note by the Secretary-General, C(69)123, 18 September 1969, OECDA, 1–4.

29. Van Lennep, Problems of Modern Society, C(69)168, 1, emphasis in original.

30. Ibid., 5–6.

31. Minutes of Council Meeting, C/M(70)1 (Prov.) (1st Revision), 13 January 1970; Council. Minutes, C/M (70)2 (1st Rev.), 20 January 1970, both OECDA.

32. Council. Problems of Economic Growth, Environment and Welfare. Proposals for Preparatory Considerations of the Problems, C(70)12, 16 January 1970; Draft Resolution of the Council setting up an ad hoc preparatory committee of the activities of the Organisation on environmental problems relating to economic growth, C(70)20, 4 February 1970; Creation of an ad hoc preparatory

committee on the activities of the Organisation on environmental problems relating to economic growth, C(70)22, 5 February 1970, all OECDA.

33. Ad Hoc Preparatory Committee on the Activities of the Organisation on Environmental Problems Related to Economic Growth. Annotated Agenda, ENV(70)5, 26 February 1970, OECDA.

34. Ad Hoc Preparatory Committee on the Activities of the Organisation on Environmental Problems Related to Economic Growth, ENV (70)11. Overall Conclusions, corrected date: 30 April 1970, OECDA.

35. Ad Hoc Preparatory Committee on the Activities of the Organisation on Environmental Problems Related to Economic Growth, ENV(70)9, 15 April 1970; Council Minutes of the 218th Meeting, 12 May 1970, C/M(70)13(Prov.), 16 June 1970, all OECDA; Long, *International Environmental Issues*, 34.

36. Margalit Fox, 'Christian Herter Jr., Longtime Public Servant, Dies at 88', *New York Times*, 1 October 2007.

37. Long, *International Environmental Issues*, 39.

38. Ibid., 34–38.

39. MacNeill to the author, 2 February 2014.

40. Ugo Bardi, *The Limits to Growth Revisited* (New York: Springer, 2011); Mauricio Schoijet, 'Limits to Growth and the Rise of Catastrophism', *Environmental History* 4(4) (1999): 515–30, 520 f.

41. Busch, 'The OECD Environment Directorate', 1.

42. Long, *International Environmental Issues*, 35–43.

43. Schmelzer, 'Crisis before the Crisis'; Long, *International Environmental Issues*, 41–44. See also the chapter by Wolfram Kaiser in this volume.

44. Informal Summery, Working Party of the Trade Committee, 10 December 1972, Folder 161714, OECDA.

45. Long, *International Environmental Issues*, 41–51.

46. OECD Council, 'Guiding Principles Concerning International Economic Aspects of Environmental Policies' C(72)128, 1972, OECDA.

47. Report of the Environmental Committee. No.91 Principes Directeurs relatifs aux aspects économiques des politiques de l'environnement sur le plan international, 23 March 1972, Folder 161714, OECDA.

48. Iris Borowy, *Defining Sustainable Development for Our Common* Future (Milton Park: Earthscan/Routledge, 2014), 30–35.

49. Council of Europe, *Resolution (68)4, Adopted by the Ministers' Deputies on 8 March 1968, Approving the 'Declaration of Principles' on Air Pollution Control* (Strasbourg, 1968).

50. Carroll and Kellow, *OECD*, 147, 214; Ruffing, 'Role of the Organization', 204.

51. Long, *International Environmental Issues*, 47; Gro Harlem Brundtland, *Grundrecht Gesundheit* (Frankfurt: Campus Verlag: 2000), 13.

52. Long, *International Environmental Issues*, 49.

53. Jim MacNeill, *Environmental Management* (Ottawa: Information Canada, 1971); Jim MacNeill, 'From Controversy to Consensus: The World Commission on Environment and Development', *Environmental Policy and Law* 37(2–3) (2007): 242–65, 249.

54. Environment Committee, Environment and Current Economic Issues, ENV(79)4, 20 February 1979, 39, OECDA.

55. OECD, Declaration on Anticipatory Environmental Policies, OECD and the Environment, Paris: OECD, 1979, http://sedac.ciesin.org/entri/texts/oecd/OECD-4.05.html, accessed 20 May 2016.

56. OECD Environment Committee, 1983 Draft Programme, 2 February 1982, ENV(81)21, OECDA; Ruffing, 'Role of the Organization', 208; MacNeill to the author, 1 February 2012.

57. MacNeill to the author, 1 February 2012.

58. Porter and Webb, 'Role of the OECD', 50

59. Long, *International Environmental Issues*, 53; MacNeill to the author, 2 February 2014.

60. Long, *International Environmental Issues*, 61.

61. OECD, *Results of the International Conference on Environment and Economics* (Paris: OECD, 1985), OECDA, 10.

62. As emphasized by MacNeill to the author, 1 February 2012.

63. Draft Resolution Recommended by the Governing Council of UNEP at its 11th session, 23 May 1982, S-1051-0014-05, United Nations Archives (UNA); Borowy, *Defining Sustainable Development*, 48–52.

64. Key Issues, WCED/84/10-1, Inaugural Meeting, Geneva, 1–3 October 1984, International Development Research Archive.

65. Borowy, *Defining Sustainable Development*, 55–161.

66. The World Commission on Environment and Development, *Our Common Future* (Geneva, 1987), 43, 45.

67. WCED, *Our Common Future*, 314.

68. Ibid., 314.

69. Ibid., 317–40.

70. UN General Assembly, A/RES/42/187, 11 December 1987, S-1051-0026-0008, UNA.

71. Borowy, *Defining Sustainable Development*, 167–74.

72. Long, *International Environmental Issues*, 72.

73. OECD, Programme de Travail de l'OCDE et Rapport de la Commission Mondiale pour d'Environnement et le Développement, 15 September 1988, Folder 161714, OECDA.

74. OECD, International Response to the Report of the World Commission on Environment and Development, SG/WCED (88)2, 25 October 1988, OECDA.

75. OECD, Activities Supportive of the Report of the WCED, SG/WCED (88)3, 25 October 1988, OECDA.

76. Statement by Jim MacNeill, Seminar on OECD and the Report of the WCED, under cover of OECD Seminar on the Report of the WCED, SG/WCED (88)4, 2 December 1988, OECDA.

77. OECD Seminar on the Report of the WCED, SG/WCED (88)4, 2 December 1988, OECDA, 9–10.

78. OECD Seminar on the Report of the WCED, 9–10.

79. Long, *International Environmental Issues*, 72.

80. The Trade Committee, the Environment Directorate, the Committee for Science and Technology, the Business and Industry Advisory Committee and the Development Assistance Committee.
81. OECD, Statement by Mr. Long, Director for Environment, on the Follow-up to the Report of the WCED, CES/89.14, 14 March 1989, OECDA.
82. Long, *International Environmental Issues*, 71.
83. OECD, WCED Report Subjects Calling for Examination and/or Action in OECD, undated, Folder 161714, OECDA; Ruffing, 'Role of the Organization', 206–7.
84. Long, *International Environmental Issues*, 72–73.
85. Carroll and Kellow, *OECD*, 219.
86. Ruffing, 'Role of the Organization', 203.
87. Among a growing body of literature on this topic, see James G. Speth, *The Bridge at the Edge of the World* (New Haven: Yale University Press, 2008), 57, 115–16.
88. Meeting of the Council at Ministerial Level, 24–25 June 2009, Declaration on Green Growth, C/MIN(2009)5/ADD1/FINAL, http://search.oecd.org/officialdocuments/displaydocumentpdf/?doclanguage=en&cote=C/MIN(2009)5/ADD1/FINAL, accessed 20 May 2016.
89. http://www.oecd.org/greengrowth/49313167.pdf, accessed 20 May 2016.
90. http://oecdwatch.org/about-us, accessed 20 May 2016.
91. http://www.oecd.org/greengrowth/investing-in-a-sustainable-future.htm, accessed 20 May 2016.

Bibliography

Bardi, Ugo, *The Limits to Growth Revisited* (New York: Springer, 2011).
Borowy, Iris, *Defining Sustainable Development for Our Common Future* (Milton Park: Earthscan/Routledge, 2014).
Brundtland, Gro Harlem, *Grundrecht Gesundheit* (Frankfurt: Campus: 2000).
Busch, Per-Olof, 'The OECD Environment Directorate. The Art of Persuasion and its Limitations', *Global Governance Working Paper* 20 (2006), http://www.glogov.org/images/doc/wp20.pdf, accessed 5 July 2016.
Carroll, Peter and Aynsley Kellow, *The OECD: A Study of Organisational Adaptation* (Cheltenham: Edgar Elgar, 2011).
Carson, Rachel, *Silent Spring* (Boston, MA: Houghton Mifflin, 1962).
Council of Europe, *Resolution (68)4, Adopted by the Ministers' Deputies on 8 March 1968, Approving the 'Declaration of Principles' on Air Pollution Control* (Strasbourg, 1968).
Fox, Margalit, 'Christian Herter Jr., Longtime Public Servant, Dies at 88', *New York Times*, 1 October 2007.
Galbraith, Kenneth, *The Affluent Society* (Boston, MA: Houghton Mifflin Harcourt, 1958).
Kaiser, Wolfram, 'Sometimes it's the Economy, Stupid! International Organisations, Steel and the Environment', in *International Organizations and Environmental*

Protection. Conservation and Globalization in the Twentieth Century, Wolfram Kaiser and Jan-Henrik Meyer (eds) (New York: Berghahn Books, 2017).

Long, Bill, *International Environmental Issues and the OECD 1950–2000* (Paris: OECD, 2000).

MacNeill, Jim, *Environmental Management* (Ottawa: Information Canada, 1971).

———. Statement, Seminar on OECD and the Report of the WCED, under cover of OECD Seminar on the Report of the WCED, SG/WCED (88)4, 2 December 1988, OECDA.

———. 'From Controversy to Consensus: The World Commission on Environment and Development', *Environmental Policy and Law* 37(2–3) (2007): 242–65.

MacNeill, Jim to Iris Borowy, 1 February 2012.

———. 2 February 2014.

Mahon, Rianne and Stephen McBride, 'Introduction', in *The OECD and Transnational Governance,* Rianne Mahon and Stephen McBride (eds) (Vancouver: UBC Press, 2008), 3–24.

———. (eds), *The OECD and Transnational Governance,* (Vancouver: UBC Press, 2008)

Martens, Kerstin and Anja P. Jakobi, 'The OECD as an Actor in International Politics', in *Mechanisms of OECD Governance: International Incentives for National Policy-Making?*, Kerstin Martens and Anja P. Jakobi (eds) (Oxford: Oxford Scholarship Online, 2010), 1–25.

McNeill, John R. *Something New under the Sun* (New York: W.W. Norton, 2000).

Meyer, Jan-Henrik, 'Appropriating the Environment: How the European Institutions Received the Novel Idea of the Environment and Made it Their Own', *KFG Working Paper* 31 (2011), Research College 'The Transformative Power of Europe', Free University Berlin, http://www.polsoz.fu-berlin.de/en/v/ transformeurope/publications/working_paper/WP_31_Meyer_neu.pdf, accessed 25 May 2016.

Mishan, Ezra, *The Costs of Economic Growth* (New York: Staples Press, 1967).

OECD, Convention on the Organisation for Economic Co-operation and Development, 14 December 1960, http://www.oecd.org/general/conventionon theorganisationforeconomicco-operationanddevelopment.htm, accessed 25 May 2016.

———. Problems of Modern Society. Note by the Secretary-General, C(69)123, 18 September 1969, OECDA, 1–4.

———. Council, Minutes of Council Meeting, C/M(70)1 (Prov.) (1st Revision), 13 January 1970, OECDA.

———. Council, Problems of Economic Growth, Environment and Welfare. Proposals for Preparatory Considerations of the Problems, C(70)12, 16 January 1970, OECDA.

———. Council, Minutes, C/M (70)2 (1st Rev.), 20 January 1970, OECDA.

———. Council, Draft Resolution of the Council setting up an ad hoc preparatory committee of the activities of the Organisation on environmental problems relating to economic growth, C(70)20, 4 February 1970, OECDA.

——. Creation of an ad hoc preparatory committee on the activities of the Organisation on environmental problems relating to economic growth, C(70)22, 5 February 1970, OECDA.

——. Ad Hoc Preparatory Committee on the Activities of the Organisation on Environmental Problems Related to Economic Growth. Annotated Agenda, ENV(70)5, 26 February 1970, OECDA.

——. List Prepared by Ad Hoc Preparatory Committee on the Activities of the Organisation on Environmental Problems Related to Economic Growth, ENV (70)3, 27 February 1970, OECDA.

——. Present Status of Work of the Committee for Research Cooperation and Other OECD Bodies Relating to the Environment, ENV (70)4, 4 March 1970, OECD archive (OECDA).

——. Ad Hoc Preparatory Committee on the Activities of the Organisation on Environmental Problems Related to Economic Growth, ENV(70)9, 15 April 1970, OECDA.

——. Ad-Hoc Preparatory Committee on the Activities of the Organisation on Environmental Problems Related to Economic Growth, ENV (70)11. Overall Conclusions, corrected date: 30 April 1970, OECDA.

——. Council Minutes of the 218th Meeting, 12 May 1970, C/M(70)13(Prov.), 16 June 1970, OECDA.

——. 'Guiding Principles Concerning International Economic Aspects of Environmental Policies' C(72)128, 1972, OECDA.

——. Report of the Environmental Committee. No. 91 Principes Directeurs relatifs aux aspects économiques des politiques de l'environnement sur le plan international, 23 March 1972, Folder 161714, OECDA.

——. Informal Summary, Working Party of the Trade Committee, 10 December 1972, Folder 161714, OECDA.

——. Declaration on Anticipatory Environmental Policies, OECD and the Environment, Paris: OECD, 1979, http://sedac.ciesin.org/entri/texts/oecd/OECD-4.05.html, accessed 20 May 2016.

——. Environment Committee, Environment and Current Economic Issues, ENV(79)4, 20 February 1979, OECDA.

——. Environment Committee, 1983 Draft Programme, 2 February 1982, ENV(81)21, OECDA.

——. *Results of the International Conference on Environment and Economics* (Paris: OECD, 1985), OECDA.

——. Activities Supportive of the Report of the WCED, SG/WCED (88)3, 25 October 1988, OECDA.

——. Programme de Travail de l'OCDE et Rapport de la Commission Mondiale pour d'Environnement et le Développement, 15 September 1988, Folder 161714, OECDA.

——. International Response to the Report of the World Commission on Environment and Development, SG/WCED (88)2, 25 October 1988, OECDA.

——. OECD Seminar on the Report of the WCED, SG/WCED (88)4, 2 December 1988, OECDA.

————. Statement by Mr. Long, Director for Environment, on the Follow-Up to the Report of the WCED, CES/89.14, 14 March 1989, OECDA.

————. Meeting of the Council at Ministerial Level, 24–25 June 2009, Declaration on Green Growth, C/MIN(2009)5/ADD1/FINAL, http://search.oecd.org/officialdocuments/displaydocumentpdf/?doclanguage=en&cote=C/MIN(2009)5/ADD1/FINAL, accessed 20 May 2016.

————. 'Green Growth at the OECD: Selected Areas of Ongoing Work', September 2011, http://www.oecd.org/greengrowth/49313167.pdf, accessed 20 May 2016.

————. 'Investing in a Sustainable Future', 2013, http://www.oecd.org/greengrowth/investing-in-a-sustainable-future.htm, accessed 20 May 2016.

————. WCED Report Subjects Calling for Examination and/or Action in OECD, undated, Folder 161714, OECDA.

————. 'OECDhistoryproject. Warden of the West. The OECD and the global political economy, 1948 to present'. See http://oecdhistoryproject.net/, accessed 5 July 2016.

OECD Watch, 'About OECD Watch', http://oecdwatch.org/about-us, accessed 25 May 2016.

Porter, Tony and Michael Webb, 'Role of the OECD in the Orchestration of Global Knowledge Networks', in *The OECD and Transnational Governance,* Rianne Mahon and Stephen McBride (eds) (Vancouver: UBC Press, 2008), 43–59.

Ruffing, Kenneth G., 'The Role of the Organization for Economic Cooperation and Development in Environmental Policy Making', *Review of Environmental Economic and Policy* 4(2) (2010): 199–220.

Schmelzer, Matthias, *Freiheit für Wechselkurse und Kapital. Die Ursprünge neoliberaler Währungspolitik und die Mont Pélerin Society* (Marburg: Metropolis, 2010).

————. 'The Crisis before the Crisis: The "Problems of Modern Society" and the OECD, 1968–1974', *European Review of History* 19(6) (2012): 999–1020.

————. 'The Hegemony of Growth: The Making and Remaking of the Economic Growth Paradigm and the OECD' (Ph.D. dissertation, European University Viadrina, Frankfurt/Oder, 2013).

————. *The Hegemony of Growth: The OECD and the Making of the Economic Growth Paradigm* (Cambridge: Cambridge University Press, 2016).

Schoijet, Mauricio, 'Limits to Growth and the Rise of Catastrophism', *Environmental History* 4(4) (1999): 515–30.

Schulz, Thorsten, 'Transatlantic Environmental Security in the 1970s? NATO's "Third Dimension" as an Early Environmental and Human Security Approach', *Historical Social Research* 35(4) (2010): 309–28.

Speth, James G., *The Bridge at the Edge of the World* (New Haven: Yale University Press, 2008).

UN, Draft Resolution Recommended by the Governing Council of UNEP at its 11th session, 23 May 1982, S-1051-0014-05, United Nations Archives (UNA).

UN General Assembly, A/RES/42/187, 11 December 1987, S-1051-0026-0008, UNA.

Van Lennep, Emiel, Problems of Modern Society. Note by the Secretary-General, C(69)168, 12 December 1969, OECDA.

Wolfe, Robert, 'From Reconstructing Europe to Constructing Globalization: The OECD in Historical Perspective', in *The OECD and Transnational Governance*, Rianne Mahon and Stephen McBride (eds) (Vancouver: UBC Press, 2008), 25–42.

Woodward, Richard, 'Towards Complex Multilateralism? Civil Society and the OECD', in *The OECD and Transnational Governance*, Rianne Mahon and Stephen McBride (eds) (Vancouver: UBC Press, 2008), 77–95.

World Commission on Environment and Development (WCED), Key Issues, WCED/84/10-1, Inaugural Meeting, Geneva, 1–3 October 1984, International Development Research Archive.

———. *Our Common Future* (Geneva, 1987).

CHAPTER 8

Towards 'Sustainable' Development

The United Nations, INGOs and the Crafting of the World Conservation Strategy

Stephen Macekura

In the years following the Stockholm Conference, national leaders, international civil servants and activists alike wrestled with the tensions between environmental protection and economic growth. In particular, the Stockholm Conference had brought to the fore of international politics debates between the developing nations of the Global South, the industrialized nations of Western Europe, the United States and Japan, and environmental activists. Many leaders from developing nations viewed environmental protection measures with suspicion. They claimed that new environmental regulations would impede their growth, would prove far too costly in the short term and would represent an unfair imposition of Western standards on their own practices. Environmental activists and development experts, frustrated at these divisions, aspired to redress them by rethinking and ultimately reframing the purpose of environmental conservation, a process that culminated in a concept called 'sustainable development'.[1]

The idea of sustainable development emerged out of two critiques of how developing nations conceived of development planning – one from environmentalists and one from the development community. Environmentalists rallied against how these plans had depleted natural resources, spurred pollution, and imperiled wild flora and fauna, whilst reformers from the developing world critiqued the same plans for neglecting the basic human needs of the poor and demanded 'alternative' approaches.[2] Although developers and environmentalists had clashed for decades, in the 1970s a transnational network of leading environmentalists engaged influential development experts in the hope of resolving old tensions between the two groups. Led by policy makers and experts such as Maurice Strong, Barbara Ward, David Munro, Ray Dasmann and Lee Talbot, this network gradually developed an alternative approach to planning that sought to redress poverty and environmental degradation. The approach, which they referred to as 'sustainable development', emerged as officials in the International Union for the Conservation of Nature (IUCN) and the World Wildlife Fund (WWF) worked with the United Nations Environment Programme (UNEP) to craft the *World Conservation Strategy* (WCS).

This chapter analyses the way in which this network of actors played key roles in reframing the international conservation agenda by defining and promoting the concept of 'sustainable development'. There are many historical concepts that resemble contemporary definitions of sustainable development. Many scholars have pointed to the conservation movement of the early twentieth century that focused on a utilitarian philosophy, stressing the 'wise use' of natural resources to serve the greatest number of people for the greatest amount of time and to guarantee open access to the most visually striking landscapes to the most people. Likewise, other observers looking for sustainability's origins have focused on scientific notions from the mid twentieth century, such as 'maximum sustained yield', which promoted a utilitarian philosophy within the bounds of the regenerative capacities of a given ecosystem. Still others have gazed further back, finding key components of 'sustainable' thinking and practice as disparate as eighteenth-century German scientific practices and medieval Christian theology.[3]

Yet this long etymology and intellectual history does not explain why or how the phrase 'sustainable development' emerged when it did in the 1970s or why the sustainability synthesis became so popular in the 1980s and beyond. This chapter highlights how and why the concept of sustainable development emerged as a way to reconcile tensions between the Global South, the industrialized North and international non-governmental organizations (INGOs). In so doing, the chapter focuses on actors, the ideas they promoted and the institutions they set up to help solidify their ideas as the basis for transnational discourse and policy making. In charting this relationship, it makes two overlapping arguments.

First, this chapter argues that the emergence of the concept of sustainable development was, at root, the result of a political process. This is similar to the argument Iris Borowy makes in her chapter. Often the concept is described in scientific terms, or is considered the result of injecting ecological science into public policy in a way that had not been present before. While the scientific dimensions are significant, to focus solely on changes in scientific discourse and practice elides an important process in which leading environmentalists, frustrated after years of attempting to persuade developing nations to adopt environmental protection, came to accommodate basic ideas from Third World intellectuals and leaders – ideas about poverty eradication, socioeconomic equality and local participation. Sustainable development has been criticized for its ambiguity, in the sense that it is so broad that it can mean almost anything. But its original advocates had a very specific definition in mind. Indeed, recounting the history of this political process that gave rise to sustainability – evident in the drafting of the WCS – illuminates that tensions between environmental protection and economic development were present in the phrase's earliest incarnations: environmentalists explicitly linked the need for environmental protection with a desire to reduce poverty, improve standards of living and sanction the right of the Global South to pursue development within an ecological framework.

Secondly, focusing on the intellectual and discursive origins of sustainable development also highlights the role of transnational networks of expertise in agenda setting. The process that generated the sustainability idea occurred through international institutions and transnational exchanges between environmentalists and development experts. UNEP, INGOs and transnational gatherings of experts offered venues in which activists and experts alike could shape a global approach to conservation. Yet while these institutions provided important sites for the discussion and diffusion of new ideas, they had few mechanisms to implement those ideas within nation states. The WCS was a global document – it was crafted in international institutions and had guidelines for all nations to follow. But whether states would adopt these global standards was – and remains – uncertain. Transnational networks and international institutions helped to form the meaning of the sustainable development idea. But they were also limited in terms of reshaping national policy decisions without direct mechanisms to force governments to comply with global standards.

INGOs and Linking Environmental Concerns to International Development

In the years following the Second World War, a global push for economic development and decolonization began to reshape international politics, as Jan-Henrik Meyer also shows in his introductory chapter. Across the world, leaders of countries devastated by the war clamoured for economic development and growth to boost short-term recovery, satisfy citizens' desires for basic needs and generate material abundance to ensure long-term prosperity. Moreover, the end of the war ushered in a wave of decolonization that, over the next three decades, caused the disintegration of European colonial rule. Dozens of new states emerged in the wake of the old empires. While representing culturally, socially, politically and economically distinct societies, the frequently European-educated leaders of these new states were also keen to follow their former colonizers' model of economic growth and development.[4]

By the end of the 1940s, officials in the United States came to view these states as critical players in a global Cold War with the Soviet Union. The U.S. government (along with Western Europe and the Soviet Bloc) began to send foreign aid to help spur growth in the hope of winning the hearts and minds of politicians and people in developing countries. Development economists and intellectuals often understood development in terms of their interpretations of Western nations' past histories, and they frequently sought to replicate similar patterns abroad. U.S. and European officials collaborated with local elites to pursue development programmes that stressed industrialization, rapid increase in agricultural production, and urbanization, all of which quickly became associated with the notion of 'modernization'.[5]

Yet as this global push for modernization began, a small but vocal group of voices began to criticize the way in which Western nations had developed and to worry about reproducing the same models for growth that had generated widespread environmental problems in the United States and Europe. Early conservation activists like William Vogt and Fairfield Osborne worried about resource exhaustion. Others, such as the British scientist Julian Huxley and American conservationist Russell Train, feared the erosion of colonial era national parks and game reserves as developing countries pushed to convert those lands for agricultural and industrial production. INGOs like the IUCN and the WWF began to lobby international institutions like the United Nations (UN) and developing countries to pursue environmental protection policies alongside their economic development plans.[6]

Throughout the early period, both environmental activists and leaders of developing nations perceived a strong tension between economic development and environmental protection. Development, many believed, was a linear process that necessitated a view of nature as a static entity, designed only for human exploitation. It demanded the removal of barriers to national action and economic growth. By contrast, the kind of environmental protection conservationists sought required limitations on human activity to preserve and protect the natural world from resource exploitation.

As environmental activists attempted to persuade developing countries to adopt conservation and preservation policies, the tensions between economic development and environmental protection appeared to be stark. Environmentalists struggled to convince developing countries to maintain colonial era protection arrangements or establish new regulations. Likewise, development experts prized plans that spurred rapid growth and they often played down the environmental consequences of economic development – if they acknowledged them at all. During the 1960s, environmentalists remained deeply worried about the future of environmental protection in the Third World, especially amid growing interest in environmental issues in the industrialized nations of Western Europe, the United States and Japan. These concerns carried over to the preparations for the Stockholm Conference, where environmentalists hoped to bring about an even greater awareness of the relationship between economic growth and environmental change.[7]

Leading up to the Stockholm Conference, development experts and environmental experts began to rethink the relationship between economic development and environmental protection. In 1970, French development economist Ignacy Sachs started to explore how developing countries could pursue economic growth while minimizing environmental damage. Working with Maurice Strong, a Canadian oil executive with a long-time interest in both conservation and international development, Sachs developed a new approach to economic growth, which he called 'ecodevelopment'.[8] The ecodevelopment concept married a concern for an equitable distribution of resources and satisfaction of basic needs with a desire to limit pollution and ensure long-term ecosystem health. Ecodevelopment emphasized decentralized and participatory planning, the elimination of poverty, appropriate technologies, self-reliance and the judicious husbandry of natural resources.[9] If planners incorporated these components into their economic strategies, Sachs and Strong believed, they could still achieve material abundance and economic growth while avoiding the catastrophic missteps associated with rapid industrialization and urbanization.[10]

Ecodevelopment gained traction during a series of international conferences in the early 1970s. As two respected, expert voices in the international development community, Sachs and Strong could draw upon a wide network of elite intellectuals, policy makers in international institutions and development officials with ties to the United States, Western European nations and many of the leading developing countries. One such ally was Barbara Ward. A prominent development expert and director of the International Institute for Environment and Development (IIED), Ward brought together Sachs and other development and environmental experts at two symposiums in 1970 and 1971, and also advised the Vatican, as Luigi Piccioni shows in his chapter in this volume.[11] Enthusiastic about these exchanges, Maurice Strong, then the executive director of the Stockholm Conference, and Ward convened a major gathering in Founex, Switzerland in the spring of 1971. They invited the developing world's leading experts associated with 'alternative development' paradigms, such as Sachs, Pakistani economist Mahbub ul Haq, Sri Lankan development expert Gamani Corea and Egyptian economist Samir Amin. They hoped to mollify the concerns of developing nations in the run-up to the Stockholm Conference by creating a policy framework that linked issues of poverty eradication, public health and environmental degradation.[12] While Founex had some influence on the Stockholm Conference, it mainly provided an institutional model for how to discuss these issues and it motivated participants to continue promoting the Founex goals.

At Stockholm, tensions between the wealthy industrialized nations and the developing world pervaded most discussions and negotiations When developed nations expressed a desire for greater environmental protection measures in policies, developing nations balked and responded with criticisms of neoimperialism and arguments that stressed the moral obligation of developed nations to increase their foreign aid. The efforts by Strong, Ward and others to reconcile these conflicts proved to be short-lived, as many observers noted that discord, suspicion and frustration dominated the conference. However, as national elites bickered over rights and responsibilities for development, Strong and Ward's network of activists and experts redoubled their efforts to find other institutional venues to promote an agenda for international conservation that would take seriously the concerns of the developing nations.[13]

After Stockholm, a transnational network of environmentalists and development reformers – many of whom had met in Founex – continued to discuss environmental and development issues in a variety of non-governmental forums. They interacted with the IUCN's General

Assembly meetings, at UN conferences on food and habitat, in personal correspondence and critically in UNEP, which was headed by Strong and became a central venue for the environment-development discussions. This networking and institution building helped to provide environmentalists, who were still viewed with deep scepticism by many in the developing world, with an opportunity to work directly with elites from the so-called Third Word.

One such meeting occurred in 1974. In that year, Ward convened a follow-up to the original Founex conference in Cocoyoc, Mexico, which participants dubbed 'Founex II'. Featuring many of the same development experts as the first meeting, the Cocoyoc conference built upon Ward's network and the growing post-Stockholm interest in linking environmental protection with the needs of the Global South.[14] In the Cocoyoc gathering's summary declaration, Ward articulated the emerging synthesis between environmental and development concerns. She called for a new international system 'capable of meeting the "inner limits" of basic human needs for all the world's people and of doing so without violating the "outer limits" of the planet's resources and environment'. Furthermore, she claimed: 'The road forward does not lie through the despair of doom-watching nor through the easy optimism of successive technological fixes. It lies through a careful and dispassionate assessment of the "outer limits," through cooperative search for ways to achieve the "inner limits" of fundamental human rights, through the building of social structures to express those rights, and through all the patient work of devising techniques and styles of development which enhance and preserve our planetary inheritance.'[15] In making such claims, Ward summarized a new line of thinking about the environment and development that linked together environmental protection, alternative development strategies and a need for socioeconomic equality between North and South.

Alongside these reform efforts, environmentalists built on Sachs and Strong's early work on ecodevelopment by continuing to rethink development planning. Central to this process was the need to include ecological consideration in economic planning, but to do so without eliminating the growth imperative altogether. In the early 1970s, the IUCN funded and coordinated multiple conferences designed to derive such ecological guidelines for developers across the world. Their efforts resulted in a 1973 book, *Ecological Principles for Economic Development*, written by IUCN biologist Ray Dasmann and Conservation Foundation official John Milton. The authors listed a series of general ecological principles for development experts to follow: 'Lack of consideration for the ecological realities of an environment can doom development efforts, with consequent waste of money and impairment in the condition of life.' Like Ward, they hoped that the injection of ecological knowledge into development plans could satisfy 'human needs and aspirations' of the present and contribute 'to the stability and productivity of the planet' in the long term.[16] Over the remainder of the 1970s, many environmentalists and development experts worked to produce a model for this new style of planning.

The IUCN and UNEP: Rethinking Environmental Protection

Efforts to rethink international development planning along ecological lines gained momentum as environmental INGOs came to incorporate many insights from the development community. As a result of the transnational exchanges and reconsideration of development planning, the environmental community began to rethink old assumptions about the purpose of conservation. For one, the IUCN began to reframe conservation as a tool to pursue economic development. For decades, the organization had focused narrowly on species protection, or carving out nature, in the form of parks and reserves, away from human use. But beginning in 1972, the IUCN made 'conservation *for* development' (emphasis added) a key theme of its activities. By the end of 1974, IUCN President Gerardo Budowski acknowledged the extent to which his organization had also begun to wrestle with the need to focus on eradicating poverty along with nature protection. There was, he claimed, 'no question of global inequities ... But ecological realties must also be faced at the same time, and action to correct the grave imbalances between human populations, resources and environment must go hand in hand with struggles against the injustices of social and economic systems'.[17]

This desire to incorporate the concerns of the developing world mirrored a broader intellectual project about the relationship between local peoples and the management of national parks and protected areas. In particular, Ray Dasmann, the IUCN's senior ecologist, began to prod environmentalists to incorporate indigenous peoples and rural communities in the management of protected spaces.[18] Working with Strong's broader network, in 1973 and 1974 Dasmann grew sympathetic to many concerns of development experts who stressed the need for environmental programmes to serve the basic human needs of the poorest populations. He chided the IUCN for ignoring local self-determination and for rejecting participatory planning.[19] Over the next few years, he spread this message, and the WWF and the IUCN slowly reformed their approaches to park management to include greater participation for local people living on or around park territory, particularly indigenous communities. For groups with a long history of casting aside and actually removing indigenous persons from national parks based on a vision of a wilderness in which human beings had no place, it was a remarkable shift.[20]

Alongside this more participatory vision of planning and management, many environmental INGOs also focused on new issues beyond a past emphasis on species protection and wildlife management. Ward's group established a new focus on urban pollution and poverty under the moniker of 'human settlements'.[21] Many INGOs took an interest in energy sources and use, particularly after the 1973 oil crisis.[22] The groups also adopted a broader focus on ecological areas that cut across national lines, particularly around deforestation in tropical rainforests and desertification in East Africa.[23] By the end of the 1970s, major environmental groups had begun to tackle a much wider variety of issues than they had at the start of the decade, with a greater emphasis on incorporating concerns from the developing world into their activities.[24]

The efforts by environmentalists to accommodate concerns from the development experts marked an important moment within the environmental community. Following the tensions evident at the Stockholm Conference, a generation of environmental activists had now come to place emphasis on the developmental needs of the Global South. In this process of intellectual change, new institutional connections – particularly those between environmental activist groups and international institutions – reinforced the transformations taking place.

As environmentalists increasingly engaged the development community in the 1970s, UNEP played a key role in facilitating such exchanges. It became a key vehicle for such connections because developing countries were well represented in it. In part, this derived from the influence that developing countries had achieved within the UN system. Their voting majority in the UN General Assembly gave them enough leverage to have the new agency placed in Nairobi – the first UN agency in the developing world. Developing countries, which also had an inbuilt majority on the UNEP Governing Council, used the organization to demand additional financial support and frame environmental discussions around themes of poverty and inequality. Though many UN officials envisioned it as an apolitical, technocratic organization, after the UNEP Governing Council's first session in 1973, it 'became clear that the work of UNEP was going to be as seriously influenced as other UN activities by the division between developing and developed worlds'.[25]

When Strong took over as executive director of UNEP in 1974, he seized the opportunity to use the agency as a venue for bringing together environmentalists and representatives from the developing world. As a friend of leading environmentalists, a respected development expert and well networked among Third World elites, Strong positioned himself as a mediator between these groups. From his arrival in Nairobi, he viewed UNEP as a 'catalyst, initiator, and co-ordinator' for linking environmental and development issues, particularly in the developing world.[26] He encouraged environmentalists to work with developing nations seeking technical support, often presenting IUCN officials with the opportunity to work as consultants.[27] Through the mid 1970s, UNEP hosted a number of regional conferences in the developing world, which brought together development economists, ecologists and officials from national planning ministries to 'show environment as a dimension of development'.[28]

Strong also hoped to strengthen ties between environmental INGOs and the UN system by encouraging officials to work in both worlds. 'I am convinced', he wrote in 1973, 'that effective environmental action depends on a much closer and more cooperative relationship between the inter-governmental and non-governmental institutions that has yet been experienced, at least internationally'.[29] Under Strong's guidance, from the mid 1970s, many officials moved between INGOs and UNEP. For instance, David Munro, a high-ranking UNEP official during the mid 1970s, became IUCN's executive director in 1978. Peter Thacher, an American diplomat who had assisted Strong during the Stockholm proceedings, served as executive director of UNEP in the late 1970s before working for various INGOs in the 1980s. Strong himself moved between different organizations. In addition to his role with UNEP, he held key positions within the IUCN, sat on the Executive Council of the WWF and unofficially advised the IIED throughout the 1970s.[30]

By working in closer connection with UNEP, IUCN officials had the opportunity to redress these objections by incorporating more of the concerns of developing nations into their overall strategy. In so doing, they began to reshape the agenda for global conservation. Their efforts to do so culminated at the end of the decade, with the publication of the WCS.

The WCS and the Rise of Sustainable Development

Environmental NGOs developed a desire for a global strategy for conservation in the mid 1970s. Maurice Strong and other officials began discussing the crafting of a strategy that was global in scope and that could apply to all nations in 1975. In 1977, the IUCN, the WWF and UNEP agreed to support a group project that would identify general strategies countries around the world that could be applied in order to pursue economic development within an ecologically sound framework. The work of drafting the main document began in earnest in 1978.

IUCN Senior Policy Advisor Robert Allen, who had authored multiple essays and books on international conservation, drafted the first version.[31] The crafting of the strategy resembled debates within the environmental community over the extent to which conservation and preservation should serve developmental goals. Allen's first draft attempted to balance the concerns of those who still believed in an older, preservationist concept of environmental protection and those who championed the 'conservation for development' concept.[32] However, the draft struggled to reconcile the two positions. Allen wrote: 'It would silence a frequent criticism if it were to be made quite explicit that ecosystem conservation must be carried out within the framework of environmental policies which work towards the elimination of social and economic injustices and which strive to raise the level of well-being of all mankind … ecosystem conservation … should make an important contribution to that end.'[33] But he did not elaborate much on that point, and the rest of the document offered basic strategies for ecosystem management with little discussion of how they could serve human needs.

Though Allen made a rhetorical gesture toward the development community, the IUCN's Lee Talbot lamented that the first draft amounted to 'basically a wildlife conservation textbook'.[34] Allen's halfhearted embrace of the 'conservation for development' message appalled many environmentalists who read the first draft. Sierra Club official Patricia Scharlin explained that: 'The draft does not yet come to grips with … the place of conservation in the global development process … and the place of people as part of ecosystems. With regard to the latter point, there still remains the old conservationist point of view that man must control nature.' She continued: 'There is still too much emphasis on protection of specific areas or species and not enough focus on the part conservation plays [in developing countries].' She even suggested the IUCN go further to help the developing countries' push for a New International Economic Order (NIEO) and global socioeconomic justice.[35] Many other environmental activists shared her opinions, especially those from newer NGOs. 'If we are to win the battle for environmental protection in the developing world', Thomas Stoehl, Jr. from the National Resources Defense Council (NRDC), an organization formed by young lawyers in the United States in 1970 to offer legal support and policy advice for environmental causes, told Robert Allen, 'environmental protection almost by definition must become an aspect of the development process.'[36]

While these officials argued that the document was far too narrow, others felt that it had, in fact, gone too far in assimilating the concerns of developing countries. 'Conservation's job is not to accommodate development', explained one WWF official in a letter to Robert Allen. In particular, Allen needed to show more concrete strategies for 'sane development' based on 'preservation of [biological] resources'. The official also claimed that the strategy should be more than a reflection of the 'IUCN's plans for itself, nor solely defined by what UNEP is requiring of IUCN'.[37]

While the ambiguity of Allen's draft led to widely differing criticisms from the environmental community, UNEP officials unequivocally rejected it for not sufficiently addressing the developmental aspirations of the developing world. Upon seeing a draft, one official exclaimed: 'This thing stinks! It is dyed-in-the-wool preservationism.' Reuben Olembo, an UNEP official who had been working to popularize the old ecodevelopment concept, wrote directly to the IUCN expressing similar dismay.[38] In a comprehensive evaluation of the early drafting process, Thomas Power, a consultant hired to analyse the draft for UNEP, found it 'regrettable and ironic' that the WCS had 'failed' to integrate developmental concerns when the IUCN itself had changed its own programme to accommodate the developing world.[39]

As the major funding source for the project, UNEP used its considerable leverage to shape the final document, and the IUCN went to great lengths to align the strategy with its partners' concerns. Discussions between Power, UNEP staff and the IUCN resulted in Allen rewriting multiple chapters.[40] Shortly after hearing the third round of criticisms, Allen flew to Nairobi from Switzerland to meet with Olembo and other staff members. After a series of discussions, Allen concluded: 'The strategy had to be put in the context of development for it to be taken seriously … It had to cover what development was, and how conservation furthered it.' With help from Olembo and David Munro, he rounded the last draft into form in late 1979 and early 1980.[41] Officials finally completed the strategy for international release in March 1980.

In its final iteration, the WCS offered general principles that could be applied to development to incorporate environmental concerns. It outlined three principles to guide all development plans: they should maintain essential ecological processes and life-support systems; preserve genetic diversity within the world's ecosystems; and ensure the sustainable utilization of species in ecosystems. By following these guidelines, planners could orient their economic policies towards satisfying the basic needs of the population without destroying the resource base on which development depended. The document referred to this compromise as a model for 'sustainable development'.[42]

The document also revealed striking evidence of the extent to which IUCN officials had accommodated developmental aspirations. In general, the document called for the 'need for global strategies both for development and for the conservation of nature'.[43] The WCS also claimed that 'humanity's relationship with the biosphere ... will continue to deteriorate until a new international economic order is achieved ... and sustainable modes of development become the rule rather than the exception'. Amid older conservation ideas, the document also endorsed the developing countries' push for global socioeconomic reform as a fundamental tenet of environmental protection. Furthermore, the document declared that conservationists had actually encouraged the incompatibility of environmental protection and development by 'too often resisting all development'. Alongside sections on managing various ecosystems, the WCS also stressed poverty eradication, local participation in conservation activities and various aspects of 'people-centered development'.[44]

In 1980, many in the environmental and development community celebrated the strategy. Mark Halle, an official who worked with UNEP, the WWF and the IUCN, hailed the WCS for bringing 'conservation solidly into the mainstream of social and economic development'. Since the document reconciled environmentalists to framing their work within the developmental needs of the poorest countries, he also called the strategy 'the most significant advance in conservation' since the birth of modern environmentalism a few decades earlier.[45] American environmental expert and official Lee Talbot looked positively upon the final version as a 'consensus between the practitioners of conservation and development'.[46]

IUCN officials released the strategy with much fanfare on 6 March 1980. The IUCN convened a public launch, in which the organization announced that it was distributing a condensed pack of materials especially designed for policy makers in national governments and international institutions around the world. It also organized national launch parties in countries where it had influential allies. Russell Train, former administrator of the U.S. Environmental Protection Agency and head of the WWF U.S. office, and Robert McNamara, former U.S. Secretary of Defense and President of the World Bank, hosted the strategy's release party in the United States. Peter Scott from the WWF U.K. office oversaw a series of similar celebratory events in the United Kingdom.[47] The UNEP General Council presented the strategy at its annual session to 'very positive reception'.[48] The launch attracted worldwide attention from major media sources in the developed and developing world, in nations such as US, China and Kenya.[49]

While the publication of the document was an important milestone, INGO officials recognized that tremendous work remained to put its ideas into policy. To encourage implementation in developing nations, the IUCN established, with financial support from UNEP, the Conservation for Development Centre (CDC). Spearheaded by Strong, the new organization provided consultative services to governments that wanted to adopt the WCS and for development assistance agencies that wanted to fund the project in line with the strategy's principles.[50]

After the publication of the WCS, 'sustainable development' moved to the top of the agenda for many environmental INGOs, experts within the development community and even many international organizations. The IUCN released numerous publications related to the strategy and rephrased much of its programming to include the use of 'sustainable development' and other ideas from the WCS.[51] The sustainability discourse predominated in discussions surrounding the ten-year follow-up to Stockholm, held in Nairobi in 1982.[52] The World Bank began using the term by the mid 1980s.[53] Sustainable development was a major theme of the 1980 *Global 2000 Report* in the United States, which had been commissioned by President Jimmy Carter in 1977 and was the closest the United States came to drafting a national conservation strategy.[54] In all these cases, sustainable development became a common phrase to describe how development concerns could take in environmental ideas and vice versa.

Talk of sustainable development became truly dominant in global development discourse following the work of the UN's World Commission on Environment and Development, better known as the Brundtland Commission, which Borowy discusses in her chapter in this book. Chaired by Norwegian Prime Minister Gro Harlem Brundtland, the group formed in 1983 to build on the conceptual work done by the WCS and offer more practical suggestions for how nations might adopt its guidelines. Between 1985 and 1987, the commission supported dozens of surveys, reports and meetings. While Brundtland nominally chaired the commission, its members – led by Strong and Canadian environmentalist Jim MacNeill – did much of the legwork. Many officials affiliated with INGOs, particularly the IIED, also helped to shape the report's content.[55]

The commission's final report, entitled *Our Common Future*, appeared in 1987. It stressed the need to strengthen environmental institutions in developing nations, increase funding for UNEP, and deepen connections between policymakers and INGOs. The report defined sustainable development as the need 'to ensure that [development] meets the needs of the present without compromising the ability of future generations to meet their own needs'. In essence, the Brundtland Commission culled together earlier ideas – from Strong and Sachs' ecodevelopment concept, to Ward's notion of inner and outer limits, to the WCS – to emphasize both contemporary short-term socioeconomic equality and long-term, intergenerational environmental equality.[56] Many nations, including the United States, endorsed this definition of sustainable development.[57]

Despite its proliferation, sustainable development discourse invited attacks for its broadness and its vagueness. Following the release of the Brundtland Commission's report, the meaning of sustainability became separated from its roots as many different actors – from corporate leaders to national elites – employed the phrase to suggest that environmental protection was compatible with continued economic growth and neoliberal development approaches. During the 1980s, many intellectuals and policy makers criticized the role of the state in economic development and turned to privatization of public goods and services, trade liberalization and deregulation to promote economic growth. Amid this rise of neoliberal thinking, many international organizations (IOs), particularly the World Bank and the International Monetary Fund (IMF), began to embrace such policies for developing nations. In turn, advocates for sustained economic growth, minimal regulations, liberal trading regimes and privatization – from writers of *The Economist* magazine to officials at leading IOs and multinational corporations – adopted the sustainability phrase to suggest that economic growth and environmental protection could be reconciled through market-based mechanisms.[58] In time, sustainable development

grew to mean many different things to different people; by the early
1990s, scholars identified over forty different definitions for the phrase.[59]
A decade after the WCS, few tied it to North–South equality, the NIEO
or any of its more radical implications related to global resource
transfers. As Borowy shows in her chapter in this book, many experts
simply took sustainable development to mean that economic growth
and environmental protection were compatible rather than that they
could be made compatible through fundamental changes in the global
economy and North–South relations.

Altogether, in the global diffusion of sustainable development as a
phrase, concept and model for developers and environmentalists to
pursue, questions of environment and development became fixed
around the notion that environmental protection measures needed to
serve economic development. Environmental protection initiatives, the
new thinking went, had to promote local participation, the satisfaction
of basic needs and the eradication of poverty. By the late 1980s, even as
the phrase came to be associated with neoliberal development practices,
it still, at root, signified a need to place environmental protection within
a developmental framework. Thus, in the end, the work of the IUCN,
the WWF and UNEP over the 1970s simply ensured that, at least in
rhetoric, worldwide conservation would be *for* development.

Conclusion

While the WCS greatly influenced strategic planning of major INGOs
and reshaped the discourse around the environment–development
relationship, it had far less influence on the policies of major states.
Although Maurice Strong led the newly created CDC to encourage
national governments to implement development plans based on the
strategy, few states and foreign aid agencies ever fully did so. Zambia,
one of the few countries to draft a full national conservation strategy
with the IUCN's help, lacked the financial capability to implement its
strategy and could not attract enough foreign aid to cover its costs.[60]
Similarly, while the CDC was able to negotiate a contract with the U.S.
Agency for International Development (AID) to fund the drafting of
national conservation strategies worldwide, it was unable to secure
funding to guarantee that the plans would be put into practice. For
instance, AID offered contracts to INGOs, such as the IIED, to help craft
conservation strategies in countries from Sri Lanka to Nepal and the

Philippines. But without a dramatic increase in funding necessary for reformulating national economic policy as the plans suggested, the strategies did not enter fully into policy.[61] And while the IUCN often encouraged its member states to craft strategies throughout the 1980s, it too lacked the ability to pay the costs of implementing the new plans.[62] In the end, without a subsequent increase in foreign aid to fund the national economic reorganization that conservation strategies often demanded, any national strategy based on the WCS was bound to fall short of full implementation.

Although the WCS was never implemented worldwide, it nonetheless represented a major transformation for the environmental movement and many environmental INGOs. Transnational exchanges with the development community significantly reshaped international environmentalism in the 1970s. The sustainability discourse resulted in part from the development community's acceptance of environmental limitations to economic growth. But it also came from environmentalists accepting issues such as poverty eradication and equality as objectives for global conservation. The concept of sustainable development was as much about incorporating developmental aspirations into environmentalism as it was about including insights from environmentalism into development thought.

Furthermore, the story of sustainable development shows the importance of embedding the history of ideas in the history of institutions. The types of institutions used to discuss the meaning of environmental protection in the meaning of development ultimately helped to shape the meaning of sustainable development. The actors who led the charge – Maurice Strong and Barbara Ward – worked in INGOs that brought them into contact with development experts and activists from around the world. They often reached out to development reformers, which brought a different set of voices into environmental discussions. Likewise, forging ties with UNEP in the drafting of the WCS meant that IUCN officials would work closely with a global network of environmental and development experts as well as officials within the UN system, many of whom had sympathies for the plight of developing countries. Accordingly, the institutional framework in which the WCS was drafted – formed by connections between international institutions and environmental groups – shaped its meaning by engaging environmental activists with development reformers and experts from the Global South.

The story of sustainable development planning shows the significance of illuminating the connections and interactions between actors and institutions in agenda setting. Rather than just highlighting the various ideas that inspired actors to reshape the intellectual agenda behind global conservation in the 1970s, the long history of the WCS reveals how specific actors working in very specific institutions led environmental reformers to infuse the concept of sustainable development with concerns and aspirations of the Global South and development reformers – even to the point of frustrating some environmental experts in the process. Today, 'sustainable development' has many meanings. But in its original formulation in the WCS, the phrase had a very direct meaning and purpose. Its architects put conservation in the service of development while acknowledging the right of the Global South to pursue development, reaffirming the need for greater local participation in environmental protection and stressing that conservation programmes should help eradicate poverty. A political process of negotiation and revision, in which environmentalists worked through INGOs and international institutions to engage development experts and elites from the developing world, shaped this particular meaning long before 'sustainability' became a popular phrase spoken the world over.

Stephen Macekura is Assistant Professor of International Studies at Indiana University, Bloomington, United States.

Notes

1. This chapter draws on material from my book, *Of Limits and Growth: The Rise of Global Sustainable Development in the Twentieth Century* (Cambridge: Cambridge University Press, 2015).
2. On the rise of 'alternative development' and the basic human needs approach to development see Victor Nemchenok, 'A Dialogue of Power' (Ph.D. dissertation, University of Virginia, 2013); and Gilbert Rist, *The History of Development: From Western Origins to Global Faith* (London: Zed Books, 2008), Chapter 9. While the alternative development reformers had many diverse objectives, they shared a concern for redressing poverty, protecting individual human needs, overcoming the legacies of colonialism and drawing global attention to North–South socioeconomic conflict. Specifically, they advocated major increases in foreign aid and technology transfers, global standards for price floors on commodities, strict regulation of transnational corporations and guaranteed access to markets of OECD member states.
3. For scholarship that addresses the meaning and origins of sustainable development concepts, see Sharachchandra M. Lele, 'Sustainable Development:

A Critical Review', *World Development* 19(6) (1991): 607–21; Charles V. Kidd, 'The Evolution of Sustainability', *Journal of Agricultural and Environmental Ethics* 5(1) (1992): 1–26; John Robinson, 'Squaring the Circle? Some Thoughts on the Idea of Sustainable Development', *Ecological Economics* 48(4) (2004): 369–84; Paul Warde, 'The Invention of Sustainability', *Modern Intellectual History* 8(1) (2011): 153–70.

4. Macekura, *Of Limits and Growth*. See also Amy L. Staples, *The Birth of Development: How the World Bank, FAO, and WHO Changed the World, 1945–1965* (Kent, OH: Kent State University Press, 2006).

5. Nick Cullather, *The Hungry World: America's Cold War Battle against Poverty in Asia* (Cambridge, MA: Harvard University Press, 2011).

6. On Vogt and Osborne, see Thomas Robertson, *The Malthusian Moment: Global Population Growth and the Birth of American Environmentalism* (New Brunswick, NJ: Rutgers University Press, 2012). On the IUCN and the WWF in the 1950s and 1960s, see Macekura, *Of Limits and Growth*, Chapters 1 and 2.

7. Macekura, *Of Limits and Growth*, Chapter 3.

8. Ignacy Sachs, 'Ecodevelopment: A Definition', *Ambio* 8(2/3) (1979): 113.

9. Ibid.

10. B. Gosovic to Maurice Strong, 19 September 1978, box 103, V. UNEP, folder 979, Maurice F. Strong Papers, Environmental Science and Public Policy Archives, Harvard University, Cambridge, MA (hereinafter Strong Papers). Under Strong's direction, UNEP produced a major research paper on the subject that it released in early 1976. 'Ecodevelopment', 15 January 1976, UNEP/GC/80. Viewed in G.X 34/26 'Co-operation between ECE and the United Nations Environment Programme (UNEP)' – Jacket 2', The United Nations Office at Geneva Registry Collection (hereinafter UNOG Registry Collection), Geneva, Switzerland.

11. Satterthwaite, *Barbara Ward and the Origins of Sustainable Development* (London: IIED, 2006) http://pubs.iied.org/pdfs/11500IIED.pdf, accessed 7 July 2016, 43; Barbara Ward, J.D. Runnalls, and Lenore D'Anjou (eds), *The Widening Gap: Development in the 1970s* (New York: Columbia University Press, 1971).

12. Maurice Strong, 'Stockholm: The Founding of IIED', in *Evidence for Hope: The Search for Sustainable Development*, Nigel Cross (ed.) (London: Earthscan, 2003), 21–23; 'Development and Environment: Founex, Switzerland, 4–12 June 1971' (Mouton, France: École Pratique des Hautes Études, 1971), 125–28. See also Frederic Lapeyre, 'Transcript of Interview of Ignacy Sachs', 9 May 2000, United Nations Intellectual History Project, *The Graduate School and University Center, City University of New York*, 36–39, oral history in author's possession.

13. Stephen Macekura, 'The Limits of Global Community: The Nixon Administration and Global Environmental Politics', *Cold War History* 11(4) (2011): 489–518.

14. Nemchenok, 'A Dialogue of Power', Chapter 2, 62–65.

15. 'The Cocoyoc Declaration', *International Organization* 29(3) (1975): 893–901, 896.

16. Raymond Fredric Dasmann, John P. Milton and Peter H. Freeman, *Ecological Principles for Economic Development* (London: John Wiley & Sons, 1973), 1–2; 'Ecology in Development Programmes: An IUCN Project', *IUCN Bulletin* 2(17) (1970), 141; D. Hackett, K. Hale and K. Smith, 'The Role of the Ecologist in Environmental Planning, Development and Management', a study for the Environmental Planning Commission of IUCN, December 1973, International Union for the Conservation of Nature Library (hereinafter IUCN Library), Gland, Switzerland.

17. Gerardo Budowski, 'Should Ecology Conform to Politics?', *IUCN Bulletin* 5(12) (1974): 44–45.

18. Macekura, 'The Limits of Global Community'.

19. Raymond Dasmann, 'Sanctuaries for Life Styles', *IUCN Bulletin* 4(8) (1973): 29.

20. Robert Allen, 'Sustainable Development and Cultural Diversity – Two Sides of the Same Coin', *IUCN Bulletin* 6(4) (1975): 13. On national parks and aboriginal removal, see, for instance, Mark Dowie, *Conservation Refugees: The Hundred-Year Conflict between Global Conservation and Native Peoples* (Cambridge, MA: MIT Press, 2009); Roderick Neumann, *Imposing Wilderness: Struggles over Livelihood and Nature Preservation in Africa* (Berkeley: University of California Press, 2002); David Anderson and Richard Grove, 'The Scramble for Eden: Past, Present, and Future in African Conservation', in *Conservation in Africa: People, Policies, and Practice*, David Anderson and Richard Grove (eds) (Cambridge: Cambridge University Press, 1987), 1–12; Bernhard Gissibl, Sabine Höhler and Patrick Kupper (eds), *Civilizing Nature: National Parks in Global Historical Perspective*, (New York: Berghahn Books, 2012).

21. David Satterthwaite, 'Setting an Urban Agenda: Human Settlements and IIED-América Latina', in Cross (ed.), *Evidence for Hope*, 122–46.

22. 'Energy and Conservation', *IUCN Bulletin* 5(6) (1974): 14; Gerald Leach, 'The Energy Programme', in Cross (ed.), *Evidence for Hope*, 96–107.

23. Duncan Poore, 'Saving Tropical Rain Forests', *IUCN Bulletin* 5(8) (1974): 29.

24. Barbara Ward to Maurice Strong, 24 January 1978, box 75, V. IIED 1977–78, Strong Papers.

25. 'United Nations Environment Programme – Future UK Policy', December 1973, National Archives of the United Kingdom, Records Pertaining to the Foreign and Commonwealth Office 55/1007.

26. 'Remarks by Maurice Strong, Executive Director, UNEP to the UNA of Great Britain and Northern Ireland', 22 July 1975, folder C. 1246, Sir Peter Markham Scott Papers, Department of Manuscripts and University Archives, Cambridge University, Cambridge, UK.

27. Kai Curry-Lindahl to Robert Hunter, 22 December 1976, box 103, V. UNEP, folder 978, Strong Papers.

28. 'Ad Hoc Expert Group Meeting on Alternative Patterns of Development and Life Styles for Africa, Nairobi, 23–25 March, 1977', G.X 34/42 UNEP/ECE Conference on Alternative Lifestyles and Development, UNOG Registry Collection.

29. Maurice Strong to Robert O. Anderson, 12 January 1973, box 73, V. IIED 73–74, Strong Papers.

30. John McCormick, *The Global Environmental Movement* (Chichester: John Wiley, 1995), 125–26.

31. 'Robert Allen', carton 8-11 World Conservation Strategy, 1979–80, Sierra Club International Program Records, BANC MSS 71/290c (hereinafter SCIPR), the Bancroft Library, University of California, Berkeley.

32. 'First Draft of a World Conservation Strategy', January 1978, 1, IUCN Library.

33. Ibid., 3.

34. Lee Talbot, 'The World Conservation Strategy', in *Sustaining Tomorrow: A Strategy for World Conservation*, Francis R. Thobideau and Hermann H. Field (eds) (Hanover, NH: University Press of New England, 1984), 15.

35. Patricia Scharlin to Robert Allen, 7 April 1978, World Conservation Strategy, carton 8-10, 1979, SCIPR.

36. Thomas Stoel, Jr., 'General Comments on First Draft of IUCN World Conservation Strategy', April 1978 in Thomas Stoel, Jr. to Robert Allen, 30 March 1978, World Conservation Strategy, carton 4-17, 1978, SCIPR.

37. Thomas Lovejoy to Robert Allen, 1 June 1978, World Conservation Strategy, carton 8-10, 1979, Sierra Club International Program Records; Jorge Morello and Thomas F. Power, Jr., 'Evaluation of Projects Contracted for Execution to the International Union for Conservation of Nature and Natural Resources as a Supporting Organisation', 11 May 1979, box 58, folder 555, Peter S. Thacher Collection, Part I, 1960–96, Environmental Science and Public Policy Archives, Harvard University, Cambridge, MA (hereinafter Thacher Papers, Part I).

38. Martin Holdgate, *The Green Web: A Union for World Conservation* (Abingdon: Earthscan, 2013), 151.

39. Thomas F. Power, Jr. to Mostafa K. Tolba, 17 May 1979, box 25, folder 229, Thacher Papers, Part I.

40. Holdgate, *The Green Web*, 151.

41. Ibid., 151.

42. *World Conservation Strategy: Living Resource Conservation for Sustainable Development* (IUCN-WWF-UNEP, 1980), 7.

43. Ibid., i.

44. *World Conservation Strategy*, part 1. On poverty, see 7–8: on participation, see 33–34; on conservation serving people-centred development, see 33–36.

45. Mark Halle, 'The World Conservation Strategy – An Historical Perspective', 30 September 1984, box 103, folder 979, Thacher Papers, Part I.

46. Talbot, 'The World Conservation Strategy', 14.

47. 'World Conservation Strategy: The Public Launch', First Memorandum, folder C.1138, Scott Papers; 'World Conservation Strategy: Launch of the Americas from the Hall of the Americas', carton 8-11 World Conservation Strategy, 1979–80, Sierra Club International Program Records.

48. Memo to David Munro, 26 April 1980, box 18, folder 168, Thacher Papers, Part I.

49. 'Environmental Consciousness-Raising', *Washington Post*, 10 March 1980, A26; John Yemma, 'Making Resources an Economic Global Priority', *Christian Science Monitor*, 6 March 1980, 4; 'East Africa: In Brief; Kenya's Interest in

Conservation', *BBC Summary of World Broadcasts*, 8 March 1980; 'Other Reports; Environmental Protection Month', *BBC Summary of World Broadcasts*, 15 March 1980.

50. Brian Johnson to Arthur Norman, Maurice Strong, Max Nicholson and Lee Talbot, 21 March 1980, box 253, VII IIED Correspondence London, folder 2386, Strong Papers; Lee Talbot to Terry Lisniewski, 23 March 1981, box 256, VII IUCN Correspondence, folder 2415, Strong Papers.

51. Lee Talbot to Terry Lisniewski, 23 March 1981, box 256, VII IUCN Correspondence, folder 2414, Strong Papers.

52. 'Nairobi Declaration', May 1982. http://www.un-documents.net/nair-dec.htm, accessed 25 May 2016.

53. A.W. Clausen, 'Sustainable Development: The Global Imperative', 1981, http://documents.worldbank.org/curated/en/1981/11/15532206/sustainable-development-global-imperative-fairfield-osborn-memorial-lecture-w-clausen-president-world-bank, accessed 25 May 2016.

54. *The Global 2000 Report to the President*, Vols. 1–3 (Washington DC: GPO, 1980).

55. Brian W. Walker, 'Time for Change: IIED, 1984–1990', in Cross (ed.), 'Evidence for Hope', 35.

56. Report of the World Commission on Environment and Development, A/RES/42/187, 11 December 1987, http://www.un.org/documents/ga/res/42/ares42-187.htm, accessed 25 May 2016. On the history of the Brundtland Commission, see Iris Borowy, *Defining Sustainable Development for Our Common Future: A History of the World Commission on Environment and Development (Brundtland Commission)* (Abingdon: Routledge, 2014).

57. The Reagan administration endorsed the concept in the most general sense, but 'refrained from any broad endorsement' of the report and thus any of its specific policy suggestions. 'Points to be Made: Environment', addendum to letter Jerry W. Leach to Colin L. Powell, 31 March 1988, 'Environment – Ozone & Acid Rain [1]', box 92365, Tyrus Cobbs Files, Ronald Reagan Presidential Library, Simi Valley, CA.

58. Steven F. Bernstein, *The Compromise of Liberal Environmentalism* (New York: Columbia University Press, 2001), 70–83.

59. David Pearce, Anil Markandya and Edward Barbier, *Blueprint for a Green Economy* (London: Earthscan, 1989), 173–85.

60. For a full elaboration of the Zambia case, see Macekura, *Of Limits and Growth*, Chapter 6.

61. 'World Conservation Strategy: A Follow-up Report', box 253, VII WWF Board, folder 2396, Strong Papers; David Runnalls to Maurice Strong, 6 May 1983, Box 253, VII IIED Correspondence Washington, Folder 2392, Strong Papers; Peter H. Freeman, David Runnals and Barbara A. Ormond, 'Mid-Term Evaluation: Environmental Planning and Management. Cooperative Agreement, April 1984', iii, USAID Document Clearinghouse, http://pdf.usaid.gov/pdf_docs/pdaas010.pdf, accessed 7 July 2016.

62. Conservation for Development Centre, 'Tanzania: Natural Resources Expertise Profile', December 1986, IUCN Library; Conservation for Development Centre, 'Kenya: Natural Resources Expertise Profile: Revised Edition', January 1986, IUCN Library; Conservation for Development Centre, 'Draft Proposal: Preparation of the National Conservation Strategy for Kenya', May 1987, IUCN Library.

Bibliography

Allen, Robert, 'Sustainable Development and Cultural Diversity – Two Sides of the Same Coin', *IUCN Bulletin* 6(4) (1975): 13.

———. 'File: Robert Allen', World Conservation Strategy, 1979–80, Sierra Club International Program Records, [SCIPR], BANC MSS 71/290ccarton 8-11, The Bancroft Library, University of California, Berkeley.

Anderson, David and Richard Grove, 'The Scramble for Eden: Past, Present, and Future in African Conservation', in David Anderson and Richard Grove (eds), *Conservation in Africa: People, Policies, and Practice* (Cambridge: Cambridge University Press, 1987), 1–12.

Bernstein, Steven F., *The Compromise of Liberal Environmentalism* (New York: Columbia University Press, 2001).

Borowy, *Defining Sustainable Development for Our Common Future: A History of the World Commission on Environment and Development (Brundtland Commission)* (Abingdon: Routledge, 2014).

Budowski, Gerardo, 'Should Ecology Conform to Politics?', *IUCN Bulletin* 5(12) (1974): 44–45.

Clausen, A.W., 'Sustainable Development: The Global Imperative', 1981, http://documents.worldbank.org/curated/en/1981/11/15532206/sustainable-development-global-imperative-fairfield-osborn-memorial-lecture-w-clausen-president-world-bank, accessed 25 May 2016.

Conservation for Development Centre, 'Kenya: Natural Resources Expertise Profile: Revised Edition', January 1986, IUCN Library.

———. 'Tanzania: Natural Resources Expertise Profile', December 1986, IUCN Library.

———. 'Draft Proposal: Preparation of the National Conservation Strategy for Kenya', May 1987, IUCN Library.

'The Cocoyoc Declaration', *International Organization* 29(3) (1975): 893–901.

Cullather, Nick, *The Hungry World: America's Cold War Battle against Poverty in Asia* (Cambridge, MA: Harvard University Press, 2011).

Curry-Lindahl, Kai to Robert Hunter, 22 December 1976, box 103, V. UNEP, folder 978, Strong Papers.

Dasmann, Raymond Fredric, John P. Milton and Peter H. Freeman, *Ecological Principles for Economic Development* (London: John Wiley & Sons, 1973).

Dasmann, Raymond Fredric, 'Sanctuaries for Life Styles', *IUCN Bulletin* 4(8) (1973): 29.

Dowie, Mark, *Conservation Refugees: The Hundred-Year Conflict between Global Conservation and Native Peoples* (Cambridge, MA: MIT Press, 2009).

'East Africa: In Brief; Kenya's Interest in Conservation', *BBC Summary of World Broadcasts*, 8 March 1980.

'Ecology in Development Programmes: An IUCN Project', *IUCN Bulletin* 2(17) (1970): 141.

'Energy and Conservation', *IUCN Bulletin* 5(6) (1974): 14.

'Environmental Consciousness-Raising', *Washington Post*, 10 March 1980, A26.

Freeman, Peter H., David Runnals and Barbara A. Ormond, 'Mid-Term Evaluation: Environmental Planning and Management. Cooperative Agreement, April 1984', iii, USAID Document Clearinghouse, http://pdf.usaid.gov/pdf_docs/pdaas010.pdf, accessed 7 July 2016.

Gissibl, Bernhard, Sabine Höhler and Patrick Kupper (eds), *Civilizing Nature: National Parks in Global Historical Perspective* (New York: Berghahn Books, 2012).

The Global 2000 Report to the President, Vols. 1–3 (Washington DC: GPO, 1980).

Gosovic, B. to Maurice Strong, 19 September 1978, box 103, V. UNEP, folder 979, Maurice F. Strong Papers, Environmental Science and Public Policy Archives, Harvard University, Cambridge, MA (hereinafter Strong Papers).

Hackett, D., K. Hale, and K. Smith, 'The Role of the Ecologist in Environmental Planning, Development and Management', a study for the Environmental Planning Commission of IUCN, December 1973, International Union for the Conservation of Nature Library (hereinafter IUCN Library), Gland, Switzerland.

Halle, Mark, 'The World Conservation Strategy – An Historical Perspective', 30 September 1984, box 103, folder 979, Thacher Papers, Part I.

Holdgate, Martin, *The Green Web: A Union for World Conservation* (Abingdon: Earthscan, 2013)

IUCN, 'First Draft of a World Conservation Strategy', January 1978, 1, IUCN Library.

IUCN-WWF-UNEP, *World Conservation Strategy: Living Resource Conservation for Sustainable Development* (IUCN-WWF-UNEP, 1980).

Johnson, Brian to Arthur Norman, Maurice Strong, Max Nicholson and Lee Talbot, 21 March 1980, box 253, VII IIED Correspondence London, folder 2386, Strong Papers.

Kidd, Charles V., 'The Evolution of Sustainability', *Journal of Agricultural and Environmental Ethics* 5(1) (1992): 1–26.

Lapeyre, Frederic, 'Transcript of Interview of Ignacy Sachs', 9 May 2000, United Nations Intellectual History Project, *The Graduate School and University Center, City University of New York*, 36–39, oral history in author's possession.

Leach, Gerald, 'The Energy Programme', in *Evidence for Hope: The Search for Sustainable Development*, Nigel Cross (ed.) (London: Earthscan, 2003), 96–107.

Lele, Sharachchandra M., 'Sustainable Development: A Critical Review', *World Development* 19(6) (1991): 607–21.

Lovejoy, Thomas to Robert Allen, 1 June 1978, World Conservation Strategy, 1979, carton 8-10, Sierra Club International Program Records; Jorge Morello and Thomas F. Power, Jr., 'Evaluation of Projects Contracted for Execution to the International Union for Conservation of Nature and Natural Resources as a Supporting Organization', 11 May 1979, box 58, folder 555, Peter S. Thacher Collection, Part I, 1960–96, Environmental Science and Public Policy Archives, Harvard University, Cambridge, MA (hereinafter Thacher Papers, Part I).

Macekura, Stephen, 'The Limits of Global Community: The Nixon Administration and Global Environmental Politics', *Cold War History* 11(4) (2011): 489–518.

———. *Of Limits and Growth: The Rise of Global Sustainable Development in the Twentieth Century* (Cambridge: Cambridge University Press, 2015).

McCormick, John, *The Global Environmental Movement* (Chichester: John Wiley, 1995).

Memo to David Munro, 26 April 1980, box 18, folder 168, Thacher Papers, Part I.

Nemchenok, Victor, 'A Dialogue of Power' (Ph.D. dissertation, University of Virginia, 2013).

Neumann, Roderick, *Imposing Wilderness: Struggles over Livelihood and Nature Preservation in Africa* (Berkeley: University of California Press, 2002).

Pearce, David, Anil Markandya and Edward Barbier, *Blueprint for a Green Economy* (London: Earthscan, 1989).

'Points to Be Made: Environment', addendum to letter Jerry W. Leach to Colin L. Powell, 31 March 1988, 'Environment – Ozone & Acid Rain [1]', box 92365, Tyrus Cobbs Files, Ronald Reagan Presidential Library, Simi Valley, CA.

Poore, Duncan, 'Saving Tropical Rain Forests', *IUCN Bulletin* 5(8) (1974): 29.

Power, Thomas F. Jr. to Mostafa K. Tolba, 17 May 1979, box 25, folder 229, Thacher Papers, Part I.

Rist, Gilbert, *The History of Development: From Western Origins to Global Faith* (London: Zed Books, 2008).

Robertson, Thomas, *The Malthusian Moment: Global Population Growth and the Birth of American Environmentalism* (New Brunswick, NJ: Rutgers University Press, 2012).

Robinson, John, 'Squaring the Circle? Some Thoughts on the Idea of Sustainable Development', *Ecological Economics* 48(4) (2004): 369–84.

Sachs, Ignacy, 'Ecodevelopment: A Definition', *Ambio* 8(2/3) (1979): 113.

Satterthwaite, David, 'Setting an Urban Agenda: Human Settlements and IIED-América Latina', in *Evidence for Hope: The Search for Sustainable Development*, Nigel Cross (ed.) (London: Earthscan, 2003), 122–46.

———. *Barbara Ward and the Origins of Sustainable Development* (London: IIED, 2006), http://pubs.iied.org/pdfs/11500IIED.pdf, accessed 7 July 2016.

Scharlin, Patricia to Robert Allen, 7 April 1978, World Conservation Strategy, 1979, carton 8-10, SCIPR.

Staples, Amy L., *The Birth of Development: How the World Bank, FAO, and WHO Changed the World, 1945–1965* (Kent, OH: Kent State University Press, 2006).

Stoel, Thomas Jr., 'General Comments on First Draft of IUCN World Conservation Strategy', April 1978 in Thomas Stoel, Jr. to Robert Allen, 30 March 1978, carton 4-17 World Conservation Strategy, 1978, SCIPR.

Strong, Maurice, 'Stockholm: The Founding of IIED', in *Evidence for Hope: The Search for Sustainable Development*, Nigel Cross (ed.) (London: Earthscan, 2003), 21–23.

Strong, Maurice to Robert O. Anderson, 12 January 1973, box 73, V. IIED 73–74, Strong Papers.

Talbot, Lee, 'The World Conservation Strategy', in *Sustaining Tomorrow: A Strategy for World Conservation*, Francis R. Thobideau and Hermann H. Field (eds) (Hanover, NH: University Press of New England, 1984), 15.

Talbot, Lee to Terry Lisniewski, 23 March 1981, VII IUCN Correspondence, folder 2414, box 256, Strong Papers.

Talbot, Lee to Terry Lisniewski, 23 March 1981, VII IUCN Correspondence, folder 2415, box 256, Strong Papers.

UN, *Development and Environment: Founex, Switzerland, 4–12 June 1971* (Mouton, France: École Pratique des Hautes Études, 1971).

———. 'United Nations Environment Programme – Future UK Policy', December 1973, Records Pertaining to the Foreign and Commonwealth Office 55/1007, National Archives of the United Kingdom.

———. 'Remarks by Maurice Strong, Executive Director, UNEP to the UNA of Great Britain and Northern Ireland', 22 July 1975, folder C. 1246, Sir Peter Markham Scott Papers, Department of Manuscripts and University Archives, Cambridge University, Cambridge, UK.

———. 'Ad Hoc Expert Group Meeting on Alternative Patterns of Development and Life Styles for Africa, Nairobi, 23–25 March, 1977', G.X 34/42 UNEP/ECE Conference on Alternative Lifestyles and Development, UNOG Registry Collection.

———. 'Nairobi Declaration', May 1982, http://www.un-documents.net/nair-dec.htm, accessed 25 May 2016.

———. 'World Conservation Strategy: A Follow-Up Report', VII WWF Board, folder 2396, box 253, Strong Papers.

David Runnalls to Maurice Strong, 6 May 1983, VII IIED Correspondence Washington, folder 2392, box 253, Strong Papers;

———. Report of the World Commission on Environment and Development, A/RES/42/187, 11 December 1987, http://www.un.org/documents/ga/res/42/ares42-187.htm, accessed 25 May 2016.

UNEP, 'Ecodevelopment', 15 January 1976, UNEP/GC/80. Viewed in G.X 34/26 'Co-operation between ECE and the United Nations Environment Programme (UNEP)' – Jacket 2', United Nations Office at Geneva Registry Collection (hereinafter UNOG Registry Collection), Geneva, Switzerland.

Walker, Brian W., 'Time for Change: IIED, 1984–1990', in *Evidence for Hope: The Search for Sustainable Development*, Nigel Cross (ed.) (London: Earthscan, 2003), 35.

Ward, Barbara to Maurice Strong, 24 January 1978, box 75, V. IIED 1977–78, Strong Papers.

Ward, Barbara, J. D. Runnalls, and Lenore D'Anjou (eds), *The Widening Gap: Development in the 1970s* (New York: Columbia University Press, 1971).

Warde, Paul, 'The Invention of Sustainability', *Modern Intellectual History* 8(1) (2011): 153–70.

'World Conservation Strategy: The Public Launch', First Memorandum, folder C.1138, Scott Papers; 'World Conservation Strategy: Launch of the Americas from the Hall of the Americas', carton 8-11 World Conservation Strategy, 1979–80, Sierra Club International Program Records.

Yemma, John, 'Making Resources an Economic Global Priority', *Christian Science Monitor*, 6 March 1980, 4.

‎⁌ CHAPTER 9

Protecting the Southern Ocean Ecosystem
The Environmental Protection Agenda of Antarctic Diplomacy and Science

Alessandro Antonello

On the morning of 1 March 1978, Richard Laws, the director of the British Antarctic Survey and a leading marine biologist, gave a scientific lecture describing the characteristics of the Southern Ocean ecosystem to the first session of a diplomatic conference on the conservation of Antarctica's marine living resources in Canberra, Australia. He gave his presentation to the delegations to this meeting – composed of both diplomats and scientists – in their scientific working group. These diplomats and scientists had talked about the Southern Ocean ecosystem for some time and had committed to its conservation at the 1977 Antarctic Treaty Consultative Meeting. Nonetheless, they were still seeking to understand the potential range of the concept's meaning as well as how to encapsulate its complexities in international treaty language.

The scientific concept of the ecosystem was an old one. However, until the 1970s, it had not generally been used in international diplomacy. Donald Logan, head of the British delegation and chair of the scientific working group that Laws was addressing, suggested that the diplomats tasked with fashioning a comprehensive regime would find it 'useful to know from the scientists what that term implies'.[1] Laws was a pre-eminent scientist of the Southern Ocean, having begun research there, and in Britain's Antarctic territories, in the late 1940s. He opened his lecture by defining the Antarctic marine ecosystem as:

> a volume of ocean with unique physical and chemical properties and all the living organisms within it, the structure of the communities they form, the dynamic functions and the biomass of the organisms

and different trophic levels, and the complex interactions of species with each other and the environment.[2]

He then conveyed details of the ecosystem's trophic levels, describing the phytoplankton, zooplankton, squids and fish, whales, seals, penguins, albatrosses and petrels. He suggested two central characteristics of an ecosystem: its diversity and its stability. Maintaining both were central objectives of management. He concluded by posing the central question all conservationists faced, namely: 'What kind of an ecosystem do we wish to conserve?' Though he did not offer a specific answer, he pointed out that, for example, if all the whales were removed, there would still be an ecosystem. His challenge in making this observation was that the negotiations must 'decide what the limit of variations that can be accepted is'.[3]

The consultative parties to the Antarctic Treaty of 1959[4] met in Canberra in early 1978 to begin negotiating an agreement for the conservation of Antarctic marine living resources. The actors interested in the Antarctic aimed for collective rules at the international level, as from the late 1960s, some parties, notably the United Kingdom and the United States, became increasingly concerned about the Soviet Union's and Japan's exploitation of Antarctic krill – the main zooplankton species that supported the great bulk of the ecosystem – as well as the apparent interest of certain international organizations and developing states in exploiting these resources. The concerned parties saw two main problems, one political and one environmental. First, they wanted to preserve both order within the Antarctic Treaty regime and the exclusivity of that order by keeping other non-Treaty actors out. The second concern related to the fundamental and practical issue of managing resource exploitation on the high seas (a global commons) to a larger extent and at a different level (that is, the ecosystem) than existing fisheries agreements.

The historical experience as well as the contemporaneous pressures of science, industry and international resources law presented the Antarctic Treaty parties with two principal options for their conservation regime. On the one hand, they could view marine resources as resources to be exploited and managed, most likely through a traditional fisheries management regime. This would include viewing only individual species to be exploited at a 'maximum sustainable yield' rather than considering the ramifications of exploitation through the ecosystem, and allocating fishing quotas.[5] On the other hand, they could govern the Southern

Ocean as a fragile and simple ecosystem to be protected as a whole. Ecological ideas were central to the environmental politics of the 1970s. The concept of the ecosystem was particularly important as the environment came to be viewed as a living machine in which all its constituent parts were interconnected and existed in a particular balance. With a productive metaphorical ambiguity, this ecological discourse and these mechanistic and increasingly abstract views of ecosystems called for humans to correct their harmful actions and suggested that they could, in fact, control and manage the balance of nature.[6]

The parties agreed on a regime with both elements of a traditional fisheries regime and a novel ecosystem-level conservation standard. Signed in May 1980, the Convention on the Conservation of Antarctic Marine Living Resources (CCAMLR) created a permanent international commission and codified principles to protect the whole ecosystem, manage its exploitation, and facilitate and promote scientific research.[7] From a diplomatic point of view, CCAMLR embedded the centrality of the Antarctic Treaty and therefore its parties in the management of Antarctic affairs. From a scientific and environmental point of view, Article II of the Convention was a milestone in international law, for it was the first fisheries and environmental treaty that provided for the conservation of an entire ecosystem.[8]

How did the Southern Ocean ecosystem become central to the environmental protection agenda of Antarctic science and diplomacy in the 1970s? This chapter argues that the Antarctic Treaty parties enshrined the ecosystem in CCAMLR because it allowed them collectively to insist on their responsibilities for and interests in the Antarctic region and exclude countries outside the Treaty framework. At the same time, the Scientific Committee on Antarctic Research (SCAR) – an international non-governmental scientific body with a close relationship to the Treaty parties – pushed for ecosystem conservation against resource development so that it could entrench its institutional standing, in intellectual and spatial terms, as the leading Antarctic scientific body. The Treaty parties and SCAR put the Southern Ocean ecosystem at the centre of their respective diplomatic and scientific agendas to marginalize the countervailing agendas of resource exploitation and the states and organizations associated with that view. By codifying the ecosystem, the Treaty parties constructed an enlarged and interconnected region to govern for themselves. Admittance to the regime of ecosystem protection and the region would be based on the acceptance of the obligations and geographies embedded in CCAMLR. The issue of institution building was not simply one of creating a new institution – that is, a regional fisheries management organization – to embody, enact and defend the principles of an ecosystem protection convention. It was also about maintaining and expanding existing institutions; in this case, both SCAR and the Antarctic Treaty system itself.

By emphasizing the links between the scientific and political visions of the ecosystem, this chapter engages with what the science studies scholar Sheila Jasanoff has called the 'idiom of co-production'. She has argued that 'we gain explanatory power by thinking of natural and social orders as being produced together' and has suggested that 'the ways in which we know and represent the world (both nature and society) are inseparable from the ways in which we choose to live in it'.[9] Concentrating on the political side of coproduction, this chapter is alert to the ways in which the apparently objective science of the ecosystem coexisted with and encouraged a particular disposition in the very structures of scientific research and in the Antarctic diplomatic regime. We must be attentive to the politics, broadly conceived, of any act of environmental protection, not simply to the apparent quality of scientific argument and reasoning.

Antarctic Actors

The issue of the conservation of Antarctic marine living resources and the Southern Ocean ecosystem was primarily discussed within the structures of the Antarctic Treaty regime. In addition to the Antarctic Treaty, this regime also included the measures passed by the Treaty parties at their periodic consultative meetings, the text of the Convention for the Conservation of Antarctic Seals of 1972,[10] and the informal relationship between the Treaty parties as a group and SCAR.

Unlike the international organizations discussed in other chapters in this book, the Antarctic Treaty was a regime according to Stephen Krasner's definition as it had 'principles, norms, rules, and decision-making procedures around which actors' expectations converge in a given issue-area'.[11] The regime did not have an autonomous or separate secretariat to set the agenda or develop its own expertise – agenda setting was done either by the parties or by SCAR, the closely tied international non-governmental organization (NGO) discussed below. Yet, the stability of the actors involved (states, scientific bodies and individuals), the single-issue area, and the periodic and structured meetings, combined with the fact that Antarctic issues were not always pressing concerns for foreign policy or defence elites, allowed a limited autonomy of action on the part of the diplomats and scientists involved.

The Antarctic Treaty was negotiated in response to three developments in the postwar period. The first concerned the territorial dispute between Argentina, the United Kingdom and Chile over their overlapping claims to the Antarctic Peninsula. The United Kingdom had made claims to these lands in the early twentieth century, based, in part, on priority of discovery. In the late 1930s, Argentina and Chile made competing claims arising from nationalist and anti-colonial politics. Tensions arising from these overlapping territories came to a head in the late 1940s and early 1950s, and even included armed encounters between British and Argentine personnel. The United Kingdom's attempts to resolve these tensions through the International Court of Justice came to nothing and the tensions remained unresolved.[12]

The second major tension arose in the late 1940s with the increasing interest of the Soviet Union in the Antarctic. Before this time, the Soviet Union had not had any Antarctic interests, yet it used a whaling fleet it had received from Germany as war reparations to expand its whaling activities into the Southern Ocean during 1946–47.[13] Soon after, when the United States tried to negotiate a resolution to the Argentine–British–Chilean territorial dispute and to include the other territorial issues in 1948, the Soviet Union, in the new context of superpower rivalry and the beginning of the Cold War, insisted that it had to be part of any international agreement.[14] The Soviet Union also began a significant programme of continental exploration and scientific stations in the early 1950s.

The final development was the International Geophysical Year (IGY) of 1957–58. This was a worldwide programme of scientific research that sought to understand the earth's geophysical phenomena through concentrated, simultaneous and synoptic observations and data collection. It had a particular focus on Antarctica. The IGY brought a significant expansion of scientific activity in Antarctica from a wide variety of states, including the Soviet Union. It was interpreted at the time as signalling the power of international cooperation that could overcome or avoid suspicions and tensions.[15]

Having been eager to resolve these geopolitical tensions since the late 1940s, the United States used the IGY as a catalyst to convene a diplomatic conference of the twelve concerned states. This conference met in October and November 1959 following eighteen months of preparatory meetings in Washington DC. As originally negotiated, the

Treaty was concerned mostly with guaranteeing free and peaceful access to Antarctica for scientific research. It also prohibited military activities, nuclear explosions and the disposal of radioactive waste. To achieve this agreement, it relied on a peculiar legal formulation relating to territorial sovereignty. Article IV stated that, by signing the Treaty, those states that claimed territory were not renouncing their claims, those states with a basis of claim were not renouncing or diminishing that basis for claim, and that states were not prejudicing their recognition or nonrecognition of other states' claims or basis of claims.

The Treaty was signed by twelve states, which were divided into two main groupings. Seven of the original signatories – Argentina, Australia, Chile, France, New Zealand, Norway and the United Kingdom – claimed territories in the Antarctic. For them, the Article IV compromise was crucially important. The remaining five parties – Belgium, Japan, South Africa, the Soviet Union and the United States – had longer histories of Antarctic exploration or had participated in the IGY, but they neither claimed territory nor recognized the territorial claims of the other seven parties. These twelve original signatories also constituted the 'consultative parties' – those parties with a right to participate in the periodic meetings envisaged by Article IX of the Treaty.

There are three important elements to keep in mind regarding the Antarctic Treaty in its first two decades. The first element is structure. Article IX of the Treaty envisaged that the Treaty's consultative parties would meet periodically to exchange information, consult 'on matters of common interest' and formulate, consider and recommend 'measures in furtherance of the principles and objectives of the Treaty', covering scientific research and cooperation, issues of jurisdiction and the 'preservation and conservation of living resources'. These meetings occurred roughly biennially, rotating among the parties.[16]

The second element was the settlement of territorial and sovereignty issues. Because of the different approaches to Antarctic territory – including claims, potential claims and the rejection of claims – the Treaty parties included the Article IV compromise to allow them to cooperate in the field of science. But Article IV did not solve the problem comprehensively, so new issues that arose had to be brought within the territorial compromise. This became especially important in the 1970s, when the issues of mineral resources and marine living resources arose. In a situation where the territorial claimant states had more to potentially gain from the Antarctic, the Article IV settlement had to be expanded.

The final aspect was the development of environmental issues. Slowly emerging from an abstract emphasis on science was a concentration on matters of nature conservation and resource conservation. The first major landmark of Treaty diplomacy was the Agreed Measures on the Conservation of Antarctic Fauna and Flora in 1964, which was followed by the Convention on the Conservation of Antarctic Seals in 1972. These two agreements covered highly specific, species-level issues. The Agreed Measures were pushed by biological scientists in the early 1960s in response to the harm being done to Antarctic fauna by the actions of geophysical scientists in large exploratory programmes. The Convention on the Conservation of Antarctic Seals was negotiated in response to a resurgent interest on the part of Norway in exploiting Antarctic seals for furs in the 1960s.

The principal international non-governmental organization (INGO) involved in the Antarctic Treaty regime was SCAR, mentioned above. Emerging out of the organizational structures of the IGY, SCAR was a committee of the International Council of Scientific Unions. Representatives of each of the countries participating in Antarctic research sat on it. It had a secretariat based at the Scott Polar Research Institute in Cambridge in the United Kingdom and a committee structure, with one committee for each of the main scientific disciplines.[17] There was a close relationship between the Treaty parties and SCAR – especially as scientists meeting within SCAR often combined government and university roles – but it was neither formally structured nor consistently productive. Though scientists working within SCAR had, for example, precipitated official interest in nature conservation (eventually leading to the 1964 Agreed Measures), lack of funds and a commitment to outward consensus between often conflicting scientific groups meant that it was not always ready or able to respond quickly or effectively to questions raised by the Treaty parties.

Other environmental INGOs gradually entered Antarctic affairs from the mid 1970s, though their influence and presence was relatively minor at the time.[18] The most notable development was the founding of the Antarctic and Southern Ocean Coalition (ASOC) in 1978. This was constituted by an international group of environmental NGOs, especially several national branches of Friends of the Earth and Greenpeace, several branches of the World Wildlife Fund (WWF, today the Worldwide Fund for Nature) and other groups, especially those based in Australia and the United States, along with some Western and Northern European countries. ASOC drew much of its intellectual and organizational force from James Barnes, an American lawyer whose advocacy began in the Washington DC lobby group, the Center for Law and Social Policy. Notably, Barnes was invited to be an advising member of the U.S. delegation at the CCAMLR negotiations. Only in the 1980s and in connection with the negotiations regarding an Antarctic minerals regime did INGOs, including ASOC and Greenpeace, make an increasingly vocal and concerted effort at influencing Antarctic affairs.

Competing Approaches: Resource Development and Ecosystem Conservation

The principal cause of scientific and diplomatic interest in the Southern Ocean ecosystem was Soviet fisheries research in the Antarctic beginning in 1962. In that year, a Soviet fishing research ship trawled the Atlantic sector of the Southern Ocean for krill and fish. It was the beginning of a study to see if Antarctic marine species could be caught and consumed, what fishing gear was needed, and part of the continuing investigation of their biology and stocks.[19] The Soviets began fishing Antarctic waters because their fleets had slowly been expanding throughout the world's oceans as part of a significant enlargement of capacity. From the mid 1950s, the Soviets used the advances in shipbuilding and other technologies that emerged from the Second World War and the early Cold War to grow an industry that they believed could contribute to the Soviet economy.[20]

Antarctic krill (*Euphausia superba*) was the main species being sought. A small crustacean that grows to a length of about six centimetres, the krill live in great swarms, sometimes tens of kilometres long and wide, and tens to hundreds of metres deep. Krill is fundamental to the life of the Southern Ocean and Antarctic ecosystems, as it is the food of whales, seals, penguins and fish. The importance of krill to the Antarctic food chain can be expressed in its species biomass, which is the largest of any animal on earth. This huge biomass made it especially attractive for exploitation.[21]

During the early stages of these exploitation activities, a scientific vision of the Antarctic marine ecosystem was also developing. Scientists had known for many decades that the cold Antarctic seas were strongly demarcated from the warmer oceans to the north by a boundary called the Antarctic convergence. They had also gained a basic knowledge of the ecological relationships among the species living in this ocean.

British marine scientists led this research in the 1920s and 1930s with the '*Discovery* Investigations' in the South Atlantic Ocean. They undertook these research cruises especially to understand whale biology and ecology to support the whaling industry, but in the process also came to provide a thorough understanding of the Southern Ocean.[22] Although many British scientists remained at the forefront of this research, U.S. scientists led the new developments in the 1960s financially supported by the National Science Foundation (NSF) and the American Antarctic programme. Due to the accretion of experience with the Southern Ocean combined with major developments in regulatory approaches to ecosystems, American scientists and officials were most capable – in intellectual and infrastructural terms – of generating and disseminating research and ideas about the ecosystem. Much research in this area was conducted aboard the USNS *Eltanin*, a floating laboratory for Antarctic marine research commissioned by the NSF in 1961, covering both physical and biological oceanography. Between 1962 and 1972, it completed fifty-five cruises throughout all areas the Southern Ocean, contributing fundamental knowledge.[23]

That the United States and the Soviet Union were at the forefront of these particular approaches to the Southern Ocean is representative of postwar oceanography. In both physical and biological oceanography and marine biology, both nations deployed great efforts in studying not just the Southern Ocean, but all the world's oceans. Driven by the countervailing, though linked, demands of Cold War competition and international cooperation, the 1950s and 1960s saw enormous growth in the scale of oceanographic research. But they set out on to the waves with different perspectives and different questions.[24] While the Americans were interested in the scientific problems of ecological productivity and the movement of the sea floor, the Soviets wanted to survey fish stocks and hydrography.

By the early 1970s, then, there were two major visions for human interaction with the Southern Ocean. One saw an ecosystem in need of conservation and the other saw the exploitation potential of its marine living resources. Though there was some overlap between these visions, each of them was held by one of the superpowers, thus resonating with the geopolitical situation. Moreover, as global politics became more concerned in the 1970s with both environmental protection and economic development for developing countries, these competing visions also took on the complexities of those discussions.

Competing Scientific Agendas and Institutions: SCAR and the FAO

From the signing of the Antarctic Treaty, SCAR was a significant influence on the parties' engagement with the environment. In the 1970s, SCAR strongly influenced the ideas of ecosystem conservation that had been developing among a number of its members, particularly American and British marine biologists and oceanographers. With the growing interest in Southern Ocean fisheries, along with other political and economic developments, other international non-governmental and intergovernmental bodies began to claim a place in the research effort. The fisheries department of the Food and Agriculture Organization (FAO) was an especially important new actor in the field. They saw the Southern Ocean as a region of great untapped potential, ripe for development that it could lead. In these changing conditions, SCAR's ability to keep its dominant position in shaping the research agenda for the Antarctic and Southern Ocean was consequential for the development of an ecosystem focus for the Southern Ocean.

The initial impetus for SCAR's actions was the work in the late 1960s of the Intergovernmental Oceanographic Commission (IOC, a body within the United Nations Educational, Scientific and Cultural Organization (UNESCO)), which had established, following a Soviet initiative, a Southern Ocean Co-ordination Group.[25] The SCAR Working Group on Oceanography discussed the IOC's initiatives in September 1970 and supported moves for a comprehensive study of the Southern Ocean.[26] Eventually, in January 1972, SCAR's Executive Committee agreed with the IOC recommendations, but requested that the Working Group on Biology also offer its opinion on the research programme, specifically on krill research.[27] That SCAR scientists refused to let the IOC retain a major stake in these developments reflected the low opinion British and American oceanographers, in particular, had of the IOC. Jacob Hamblin has argued that the IOC became a site of superpower rivalry during its first decade in the 1960s, with U.S. delegations pushing for concentrated and problem-led studies of specific ocean regions, while Soviet delegations pushed for data-driven surveys of the whole 'world ocean'. Hamblin has further demonstrated the low opinion Western oceanographers had for Soviet approaches to oceanography (and science in general), seeing it as 'old-fashioned' and unconcerned with 'problem-based studies'.[28] After a decade of lacklustre scientific work, Western scientists were turning away from the IOC.

In August 1972, the Working Group on Biology established a Subcommittee on Marine Resources chaired by the leading American marine biologist and oceanographer Sayed el-Sayed. The working group was dominated by British and American scientists, including el-Sayed, Laws, Inigo Everson, George Llano, Bruce Parker and Don Siniff. Other important participants were the New Zealander George Knox, the French Jean Prevost and Jean-Claude Hureau, and the Australian Donald Tranter, with no Japanese representative and only one Soviet member, Vyacheslav A. Zemsky. Their agreed collective position (reached through consensus) at this time was quite clear: 'any future development in the exploitation of these resources [krill, *inter alia*] should be viewed in the context of the total ecosystem in which krill plays a key role'.[29]

After 1972, SCAR's position slowly shifted from providing advice based on existing research to planning new research that could provide the foundations for more comprehensive advice. The Subcommittee on Marine Resources did not meet again until May 1974, when the group basically reiterated their existing positions, with an emphasis on biological questions.[30] Following the Antarctic Treaty Consultative Meeting of June 1975, the SCAR Executive upgraded the status of the Subcommittee to a 'Group of Specialists', with el-Sayed continuing as convenor. The terms of reference for this group were developed from their existing intellectual foundations of assessing and developing the state of knowledge of the ecosystem, and additionally as the body to liaise with the Scientific Committee on Oceanographic Research, the IOC and the FAO. They also had to respond to the recommendations of the Treaty consultative meetings.[31] The change of name was at least in part politically motivated. George Hemmen, Executive Secretary of SCAR, had mooted the new designation to trumpet SCAR's greater interest in the topic and to enhance the group's influence against the various other interested bodies – the IOC, the FAO, the United Nations Environment Programme (UNEP) and others.[32] It was this Group of Specialists that pushed for concrete research plans and thus moved SCAR's position away from simply advising based on existing research to a position where new research was necessary for its position in Antarctic affairs.

The beginning of the period of new, future-oriented, research was a conference at Woods Hole in the United States in August 1976. Fifty-nine scientists met to review knowledge of Southern Ocean living resources and to propose a coordinated international scientific programme to extend that knowledge. No Soviet scientists attended this meeting, which agreed to a proposal for a scientific study entitled 'Biological Investigations of Marine Antarctic Systems and Stocks' (BIOMASS).[33] The vision and hope for BIOMASS was substantial. The Woods Hole meeting agreed that 'the principal objective of the BIOMASS programme is to gain a deeper understanding of the structure and dynamic functioning of the Antarctic marine ecosystem as a basis for the future management of potential living resources'.[34] Understanding the ecosystem was therefore intimately linked with future potential exploitation. Krill research was certainly a substantial element of the programme, but other proposals, including ecosystem modelling and research on all members of the ecosystem, were advanced. Detailed plans for implementation were also drafted for a planned duration of the research of one decade.

The Woods Hole meeting was not, however, a simple triumph of an ecosystem-focused conservation ethic. In the dynamic political and economic situation of the 1970s, control of the research agenda for the Southern Ocean was more fraught than the outward agreement suggests. Influence even in the ostensibly objective processes of science was important. There were two broad groups present at Woods Hole: conservation-minded biologists, especially ecologists; and exploitation-oriented fisheries scientists and officials. George Llano, a biologist in the U.S. Division of Polar Programs, noted that the exploitation-oriented fisheries scientists were interested in the 'technico-economic problems of Southern Ocean fisheries'. 'Basically', Llano continued, 'the latter's principal interest was where krill swarming occurs; how large are the stocks and how much can be taken.'[35] That the BIOMASS programme should tend to favour the conservation and ecosystems approach, while still including aspects of the fisheries interests, shows that the resolution of these ideas was of some consequence. It suggests that, even in the international scientific community, an ecosystem-dominated approach was not preordained, however objective it seemed as a scientific category.

Owing to the development of a new law of the sea (which included the creation of two hundred-mile exclusive economic zones), the FAO fisheries department took an active interest in what it considered underdeveloped fisheries throughout the world. While not quite a vacuum, the lack of a regime for the Southern Ocean gave technocrats from the FAO fisheries department an opportunity to add another region to their bailiwick. Shaping the BIOMASS programme in favour of research relating directly to development and exploitation, rather than fundamental understanding of the stocks and systems, was one way of achieving their ends. For example, Llano noted that John Gulland, a longtime FAO official and pre-eminent fisheries scientist, 'deflected the discussion from scientific questioning'.[36] Indeed, Gulland and his FAO colleagues would continue to have quite different scientific and research needs and outlooks than SCAR. In August 1978, Louis DeGoes of the U.S. National Research Council referred to Gulland and Sidney Holt as the 'FAO "Mafia"', whose intentions were to control BIOMASS planning.[37]

Gulland and his FAO colleagues were clear in their research direction as fisheries scientists working within a resource development organization. Writing to el-Sayed in September 1977, Gulland wrote of BIOMASS: 'our interest in FAO is more in developing the basis for future management than in contributing to the general understanding of the world ocean'. He went on to note that 'these of course are not independent and certainly not in conflict'. He stated that 'I think it is important to realise that extremely detailed knowledge of a resource is neither a necessary nor a sufficient condition for good management', for 'participants in exploitation must have a willingness to take appropriate action', and with that willingness, good management might be successful. Gulland wanted to know more about krill stocks and thought that krill biology and ecosystems modelling, while scientifically interesting, did not contribute to management.[38]

Within the Antarctic scientific community, there was both scientific and administrative disagreement. Mary Alice McWhinnie, a leading krill biologist based at DePaul University in Chicago, had an interesting view of this meeting and the plans emerging from it. She wondered in an impassioned letter to DeGoes in November 1976 how useful any scientific programme like BIOMASS would be in the face of a determination to fish krill on the part of the Soviet Union, Poland, West Germany and Japan, or even the United Kingdom. Indeed, by McWhinnie's account, it took everything in her and el-Sayed's power to steer the research priorities away from the development of better ways to exploit the marine resources.[39] Llano noted that 'there was tendency to drift off into discussions on tonnage and catches'.[40] In addition to these scientific issues, SCAR's Executive Secretary George Hemmen and the US polar administrator Louis DeGoes sought to create a kind of bureaucratic hegemony, building SCAR's position at the expense of other organizations.

Pursuing BIOMASS as a major international research programme and advancing the ecosystem as a whole within it was an important move to entrench SCAR more fully in Antarctic affairs. The ecosystem was, in short, a very useful thing for SCAR scientists to emphasize in the 1970s. Yet, SCAR's ecosystem-minded scientists could only advance their position against more resource-minded colleagues in other organizations so far. What SCAR needed was for the diplomats of the Treaty parties to also take up the ecosystem position.

The Ecosystem and the Strength of the Antarctic Treaty Regime

Though there had been a sense among some Treaty party diplomats and officials as early as 1968 that the exploitation of marine living resources would be a topic for discussion at a future consultative meeting, it was not until 1975 that the matter came onto the agenda. The ecosystem was not present in this discussion, however. Instead, the discussions focused on the traditional fisheries and resource aspects of the issue.[41] The meeting's eventual recommendation did not actually mention the ecosystem, calling only for the parties to 'initiate or expand' their scientific programmes, especially to contribute 'to the development of effective measures for … conservation'.[42]

The discussions at the 1975 consultative meeting in Oslo were therefore brief. Yet, if the outward agreement seemed insubstantial, contemporaneous world events provided a crucial stimulus for carrying the discussions over. The meeting occurred weeks after news – broken by Greenpeace activists in their first anti-whaling campaign – that the Soviet whaling fleet had been operating in the North Pacific outside of quotas set by the International Whaling Commission.[43] Though the Antarctic Treaty regime did not include whales on its agenda because of the existence of the International Whaling Commission, the fate of whales suffused the tone of the meeting. The New Zealand representative implored his colleagues to 'bear in mind what has happened to the whale'.[44] Thus, the fate of individual species was already at the heart of the meeting, though not the larger ecosystem.

The mid 1970s were also the early years of the Third United Nations Law of the Sea Conference (UNCLOS), and fisheries formed an important part of these discussions. In these wider fisheries negotiations, the same two pressures – conservation and development – were also present. A U.S. memorandum of March 1975 updating President Gerald Ford on the negotiations noted that there was 'a clear trend in the Conference for broad coastal state control over fisheries'. However, this was 'subject to duty to conserve and ensure full utilisation of such resources'.[45] The tensions between exploitation and conservation and between resources and ecosystem were manifest and difficult to escape. In these circumstances, the transition to an ecosystem approach required a major intellectual, political and legal jump.

The other main element of the UNCLOS negotiations was that it encouraged those states that had not hitherto paid any attention to Antarctica and the Southern Ocean to see it in a new light. One of the impulses for negotiating a new law of the sea was the call in 1967 by the Maltese diplomat Arvid Pardo to treat the seabed and areas beyond national sovereignty as a 'common heritage of mankind'. Any riches on the sea floor, which seemed possible in the late 1960s, should be divided among all countries, but with preference given to developing countries.[46] With the uncertainty over Antarctic sovereignties, the Treaty parties, both claimants and nonclaimants, were concerned that this new concept, and the UNCLOS negotiations after 1974, might bring new actors into what they saw as their area.

Some diplomats from the developing world did take an interest in Antarctica. Just like SCAR, the Treaty parties also had to deal with FAO interests. Between 1975 and 1978, developing states on the FAO Fisheries Committee began to talk of exploiting Antarctic fish resources. What disturbed the Treaty parties was the tenor of these interventions. Several representatives of developing countries spoke of the need of developing Antarctic marine resources 'for the benefit of all mankind'.[47] This implied, in the minds of several Treaty parties, interference with their particular relationship with the Antarctic and a threat to dismantle their Treaty. Because of the many other issues on the FAO fisheries agenda, the Antarctic did not remain there. Few of the developing nations pursued it with any vigour after 1978. Yet it is instructive to consider that, for the Treaty parties, the 'natural' boundaries of the Southern Ocean ecosystem could be used to bolster their position in the South.

At Oslo in 1975, the parties had not yet had the benefit of major SCAR interventions into the subject. By 1977, though, SCAR's moves in establishing BIOMASS, as well as the longer scientific and intellectual developments, had created a new context for the discussions of the Treaty parties at the ninth consultative meeting in London. This meeting's resolution contained a basic settlement of ideas that was to guide further negotiations. In a substantial resolution, the parties agreed to three major points: first, that they should cooperate as much as possible in scientific research, particularly through the BIOMASS programme; second, that they would observe basic interim guidelines for living resources conservation, particularly to take care in harvesting species without 'jeopardizing the Antarctic marine ecosystem as a whole'; and, finally, that the Antarctic Treaty parties should conclude a 'definitive regime' for the conservation of marine living resources by the end of 1978, specifically including in such a regime, among other political considerations, provision 'for the effective conservation of the marine living resources of the Antarctic ecosystem as a whole', extending north of sixty degrees south latitude 'where that is necessary for the effective conservation of species of the Antarctic ecosystem'.[48]

As agreed at London, the consultative parties convened for a special consultative meeting to negotiate a convention for a definitive regime in Canberra in Australia between 27 February and 16 March 1978. Despite the outward agreement at London, though, there were still divisions between those parties who demanded a conservation agreement for the whole ecosystem and those who wanted an agreement that regulated the exploitation of marine living resources. The Soviet delegation in particular – though joined by the Japanese delegation – was the most outspoken on the need for a convention to cover and regulate exploitation. The Soviets were particularly incensed that these living resources might be left untouched. In one intervention, the delegate stated that if the resources were not utilized, 'then it is a great loss for mankind'.[49]

Yet there were many delegations willing to speak up against the Soviets. The British and Americans were the strongest opponents. Robert Brewster, the head of the U.S. delegation, implored the participants that 'everything that we have learned about the living organisms found in the Antarctic marine environment convinces us that effective conservation requires an approach which is not limited to individual species, but which treats the ecosystem as a whole'.[50] Donald Logan, head of the British delegation, was concerned that the proposed title of the convention – referring to 'marine living resources' – was misleading. Instead, it should refer to the ecosystem. When speaking to the first chairman's draft, he also protested that the conservation standards were not prominent enough in the text, 'relegating' them to Article 17. They deserved greater prominence.[51] In addition to the United States and the United Kingdom, many of the other parties were also enthusiastic for a strong conservation approach based on the ecosystem, including Australia, New Zealand, Norway, Argentina and Chile. As a result, the conservation standard eventually gained a more prominent place at the top of the second chairman's draft.

The supporters of the whole-ecosystem conservation approach believed that it was scientifically prudent, but also useful for them. The United States, for example, had obvious interests in constraining Soviet activities in the Antarctic. The spatial and discursive aspects of speaking of the ecosystem might just allow it to succeed in that policy. The claimants also hoped that such an environmental turn might help to further embed their positions in the Antarctic. At the same time, the spectre of 'Third World' interest in the Antarctic, and the associated interest of the FAO, threatened to disrupt the Antarctic Treaty regime that had provided for good relations between the parties and for productive scientific and environmental work.

The ecosystem had an uneasy place in the convention draft that emerged from the first session of the special consultative meeting. The division between those parties that wanted the convention to see marine living resources and those that wanted an ecosystem standard meant that the draft text, and its Article II, did not fulfil scientific expectations. It took further work at the second session of the special consultative meeting in Buenos Aires a few months later to embed a compromise position between conservationist and fishing positions. The final version of Article II balanced the ecosystem and resource views of the Southern Ocean. Many environmentalists and conservationists like James Barnes and the ASOC were unhappy about the article's compromise character, and many scientists, including Laws, felt it inadequate.[52] Nevertheless, the ecosystem had a prominent place in it.

Conclusion

CCAMLR was finally agreed in May 1980. Putting the ecosystem at its heart was only a first step in the process of effective environmental protection and resource management. Coming into force in 1982, the CCAMLR's first decade showed that, even though the ecosystem was at the centre of the text, there was still sustained effort required to put it at the heart of the agenda of the newly established Commission. There remain differing agendas based in the same two blocs of parties in the contemporary politics of CCAMLR: one group of parties insists that it must enact its original promise of whole-ecosystem conservation, while another group of parties argues that it must balance the needs of fisheries exploitation with conservation of fish stocks.[53] The debates and negotiations of the last few years on the establishment of marine protected areas have seen the conservationist states – especially Australia, France, New Zealand the United States – pitted against the fishing states – particularly Russia and the Ukraine.[54] The Southern Ocean ecosystem was not targeted for conservation and protection by the Antarctic Treaty parties because such an action was best-practice environmental protection or the most rational thing to do. The ecosystem was embedded at the centre of the environmental protection agenda of Antarctic diplomacy and science because it was central to maintaining the position of SCAR and the Treaty parties. It was central to the Antarctic scientific and political agenda because its meanings and spatiality could enhance the positions of some actors and their objectives at the expense of others. Thus, the history of CCAMLR's negotiation and the emphasis on the ecosystem demonstrates that we need to be attentive not simply to the environmental purposes of protection agendas, institutions and efforts; instead, we need to understand their political purposes, too.

Alessandro Antonello is a McKenzie Postdoctoral Fellow in the School of Historical and Philosophical Studies, University of Melbourne, Australia.

Notes

1. Transcribed from 'English, Monday 27/2/1978 – Thursday AM, 2/3/1978: Interpreter floor tapes used for the consultative meeting on Antarctica, Canberra', A10734, 2, National Archives of Australia (NAA), Sydney.
2. Ibid.
3. Ibid.
4. *United Nations Treaty Series* 402 (1961): 71–102.
5. D.H. Cushing, *The Provident Sea* (Cambridge: Cambridge University Press, 1988), Chapters 12 and 15; and Carmel Finley, *All the Fish in the Sea: Maximum Sustainable Yield and the Failure of Fisheries Management* (Chicago: University of Chicago Press, 2011). The Antarctic Treaty parties also had the long experience of international whaling treaties to consider: Mark Cioc, *The Game of Conservation: International Treaties to Protect the World's Migratory Animals* (Athens, OH: Ohio University Press, 2009), Chapter 3, 104–147.
6. Donald Worster, *Nature's Economy: A History of Ecological Ideas*, 2nd edn (Cambridge: Cambridge University Press, 1994); Sharon E. Kingsland, *The Evolution of American Ecology, 1890–2000* (Baltimore: Johns Hopkins University Press, 2005), Chapter 7; Chunglin Kwa, 'Representations of Nature Mediating between Ecology and Science Policy: The Case of the International Biological Programme', *Social Studies of Science* 17(3) (1987): 413–42.

7. *United Nations Treaty Series* 1329 (1983): 48–106.
8. An early and standard account of the CCAMLR negotiations is James Barnes, 'The Emerging Convention on the Conservation of Antarctic Marine Living Resources: An Attempt to Meet the New Realities of Resource Exploitation in the Southern Ocean', in *The New Nationalism and the Use of Common Spaces: Issues in Marine Pollution and the Exploitation of Antarctica*, Jonathan I. Charney (ed.) (Totowa: Allanheld, Osmun, 1982), 239–86; an early legal and political interpretation of CCAMLR is David M. Edwards and John A. Heap, 'Convention on the Conservation of Antarctic Marine Living Resources: A Commentary', *Polar Record* 20(127) (1981): 353–62.
9. Sheila Jasanoff, 'The Idiom of Co-production', in *States of Knowledge: The Co-production of Science and Social Order*, Sheila Jasanoff (ed.) (London: Routledge, 2004), 1–12, 2.
10. *United Nations Treaty Series* 1080 (1978): 176–212.
11. Stephen D. Krasner, 'Structural Causes and Regime Consequences: Regimes as Intervening Variables', *International Organization* 36(2) (1982): 185–205, 185.
12. Klaus Dodds, *Pink Ice: Britain and the South Atlantic Empire* (London: I.B. Tauris, 2002), Chapters 3–6; Adrian Howkins, 'Frozen Empires: A History of the Antarctic Sovereignty Dispute between Britain, Argentina, and Chile, 1939–1959' (Ph.D. dissertation, University of Texas, 2008).
13. Irina Gan, '"The First Practical Soviet Steps Towards Getting a Foothold in the Antarctic": The Soviet Antarctic Whaling Flotilla *Slava*', *Polar Record* 47(1) (2011): 21–28.
14. David Day, *Antarctica: A Biography* (North Sydney: Knopf, 2012), Chapter 18.
15. Peter J. Beck, *The International Politics of Antarctica* (London: Croom Helm, 1986), Chapter 3; Dian Olson Belanger, *Deep Freeze: The United States, the International Geophysical Year, and the Origins of Antarctica's Age of Science* (Boulder: University Press of Colorado, 2006); and Klaus Dodds, 'Assault on the Unknown: Geopolitics, Antarctic Science, and the International Geophysical Year (1957–8)', in *New Spaces of Exploration: Geographies of Discovery in the Twentieth Century*, Simon Naylor and James R. Ryan (eds) (London: I.B. Tauris, 2010), 148–72.
16. Beck, *International Politics of Antarctica*, 149–62.
17. David W.H. Walton, Peter D Clarkson and Colin P. Summerhayes, *Science in the Snow: Fifty Years of International Collaboration through the Scientific Committee on Antarctic Research* (Cambridge: Scientific Committee on Antarctic Research, 2011). This is a commissioned and largely celebratory history of SCAR.
18. Lee Kimball, 'The Role of Non-governmental Organizations in Antarctic Affairs', in *The Antarctic Legal Regime*, Christopher C. Joyner and Sudhir K. Chopra (eds) (Dordrecht: Martinus Nijhoff, 1988), 33–64.
19. R.N. Burukovskiy (ed.) *Soviet Fishery Research on the Antarctic Krill* (Washington DC: Joint Publications Research Service, 1967).

20. Paul R. Josephson, *Industrialized Nature: Brute Force Technology and the Transformation of the Natural World* (Washington DC: Island Press, 2002), 200–1 and 219–25.

21. James Marr, 'The Natural History and Geography of the Antarctic Krill (*Euphausia Superba* Dana)', *Discovery Reports* XXXII (1962): 33–464; John Mauchline and Leonard R. Fisher, *The Biology of Euphausiids* (London: Academic Press, 1969).

22. Peder Roberts, *The European Antarctic: Science and Strategy in Scandinavia and the British Empire* (New York: Palgrave Macmillan, 2011), Chapter 1; D. Graham Burnett, *The Sounding of the Whale: Science & Cetaceans in the Twentieth Century* (Chicago: University of Chicago Press, 2012), Chapter 2.

23. K.G. Sandved, 'USNS *Eltanin*: Four Years of Research', *Antarctic Journal of the United States* 1(4) (1966): 164–74; G.E. Fogg, *A History of Antarctic Science* (Cambridge: Cambridge University Press, 1992), 213–41.

24. Jacob Darwin Hamblin, *Oceanographers and the Cold War: Disciples of Marine Science* (Seattle: University of Washington Press, 2005), Chapters 5–6.

25. U.S., Position Paper for International Coordination Group on the Southern Ocean, First Session, Brussels, 23–26 November 1970, Intergovernmental Oceanographic Commission, Division of Polar Programs, Central Subject Files, 1969–75, RG 307, United States National Archives, College Park (NACP).

26. *SCAR Bulletin* 39 (September 1971): 985–86.

27. *SCAR Bulletin* 40 (January 1972): 185.

28. Hamblin, *Oceanographers and the Cold War*, 201, and, more generally, 169–202.

29. *SCAR Bulletin* 43 (January 1973): 634–35.

30. *SCAR Bulletin* 49 (January 1975): 440–41.

31. *SCAR Bulletin* 51 (September 1975): 714.

32. G.E. Hemmen to R.M. Laws, 30 May 1975 and Hemmen to T. Gjelsvik, 11 June 1975, SCAR Secretariat files, Cambridge, United Kingdom (SCAR Sec.).

33. Scientific Committee on Antarctic Research et al., *Biological Investigations of Marine Antarctic Systems and Stocks (BIOMASS), Volume 1 Research Proposals* (Cambridge: Scientific Committee on Antarctic Research and Scientific Committee on Oceanic Research, 1977); *SCAR Bulletin* 55 (January 1977): 419–26.

34. Scientific Committee on Antarctic Research et al., *Biological Investigations of Marine Antarctic Systems and Stocks (BIOMASS), Volume 1 Research Proposals*, 5.

35. G.A. Llano, Memorandum, 17 September 1976, Division of Polar Programs, Records of the Program Manager for Biology and Medicine, Memorandums 1976, RG 307, NACP.

36. Ibid.

37. L. DeGoes to G.A. Knox, 11 August 1978, SCAR Sec.; Hemmen seemed equally disturbed by the prospect of FAO leadership of BIOMASS: Hemmen to DeGoes, 12 September 1978 (SCAR Sec.) and Hemmen to B.B. Parrish, 5

January 1978, 243/244/01 Vol 1, British Antarctic Survey Archives (BAS), Cambridge, United Kingdom.

38. J.A. Gulland to S.Z. el-Sayed, 6 September 1977, 243/244/01 Vol 1, BAS.
39. M.A. McWhinnie to L. DeGoes, 23 November 1976, National Science Foundation, Division of Polar Programs, Central Subject Files, 1976–87, RG 307, NACP.
40. G.A. Llano, Memorandum, 17 September 1976.
41. See J.P. Lonergan's notes on the Eighth Antarctic Treaty Consultative Meeting, B1387, 1991/688 Part 1, NAA, Hobart.
42. *Report of the Eighth Consultative Meeting, Oslo, 9–20 June 1975* (Oslo: Ministry of Foreign Affairs, 1976), Recommendation VIII–10, 40.
43. Andrew Darby, *Harpoon: Into the Heart of Whaling* (Crows Nest: Allen & Unwin, 2007), 104; Frank Zelko, *Make it a Green Peace! The Rise of Countercultural Environmentalism* (Oxford: Oxford University Press, 2013), Chapter 9; Kurkpatrick Dorsey, *Whales and Nations. Environmental Diplomacy on the High Seas* (Seattle: University of Washington Press, 2013).
44. 'Our Men in Oslo Fighting to Save the Krill', *Evening Post* (Wellington), 12 June 1975, AATJ, 7428, 11/1/9 Part 1, Archives New Zealand (ANZ), Wellington.
45. Henry Kissinger, Memorandum from the President's Assistant for National Security Affairs (Kissinger) to President Ford, Washington, Foreign Relations of the United States, Volume E–3, Documents on Global Issues, 1973–76, doc 17, https://history.state.gov/historicaldocuments/frus1969-76ve03/d17, accessed 7 July 2016.
46. United Nations General Assembly, First Committee, Official Records, 22nd Session, 1 November 1967, A/C.1/PV.1516
47. FAO, 'Report of the Eleventh Session of the Committee on Fisheries', *FAO Fisheries Reports* 196 (1977); FAO, 'Report of the Twelfth Session of the Committee on Fisheries', *FAO Fisheries Reports* 208 (1978); Australian Embassy Rome, Cablegram RO6213 to Canberra, 16 June 1978, B1387, 1996/877 Part 1, NAA, Hobart; New Zealand Embassy Rome, Telegram 438 to Wellington, 22 June 1978, ABHS, 950, 208/1/10/1 Part 4, ANZ.
48. *Report of the Ninth Consultative Meeting, London, 19 September–7 October 1977* (London: Foreign and Commonwealth Office, 1977), Recommendation IX–2, 13–16.
49. Transcribed from 'English, Monday 27/2/1978 – Thursday AM, 2/3/1978: Interpreter Floor Tapes Used for the Consultative Meeting on Antarctica, Canberra', A10734, 2, NAA, Sydney.
50. Ibid.
51. Transcribed from 'English, Thursday, 3:00pm, 2/3/1978 – Thursday 4:30pm, 9/3/1978: Interpreter Floor Tapes Used for the Consultative Meeting on Antarctica, Canberra', A10734, 4, NAA, Sydney.
52. Richard Laws, 'Difficulties and Ambiguities in the Wording of Article II', 27 March 1980, FCO 7/3836, National Archives of the United Kingdom, Kew.
53. Veronica Ward, 'Sovereignty and Ecosystem Management: Clash of Concepts and Boundaries?', in *Greening of Sovereignty in World Politics*, Karen Litfin

(ed.) (Cambridge, MA: MIT Press, 1998), 79–108, 95; 'Experiences from the Convention on Living Resources (Section 1)' in *The Antarctic Treaty System in World Politics*, Arnfinn Jørgensen-Dahl and Willy Østreng (eds) (Houndmills: Macmillan, 1991), 23–76.
54. Daniel Cressey, 'Shock as Antarctic Protection Plans Scuppered', *Nature News*, 16 July 2013, http://www.nature.com/news/shock-as-antarctic-protection-plans-scuppered-1.13401, accessed 25 May 2016; and Daniel Cressey, 'Third Time Unlucky for Antarctic Protection Bid', *Nature News*, 1 November 2013, http://www.nature.com/news/third-time-unlucky-for-antarctic-protection-bid-1.14085, accessed 25 May 2016.

Bibliography

Antarctic Treaty, *United Nations Treaty Series* 402 (1961): 71–102.
Australian Embassy Rome, Cablegram RO6213 to Canberra, 16 June 1978, B1387, 1996/877 Part 1, National Archives of Australia (NAA), Hobart.
Barnes, James, 'The Emerging Convention on the Conservation of Antarctic Marine Living Resources: An Attempt to Meet the New Realities of Resource Exploitation in the Southern Ocean', in *The New Nationalism and the Use of Common Spaces: Issues in Marine Pollution and the Exploitation of Antarctica*, Jonathan I. Charney (ed.) (Totowa: Allanheld, Osmun, 1982), 239–86.
Beck, Peter J., *The International Politics of Antarctica* (London: Croom Helm, 1986).
Belanger, Dian Olson, *Deep Freeze: The United States, the International Geophysical Year, and the Origins of Antarctica's Age of Science* (Boulder: University Press of Colorado, 2006).
Burnett, D. Graham, *The Sounding of the Whale: Science & Cetaceans in the Twentieth Century* (Chicago: University of Chicago Press, 2012).
Burukovskiy, R.N. (ed.), *Soviet Fishery Research on the Antarctic Krill* (Washington DC: Joint Publications Research Service, 1967).
Cioc, Mark, *The Game of Conservation: International Treaties to Protect the World's Migratory Animals* (Athens, OH: Ohio University Press, 2009).
Convention for the Conservation of Antarctic Seals, *United Nations Treaty Series* 1080 (1978): 176–212.
Convention on the Conservation of Antarctic Marine Living Resources, *United Nations Treaty Series* 1329 (1983): 48–106.
Cressey, Daniel, 'Shock as Antarctic Protection Plans Scuppered', *Nature News*, 16 July 2013, http://www.nature.com/news/shock-as-antarctic-protection-plans-scuppered-1.13401, accessed 25 May 2016.
Cressey, Daniel, 'Third Time Unlucky for Antarctic Protection Bid', *Nature News*, 1 November2013,http://www.nature.com/news/third-time-unlucky-for-antarctic-protection-bid-1.14085, accessed 25 May 2016.
Cushing, D.H., *The Provident Sea* (Cambridge: Cambridge University Press, 1988).
Darby, Andrew *Harpoon: Into the Heart of Whaling* (Crows Nest: Allen & Unwin, 2007).
Day, David, *Antarctica: A Biography* (North Sydney: Knopf, 2012).

DeGoes, L. to G.A. Knox, 11 August 1978, SCAR Secretariat files, Cambridge, United Kingdom.

Dodds, Klaus, 'Assault on the Unknown: Geopolitics, Antarctic Science, and the International Geophysical Year (1957-8)', in *New Spaces of Exploration: Geographies of Discovery in the Twentieth Century*, Simon Naylor and James R. Ryan (eds) (London: I.B. Tauris, 2010), 148-72.

Dodds, Klaus, *Pink Ice: The United Kingdom and the South Atlantic Empire* (London: I.B. Tauris, 2002).

Dorsey, Kurkpatrick, *Whales and Nations: Environmental Diplomacy on the High Seas* (Seattle: University of Washington Press, 2013).

Edwards, David M. and John A. Heap, 'Convention on the Conservation of Antarctic Marine Living Resources: A Commentary', *Polar Record* 20(127) (1981): 353-62.

'English, Monday 27/2/1978 – Thursday AM, 2/3/1978: Interpreter Floor Tapes Used for the Consultative Meeting on Antarctica, Canberra', A10734, 2, National Archives of Australia (NAA), Sydney.

'English, Thursday, 3:00pm, 2/3/1978 – Thursday 4:30pm, 9/3/1978: Interpreter Floor Tapes Used for the Consultative Meeting on Antarctica, Canberra', A10734, 4, National Archives of Australia (NAA), Sydney.

FAO, 'Report of the Eleventh Session of the Committee on Fisheries', *FAO Fisheries Reports* 196 (1977).

FAO, 'Report of the Twelfth Session of the Committee on Fisheries', *FAO Fisheries Reports* 208 (1978).

Finley, Carmel, *All the Fish in the Sea: Maximum Sustainable Yield and the Failure of Fisheries Management* (Chicago: University of Chicago Press, 2011).

Fogg, G.E., *A History of Antarctic Science* (Cambridge: Cambridge University Press, 1992).

Gan, Irina, '"The First Practical Soviet Steps Towards Getting a Foothold in the Antarctic": The Soviet Antarctic Whaling Flotilla *Slava*', *Polar Record* 47(1) (2011): 21-28.

Gulland, J.A. to S.Z. el-Sayed, 6 September 1977, 243/244/01 Vol. 1, British Antarctic Survey Archives (BAS), Cambridge, United Kingdom.

Hamblin, Jacob Darwin, *Oceanographers and the Cold War: Disciples of Marine Science* (Seattle: University of Washington Press, 2005).

Hemmen to B.B. Parrish, 5 January 1978, 243/244/01 Vol. 1, British Antarctic Survey Archives (BAS), Cambridge, United Kingdom.

Hemmen to DeGoes, 12 September 1978, SCAR Secretariat files, Cambridge, United Kingdom.

Hemmen, G.E. to R.M. Laws, 30 May 1975 and Hemmen to T. Gjelsvik, 11 June 1975, SCAR Secretariat files, Cambridge, United Kingdom.

Howkins, Adrian, 'Frozen Empires: A History of the Antarctic Sovereignty Dispute between Britain, Argentina, and Chile, 1939-1959' (Ph.D. dissertation, University of Texas, 2008).

Jasanoff, Sheila, 'The Idiom of Co-production', in *States of Knowledge: The Co-production of Science and Social Order*, Sheila Jasanoff (ed.) (London: Routledge, 2004), 1–12.

Jorgensen-Dahl, Arnfinn and Willy Ostreng (eds), *The Antarctic Treaty System in World Politics* (Houndmills: Macmillan in association with the Fridtjof Nansen Institute, 1991).

Josephson, Paul R., *Industrialized Nature: Brute Force Technology and the Transformation of the Natural World* (Washington DC: Island Press, 2002).

Kimball, Lee, 'The Role of Non-governmental Organizations in Antarctic Affairs', in *The Antarctic Legal Regime*, Christopher C. Joyner and Sudhir K. Chopra (eds) (Dordrecht: Martinus Nijhoff, 1988), 33–64.

Kingsland, Sharon E., *The Evolution of American Ecology, 1890–2000* (Baltimore: Johns Hopkins University Press, 2005).

Kissinger, Henry, Memorandum from the President's Assistant for National Security Affairs (Kissinger) to President Ford, Washington, *Foreign Relations of the United States*, Volume E–3, Documents on Global Issues, 1973–76, doc 17, https://history.state.gov/historicaldocuments/frus1969-76ve03/d17, accessed 7 July 2016.

Krasner, Stephen D., 'Structural Causes and Regime Consequences: Regimes as Intervening Variables', *International Organization* 36(2) (1982): 185–205.

Kwa, Chunglin, 'Representations of Nature Mediating between Ecology and Science Policy: The Case of the International Biological Programme', *Social Studies of Science* 17(3) (1987): 413–42.

Laws, Richard, 'Difficulties and Ambiguities in the Wording of Article II', 27 March 1980, FCO 7/3836, National Archives of the United Kingdom, Kew.

Llano, G.A., Memorandum, 17 September 1976, Division of Polar Programs, Records of the Program Manager for Biology and Medicine, Memorandums 1976, RG 307, United States National Archives, College Park (NACP).

Lonergan, J.P., Notes on the Eighth Antarctic Treaty Consultative Meeting, B1387, 1991/688 Part 1, National Archives of Australia (NAA), Hobart.

Marr, James, 'The Natural History and Geography of the Antarctic Krill (*Euphausia Superba* Dana)', *Discovery Reports* XXXII (1962): 33–464.

Mauchline, John and Leonard R. Fisher, *The Biology of Euphausiids* (London: Academic Press, 1969).

McWhinnie, M.A. to L. DeGoes, 23 November 1976, National Science Foundation, Division of Polar Programs, Central Subject Files, 1976–87, RG 307, United States National Archives, College Park (NACP).

New Zealand Embassy Rome, Telegram 438 to Wellington, 22 June 1978, ABHS, 950, 208/1/10/1 Part 4, Archives New Zealand (ANZ), Wellington.

'Our Men in Oslo Fighting to Save the Krill', *Evening Post* (Wellington), 12 June 1975, AATJ, 7428, 11/1/9 Part 1, Archives New Zealand (ANZ), Wellington.

Report of the Eighth Consultative Meeting, Oslo, 9–20 June 1975 (Oslo: Ministry of Foreign Affairs, 1976), Recommendation VIII–10, 40.

Report of the Ninth Consultative Meeting, London, 19 September–7 October 1977 (London: Foreign and Commonwealth Office, 1977), Recommendation IX–2, 13–16.

Roberts, Peder, *The European Antarctic: Science and Strategy in Scandinavia and the British Empire* (New York: Palgrave Macmillan, 2011).

Sandved, K.G., 'USNS *Eltanin*: Four Years of Research', *Antarctic Journal of the United States* 1(4) (1966): 164–74.

SCAR Bulletin 39 (September 1971): 985–86.

SCAR Bulletin 40 (January 1972): 185.

SCAR Bulletin 43 (January 1973): 634–35.

SCAR Bulletin 49 (January 1975): 440–41.

SCAR Bulletin 51 (September 1975): 714.

SCAR Bulletin 55 (January 1977): 419–26.

Scientific Committee on Antarctic Research et al., *Biological Investigations of Marine Antarctic Systems and Stocks (BIOMASS), Volume 1 Research Proposals* (Cambridge: Scientific Committee on Antarctic Research and Scientific Committee on Oceanic Research, 1977).

United Nations General Assembly, First Committee, Official Records, 22nd Session, 1 November 1967, A/C.1/PV.1516.

U.S., Position Paper for International Coordination Group on the Southern Ocean, First Session, Brussels, 23–26 November 1970, Intergovernmental Oceanographic Commission, Division of Polar Programs, Central Subject Files, 1969–75, RG 307, United States National Archives, College Park (NACP).

Walton, David W.H., Peter D Clarkson and Colin P. Summerhayes, *Science in the Snow: Fifty Years of International Collaboration through the Scientific Committee on Antarctic Research* (Cambridge: Scientific Committee on Antarctic Research, 2011).

Ward, Veronica, 'Sovereignty and Ecosystem Management: Clash of Concepts and Boundaries?', in *Greening of Sovereignty in World Politics*, Karen Litfin (ed.) (Cambridge, MA: MIT Press, 1998), 79–108.

Worster, Donald *Nature's Economy: A History of Ecological Ideas*, 2nd edn (Cambridge: Cambridge University Press, 1994).

Zelko, Frank, *Make it a Green Peace! The Rise of Countercultural Environmentalism* (Oxford: Oxford University Press, 2013).

CHAPTER 10

Controlling the Agenda
Science, Policy and the Making of the Intergovernmental Panel on Climate Change

David G. Hirst

Writing to the U.S. Secretary of State George Schultz in 1986, the Executive Director of the United Nations Environment Programme (UNEP), Mostafa Tolba, urged 'the U.S. to take appropriate policy actions' on climate change.[1] Tolba, 'flush with the success of negotiating the Vienna Convention on Ozone, felt that the time was ripe to repeat the ozone "miracle" for climate'.[2] The delegates at the 1985 Villach Conference on the 'Assessment of the Role of Carbon Dioxide and of Other Greenhouse Gases in Climate Variations and Associated Impacts' had recommended political action on climate change. As a result, Tolba sought to raise the political status of the issue. As one of the principal architects of the Villach Conference, he was particularly clear in his political ambitions. He wanted to see negotiations begin on a global climate convention along the lines of that recently negotiated for ozone depleting chemicals (Vienna Convention, 1985). He also strongly believed that the negotiations of such a convention should be guided by the advice of a small scientific advisory panel that he wanted to see report directly to UNEP. In 1988, just two years after his letter to Schultz, the Intergovernmental Panel on Climate Change (IPCC) was jointly established by the World Meteorological Organization (WMO) and UNEP.

The significance of the Villach Conference in the history of the science and politics of climate change has been explored by several other scholars and journalists. *New Scientist* environmental correspondent Fred Pearce, for example, hailed the meeting as 'the week the climate changed'.[3] Additionally, Wendy Franz and Shardul Agrawala highlight the Villach meeting respectively as an important departure point in connecting the science and policy of climate change emphasizing the urgency of action,[4] and representing the earliest case of international consensus on the seriousness of the issue.[5] These accounts principally focus on how the Villach meeting contributed to the rising political importance of climate change. However, they fail to address the scientific and political legacy of Villach – how the meeting has subsequently shaped the political discourse.

In this chapter I will address both how and why Villach took on such importance in shaping the subsequent science–policy discourse around climate change. I will show how the actions and decisions taken at Villach affected the subsequent discussions of climate change. I will specifically argue that decisions taken at Villach, emulating the mechanisms pursued in the coincident ozone negotiations, committed a group of scientists and international administrators to a linear 'scientized' approach to climate change policy making. Furthermore, I will demonstrate that though the IPCC's establishment did raise the political status of climate change, its creation was far from the realization of Tolba's early ambitions. Indeed, by tracing the emergence of climate change as a salient international political issue at the Villach meeting, I will show how the scientists and international administrators framed climate change as an urgent political issue and how the ensuing tussle for control resulted in the U.S. government's proposal for a specifically *intergovernmental* assessment mechanism. The U.S. proposal was intended as a means of overcoming the overly technocratic approaches recommended by the Villach delegates. Significantly, the intergovernmental proposal effectively repoliticized climate change.

In this chapter I will first detail the historical context of the international assessment held at Villach in 1985. I will show that between 1980 and 1985, an important change in emphasis occurred. From recommending more money for more research to recommending political intervention along the lines of a global climate convention, the scientists in their advisory capacity raised the political stakes on climate change. Secondly, I explore how the ways in which the Villach delegates raised the political profile of climate change framed the issue. I argue that, influenced by the coincident ozone negotiations, the Villach recommendations have subsequently committed both scientists and politicians to a highly scientized linear approach to decision making on climate change. Finally, I track the proposals and counterproposals that eventually culminated in the final and formal decision to establish the IPCC in June 1988. Key individuals, it turns out, were very important in shaping the initial structures and remit of the assessment mechanism.

More Money, More Research

By 1985, research into the earth's climate had been going on for several decades. The main focus of this research centred on improvements to climate models, which enhanced the predictive tools available to meteorologists. Out of this basic research, concerns emerged around the possibility for dramatic changes to the earth's climate. The first was that anthropogenic CO_2 emissions would contribute to an enhanced greenhouse effect, warming the planet by somewhere between 1 and 5.5°C. This was briefly followed by concerns in the 1970s that aerosols produced in the course of burning coal would lead to a dramatic global cooling, bringing on another ice age. But, by the late 1970s and into the early 1980s, the prevailing concern was once again squarely focused on the warming effect of the increasing concentration of CO_2 emissions. At the Villach meeting in 1985, the delegates considered these concerns to be sufficiently worrying to require recommendations for political action. Their principal recommendation then was for the creation of an advisory panel that would consider the merits and shortcomings of various courses of political action. Guided by the use of scientific advice and science advisors in the successfully negotiated Vienna Convention for the Protection of the Ozone Layer, the Villach delegates sought to emulate analogous mechanisms.

In order to appreciate the significance of the 1985 Villach Conference, it is necessary to investigate the historical context from which it emerged. The World Climate Programme (WCP), under which the Villach Conference was convened, had itself only recently been established as a result of the historic World Climate Conference (WCC) held in Geneva in 1979. Stimulated by concerns over the possible threat posed by climate change, the WCC conclusions simply agreed upon the need for more research to address the widespread uncertainties rather than advocating specific policies. As a result of these uncertainties and widespread hesitancy to advocate any political intervention, the WCC formally backed the proposal for the establishment of the WCP to carry out more research.

The WCP was designed to bring together the WMO, UNEP and the International Council of Scientific Unions (ICSU) to begin a collaborative effort to monitor, assess and develop a greater understanding of climate change. Under the auspices of this programme, several meetings were convened to examine and update the sponsoring agencies on the current state of the climate. The first was held in Villach in Austria in 1980 and the second – the principal focus of this chapter – in the same town five years later in 1985.

The first of these two meetings was substantially different in both size and purpose from the WCC. It was much smaller and more focused – but nonetheless it repeated the same general message. With the delegates attending the meeting as representatives of their governments, the conclusions and recommendations again called for more research. The final conference statement suggested that the potentially serious impact of climate change was sufficiently probable as to require an international commitment to a programme of cooperation in research to reduce uncertainties.[6] The delegates accordingly outlined five areas they believed required further attention in order to improve existing knowledge and reduce uncertainty. For the delegates at this meeting, their concerns about scientific uncertainty provided an effective and continuing mandate for demanding more money for more research.[7]

Bert Bolin, a Swedish professor of meteorology at Stockholm University and later the first chairman of the IPCC, left the Villach assessment in 1980 extremely frustrated. He felt that he and his fellow participants had not been able to comprehensively assess the issue due to a lack of time. The meeting had lasted only a week and upon departing several of the participants began discussing the prospect of yet another scientific assessment. The final Villach Conference statement merely reiterated the conclusions drawn a year earlier. For Bolin, simply calling on governments and international organizations to sponsor yet more research was not enough, as he felt that the time was ripe for some sort of real political action. He was confident that the negative effects of climate change were of such concern that he could not stand idly by without at least warning of the probable threat. Therefore, during these discussions, Bolin continually expressed his view that the assessment ought to be 'wider in scope, greater in depth and more international'.[8] He was looking at how he and his fellow delegates could maximize their influence, whilst raising the political profile of climate change.

Nothing more materialized out of these discussions until, in 1982, Tolba travelled to Stockholm to discuss with Bolin a UNEP project to carry out a more extensive assessment of climate change under the patronage of the World Climate Impacts Programme (WCIP). What transpired entailed a project generously funded by UNEP, with input from the WMO and an agreement for ICSU to publish the assessment.[9] The publication accordingly fell under the auspices of the Scientific Committee on Problems of the Environment (SCOPE), which was (and still is) a subsidiary body of ICSU established in 1969 as an international scientific non-governmental organization to identify and undertake analyses of emerging environmental issues that are caused by or impact upon humans and the environment.

Led by Bolin, the editors of the SCOPE report commissioned chapters evaluating the likely emissions of CO_2 into the atmosphere due to future energy demands and projecting its future atmospheric concentrations. In both cases, the conclusions cited a high degree of uncertainty. They suggested that it was not very meaningful to attempt precise projections of energy use beyond thirty to forty years, hence upper and lower bounds were estimated. The report also reviewed how the climate may change as it considered what General Circulation Models (GCMs) of the atmosphere were telling them about observed changes in the global atmospheric temperature record and projections for future trends. Overall, the report essentially mirrored the national and international assessments that had preceded it.[10] For instance, the predicted Equilibrium Climate Sensitivity (ECS) – the surface temperature warming projected in response to a doubling of the atmospheric concentration of CO_2 – according to the SCOPE report would be between 1.5°C and 5.5°C. Moreover, the executive summary warned that: 'The possible problem of a change in climate due to the emissions of greenhouse gases should be considered as one of today's most important long-term environmental problems.'[11]

Significantly, this report illustrated that, across the various assessments conducted in the 1980s, there was very little divergence in the model predictions of changes in global average temperature as a result of a doubling in the atmospheric concentration of CO_2. In one table, the report clearly showed that across a six year period (1979–85), the predicted climate sensitivity of 1.5–5.5°C was in the same range as all the other assessments reviewed. Jerome van der Sluijs et al. suggest that this observed stability was a product of the pragmatic value in enabling communication between the two social worlds of science and policy.[12] The relative stability operated as an 'anchoring device' in scientific assessments of climate change, functioning as a means of managing uncertainty by concealing the underlying scientific drift. This interpretative flexibility allowed climate modellers the scope to translate and adapt the meaning of the stable range. They thus argue that 'the dominant discourse, in postulating transfer of separately predetermined knowledge from science to policy, may conceal a more complex indeterminate process of mutual validation between these two worlds'.[13] As such, consensus, they argue, is more complex and multidimensional than a simple agreement based on shared beliefs and uniform interpretations.

In communicating the consensus through these scientific assessments of climate change, the authors of the SCOPE report were thus able to construct the problem and also develop and present solutions to climate change. Furthermore, it is notable that the SCOPE report dealt with scientific uncertainty in a considerably different way from the first Villach meeting in 1980. While the predicted temperature change of 1.5–5.5°C was the same in both assessments, the conclusions drawn from them were substantially different. The SCOPE assessment used the inherent scientific uncertainty as a justification for action rather than inaction. The flexibility in possible communications around a stable consensus allowed varying rhetorical justifications of uncertainty. For example, in the assessment summary the editors warned that while 'some of the uncertainties will still exist in the future, despite intensive research on the individual topics [the] *prevailing uncertainty does not mean that the problem can or should be dismissed* [emphasis in original]'. The summary went on to argue that 'it is necessary to examine the characteristics of these uncertainties and assess what can be said about future changes and to consider if and when some actions are needed in view of such possible changes'.[14] This statement repeated the need for money to reduce the uncertainties whilst utilizing uncertainty as a rationale for political action.

Villach: Setting the Agenda

The findings of the SCOPE assessment project were not published until 1986. However, they were made available in advance of publication to the delegates at the Villach Conference in 1985, where they served as the background scientific document facilitating discussions. At Villach in 1985, eighty-five delegates from twenty-nine developed and developing countries met for one week, from 9–15 October 1985, to discuss the role of CO_2 and other greenhouse gases (GHGs) on climate change. Alongside the efforts to provide a new worldwide consensus on the existing knowledge of climate change,[15] the delegates were loudly extolling the fact that climate change was a contemporary problem that mandated a solution there and then. Commenting on the changes predicted in the SCOPE assessment, Bolin has suggested that by the time of the Villach meeting in 1985, the 'threat of climate change became considerably more alarming'.[16]

The executive summary of the Villach report contained five recommendations: first, governments and international organizations should take into account the Villach findings when formulating their policies on social and economic development, environmental programmes and control of emissions; secondly, public information efforts publicizing climate change should be stepped up; thirdly, despite the remaining uncertainties, the scientists and policy makers should begin collaborations to explore the effectiveness of alternative policies; fourthly, continued and strengthened support of scientific research into the various aspects of the problem; fifthly, UNEP, the WMO and ICSU should establish a small task force on greenhouse gases to provide periodic assessment of the science, encourage research, advise governments on future actions and initiate consideration of a global convention.[17]

All five recommendations opened the door to a greater influence for scientific advice in the formulation of wide array of policies and represented a significant departure from the earlier 1980 Villach assessment. Franz has shown that this policy advocacy was not due to substantially new scientific findings; instead, she argues that it was because of the absence of domestic political constraints that the scientists who attended the conference did so in their personal capacities, not as representatives of their governments.[18] Moreover, the delegates were asked to 'shed their national policy perspectives' and address the global issues in as comprehensive a way as possible.[19] The freedom from national constraints at Villach in 1985 allowed the scientists the opportunity and space in which they felt able and compelled to express political views. Reflecting the increased sense of freedom and with it duty, the ICSU representative at Villach, James Dooge, an Irish politician and hydrology expert, captured the conference mood as he argued that: 'The two broad general lines of action, one concerned with science, the other with policy, are not antagonistic or incompatible … It is for this Conference to state how such an interchange can best take place.'[20]

Building upon Franz's arguments, I want to make two further points regarding the 1985 Villach meeting. First, the freedom from national constraints cited as a significant contributing factor to the conspicuous political advocacy at Villach was primarily facilitated by Tolba and UNEP. This is most evident in the decision to invite the delegates as individuals, reflecting Bolin's ambitions for a more international meeting – outlined as he left the earlier Villach meeting in 1980. Tolba, following up on those discussions, instigated the initial momentum towards the second Villach conference by commissioning Bolin to compile the assessment report ultimately published by SCOPE. Tolba's influence thus importantly highlights the role advocacy played in the emergence of the international political salience of climate change. This deliberate tactic, which had been previously deployed by Tolba and UNEP at other expert assessments on pollution in the Mediterranean,[21] was a calculated ploy to enable the scientists an opportunity to speak more freely. However, it also allowed the sponsoring agencies to select the experts they wanted to attend the meeting. In conjunction with Tolba and UNEP's stated ambitions favouring political intervention on climate change with a global convention, the delegates at Villach were encouraged to do likewise. In the course of the preparations for the meeting, Tolba persuaded the other agencies to internationalize the conference by inviting the eighty-nine delegates to attend the meeting in their own individual capacity, not as governmental representatives.

The second point regarding the 1985 Villach Conference concerns the way in which the delegates' recommendations performed considerable boundary work,[22] ensuring a future role for scientists in any decision making regarding climate change. 'Boundary work' describes the processes through which scientists demarcate science and nonscience. The social construction of these boundaries between science and nonscience are drawn and redrawn in different ways to best achieve the demarcation of the scientists' claims to authority or resources.[23] The 1985 Villach recommendations presented a specifically technocratic approach to policy making, defining an intrinsic role for scientists and expert knowledge in international policy making on climate change. The delegates presented uncertainty as manageable, but also deployed it as a rationale for action. Here, the boundary work performed was less to do with the protection of the autonomy of science from political interference and more to do with shaping the political landscape along technocratic lines. Whereas the Villach meeting in 1980 presented climate change as a possible issue, the 1985 Villach meeting utilized the interpretative flexibility around a relatively stable consensus to communicate the problem as probable and requiring urgent action.

The last of the recommendations articulated by the delegates of the second Villach conference was perhaps the most crucial for the subsequent establishment of the IPCC. It called upon UNEP, the WMO and ICSU to 'establish a small task force on greenhouse gases to provide periodic assessment of the science, encourage research, advise governments on future actions, and initiate consideration of a global convention.'[24] This was a conspicuous example of Tolba's influence, as he actively sought out a technocratic approach to international policy making on climate change. In addressing the delegates, he notably advocated the establishment of an 'international coordinating committee on greenhouse gases … [that could] issue statements to governments, international organizations and the public at large regarding the need for particular actions and the options open in response to a potential warming of the global climate'.[25]

So shortly after the conclusion of the second Villach meeting in line with this recommendation, the Advisory Group on Greenhouse Gases (AGGG) was established by the WMO, UNEP and ICSU. The AGGG was convened as a standing advisory panel comprised of experts in a position to provide the best available information to governments and international organizations. Each sponsoring organization appointed their choice of two scientists to the group. Ken Hare, Bert Bolin, Georgy Golitsyn, Gordon Goodman, Mohammed Kassas, Syukuro Manabe and Gilbert White made up the panel, whilst the Villach Conference Chairman Jim Bruce served as first secretary to the group.

The plans and recommendations for the AGGG closely resembled the Coordinating Committee on the Ozone Layer (CCOL). The CCOL had served a similar role to the one envisaged by Tolba for the AGGG in the climate regime. The CCOL was established as an international scientific assessment panel. Led by Professor Robert Watson, the director of the National Aeronautics and Space Administration (NASA) 'Mission to Planet Earth' project, CCOL reported directly to UNEP providing periodic scientific assessments of the ozone problem.[26] Meeting eight times between 1977 and 1986, it reported its findings in the *Ozone Layer Bulletin*.[27] Comprising scientists with expertise in atmospheric chemistry and modelling, the group met to report on the latest research and prepare the ground for political decision making. Watson has suggested that the scientific panel effectively had three tasks: first, to assess the science of the ozone layer; secondly, to report on the implications of changing ozone-UV radiation for human health; and, thirdly, to evaluate the various mitigation techniques and technologies.[28]

Tolba wished to make the AGGG the principal source of legitimation in climate negotiations just as the CCOL was in the ozone negotiations. Having witnessed firsthand how science and scientific advice could be used to legitimize political negotiating positions in the ozone negotiations, this advisory panel seemed a natural choice for Tolba. Moreover, in his personal account of the history of the ozone negotiations, Richard Benedick, chief negotiator for the U.S. in the ozone negotiations, emphasizes the influential role played by science, scientists and scientific assessments. He concedes that though science alone was not enough to bring about the success in the ozone negotiations, the scientific consensus was a key tool.[29] Thus, the analogous organization of the AGGG – reminiscent of the CCOL – both reflected the technocratic ambitions of Tolba and mirrored the international policy making during the ozone negotiations.

Tolba's vision was for a scientific assessment panel along the lines of that for the ozone layer, to provide input into what he hoped would be imminent negotiations of a global climate convention.[30] According to James Losey, a senior staff officer with the U.S. Environmental Protection Agency's (EPA) International Activities Office (1980–87), for some environmentalists at UNEP and the EPA, the ozone issue was nested within the larger and more complex climate issue, and an agreement on the former could be used as a springboard for dealing with the latter.[31]

The U.S. Response: Department of Energy and EPA Converge

As outlined at the beginning of this chapter, following the 1985 Villach Conference, Tolba wrote to the U.S. Secretary of State George Schultz in 1986 'urging the U.S. to take appropriate policy actions' on climate change.[32] In the ensuing discussions within and between U.S. government agencies, the proposal for a new *intergovernmental* assessment mechanism emerged. In these meetings, U.S. officials were responding to the Villach framing of climate change as an urgent threat demanding immediate action. However, the highly technocratic solutions proposed by the Villach scientists and administrators were unpalatable to various U.S. officials. Therefore, to overcome this framing of climate change, U.S. officials began developing an intergovernmental framework. They were concerned about out-of-control scientists – or 'loose cannons'[33] – setting the agenda. Advocating a new intergovernmental assessment instead, they hoped to position their own experts in prominent positions to constrain the emergent advocacy from this group of scientists and in particular UNEP officials.

Upon receiving Tolba's letter, Schultz forwarded it to the U.S. National Climate Program Office (NCPO) seeking a response. Its director, Alan Hecht, took Tolba's letter to the interagency National Climate Program (NCP) Policy Board that debated the content of the letter.[34] During these discussions, the U.S. Department of Energy (DoE) raised the prospect of another different type of scientific assessment on climate change.

In contrast to the earlier assessments in Villach in 1980 and 1985, the NCP Policy Board ultimately came to the conclusion that the new assessment process should be intergovernmental. There were two principal reasons for this decision. First, the representatives of the U.S. DoE were dissatisfied that the Villach meeting had been beyond the oversight of government, consistent with Tolba's intentions. This concern was further exacerbated by the establishment of the AGGG, reporting to the WMO, UNEP and ICSU because it was similarly beyond governmental control. Secondly, there was substantive disagreement with the report's conclusions because, for the majority of the U.S. federal agencies represented on the NCP Policy Board, the economic argument for a global convention had not been made strongly enough to support costly actions.[35] The DoE argued that the report was a UNEP document and that if governments were indeed to sign up to a convention, as suggested, then they themselves would have to be responsible for the production of reports, not groups of scientists.[36]

The stance adopted by the DoE was noticeably framed by a climate assessment it had commissioned earlier in 1983, entitled *Changing Climate*.[37] This report, produced by the National Academy of Sciences (NAS), was conducted with governmental oversight and was therefore viewed as having greater legitimacy than the 1985 Villach report. Moreover, *Changing Climate*, chaired by William A. Nierenberg, director of the Scripps Institute of Oceanography, found in the 'CO_2 issue reason for concern, but not panic'. This significantly contradicted the 1985 Villach report, which overtly advocated movement towards a global convention and emissions controls. Therefore, the DoE was in possession of a report it viewed as more credible, which also suggested that any action was premature, instead recommending '[more] research, monitoring, vigilance, and an open mind'.[38]

The findings of the NAS report framed the climate change issue for the DoE very much in economic rather than scientific terms. Naomi Oreskes et al. argue that this report was an orchestrated attempt by its chairman William Nierenberg to reframe the climate change issue, 'challenging the emerging consensus view on global warming'.[39] Oreskes and Eric Conway identify Nierenberg, in particular, as one of a triumvirate of scientists whom they call 'merchants of doubt' – merchants with extensive links to the George C. Marshall Institute, an American conservative thinktank, responsible for questioning the underlying scientific consensus on climate change for politically motivated reasons.[40]

Through a combination of the contrasting findings and questions over the legitimacy of the 1985 Villach assessment, the DoE proposed a new intergovernmental assessment, which would bring the issue back in to the hands of governments. Indeed, the executive summary to the *Changing Climate* report stated that:

> We do not believe ... that the evidence at hand about CO_2-induced climate change would support steps to change current fuel-use patterns away from fossil fuels. Such steps may be necessary or desirable at some time in the future, and we should certainly think carefully about costs and benefits of such steps; but the very near future would be better spent improving our knowledge ... than in changing fuel mix or use.[41]

Two points in this statement appear to have resonated with the DoE representatives when discussing the 1985 Villach report during the NCP Policy Board meeting. First, the economic argument for changes to be made to CO_2 emissions had not yet been made. Secondly, time and resources would be better spent on improving knowledge of processes contributing to the creation of GHGs. This second point is reflected in Alan Hecht and Dennis Tirpaks' claims that the DoE representatives saw the creation of a new intergovernmental assessment mechanism as a means of buying time before engaging the policy implications.[42] An intergovernmental assessment offered the DoE an opportunity to intervene against a move towards a global convention, whilst crucially making the U.S. government appear to be taking a proactive approach.

The DoE were not the only U.S. governmental agency that wanted a new intergovernmental assessment. The EPA, for different reasons, agreed on the need for such an assessment. In contrast to the hostility with which the DoE received the 1985 Villach findings, the EPA enthusiastically endorsed them. Much like the DoE, the EPA also commissioned a report in 1983 entitled *Can We Delay a Greenhouse Warming?* This report evaluated the scale and dimension of the climate problem.[43] In contrast to *Changing Climate*, it suggested that there was a pressing need for action to combat the threat posed by climate change. The authors of the EPA report found that 'global greenhouse warming was neither trivial nor just a long-term problem'.[44] The 1985 Villach findings therefore reinforced the conclusions of the EPA report and validated the recommendation for a global convention. So, the EPA's decision to back a call for a new intergovernmental assessment was not intended as blocking political progress towards a global convention. Instead, EPA officials saw in the proposal the next logical step towards their desired ends of a global convention.[45] As it would involve wider international participation in the processes, EPA officials, like those from the DoE, considered the intergovernmental structure important, but for entirely different reasons. Despite this divergent reasoning, both U.S. authorities' demands strengthened calls for a new intergovernmental assessment mechanism.

Proposing a New Intergovernmental Assessment Mechanism

Following on from the domestic interagency discussions, the State Department, the NCPO and the head of the National Weather Service ensured that the issue of climate change was placed on the agenda of the upcoming World Meteorological Congress. This congress gathers the WMO national representatives every four years 'to determine general policies for the fulfilment of the purposes of the Organisation'.[46] To Tolba's distress, at the meeting in May 1987, many governments agreed that the AGGG was inadequate as a basis, going forward, for informing governments.[47] Ultimately, 'the gulf between science and policy could not have been wider'.[48] The widespread dissatisfaction with the AGGG offered the U.S. delegation an opportunity to raise the prospect of a new intergovernmental assessment. Crucially, their proposal appeared to create the kind of close connection between scientists and policy makers and financial backing that the delegations considered to be absent in the AGGG design.

The U.S. delegation arrived at the meeting with a clear intention of supporting a new intergovernmental assessment mechanism. However, at this point, they were reluctant to be too forceful in proposing a resolution. In fact, the call for an intergovernmental assessment mechanism came from Gladys Ramothwa, the principal delegate of Botswana. She raised the issue with an 'impassioned call to the Congress' for the WMO to produce a document that would answer her government's questions relating to climate change.[49] She was supported by several developing countries, which also called on the WMO to produce a definitive statement to brief their governments.[50] This led to a lengthy debate that culminated in a Congress resolution requesting the WMO

Executive Council to 'arrange appropriate mechanisms to undertake further development of scientific and other aspects of greenhouse gases'.[51] Sensing an opportunity, the U.S. delegation were able to capitalize on Ramothwa's concerns by making their argument for an intergovernmental assessment mechanism. While this resolution only pointed the way for the mechanism, the debate produced a clear vision of what it should entail. Thus, the U.S. delegation achieved their aims without exerting any political pressure.

The thirty-ninth session of the WMO Executive Council assembled immediately after the conclusion of the Congress to implement its decisions. At this meeting, Richard Hallgren, head of the U.S. Weather Bureau and the U.S. permanent representative to the WMO, introduced the concept of an intergovernmental mechanism.[52] He outlined a mechanism that would see an intergovernmental group of expert scientists nominated by their government produce a consensus assessment of what was known and what might be done about climate change. The assessment would result in a document drawn up by approximately fifteen to twenty experts from the countries involved in climate change research, including the United States, the United Kingdom, France, West Germany and Australia.[53] At the conclusion of this meeting, the WMO Deputy Secretary-General Jim Bruce travelled to the meeting of UNEP's Governing Council. Here Bruce requested that UNEP contribute to an 'ad hoc [emphasis in original] intergovernmental mechanism to carry out internationally co-ordinated scientific assessments of the magnitude, timing, and potential impact of climate change'.[54] UNEP's Governing Council responded positively and urged Tolba to work in cooperation with Bruce. Together, they began to negotiate the terms of reference, the modalities, structure and timing of the first meeting.

In addition to UNEP and the WMO, the United States still had a large part to play thanks to a large extent to the actions of Bo Döös, a Swedish meteorologist and former WCP director who in 1987 was working in the NCP office.[55] Döös began discussing the proposed assessment mechanism with his friend and fellow meteorologist Eugene Bierly, the director of the Atmospheric Sciences Division at the National Science Foundation (NSF). Their discussions centred on how they might utilize structures deployed in the WMO/ICSU Global Atmospheric Research Program (GARP) as a template for the new 'mechanism', since its administration had entailed intergovernmental coordination too.[56] Döös had been intimately involved with the planning and implementation of the GARP Global Weather Experiment in 1978–79 as the Director of the Joint WMO/ICSU Planning Staff. The pair saw several similarities between the proposed mechanism and the Global Weather Experiment, both in terms of the challenges and opportunities they presented. They both relied on planned activities with a solid scientific base maintaining flexibility and independence within the intergovernmental structures.[57]

Following these discussions, Döös returned to the NCPO, where he transmitted the content of the conversation he had with Bierly to his boss Alan Hecht. Hecht agreed with the core arguments and, owing to Döös' experiences within GARP, asked him to draft a proposal. Evidently taking inspiration from his experience on the GARP Global Weather Experiment Intergovernmental Panel, Döös accordingly outlined an assessment mechanism he felt could overcome the inherent difficulties in integrating a range of views and promoting international cooperation.

Subsequent discussions between Döös, Hecht and the government agencies in the NCP Policy Board, principally involving the EPA, the State Department and the DoE (covering foreign policy, energy security and environmental protection), resulted in a six-page document outlining the activities, initial responsibilities and composition of the ad hoc intergovernmental mechanism.[58] The draft proposal was circulated to the various interested agencies that agreed to the principles outlined by Döös.[59] Hecht and Döös outlined a U.S. strategy on climate change that involved taking a lead on the issue, balancing environmental protection, economic growth and energy policy.[60] Although the decision to include or exclude points in this policy were taken by Hecht and Döös, the draft proposal broadly reflected U.S. government preferences.

The Hecht–Döös draft proposal introduced the issue by way of citing the resolutions passed by the governing bodies of the WMO and UNEP earlier in 1987. It went on to stress that there was a 'growing international concern' over the climate change issue that had been compounded by the several national and international assessments, in particular the 'major international assessment' conducted in Villach in 1985.[61] The proposal went on to conclude that the decisions of the WMO and UNEP governing bodies 'reflect the need for an *orderly process* to ensure that research, monitoring and impact assessment studies proceed together, and that internationally agreed assessments should be prerequisites for legal or regulatory activities [emphasis in original]'.[62] This document therefore recommended that, in light of the 'wide spectrum of problem areas', the panel should coordinate its activities into a dual stream.[63] Assessment of a scientific base should lead to an assessment of the socioeconomic effects and societal responses to climate change. There was an overriding sense that this new assessment would provide the answers to any questions policy makers might have. As such, the proposal replicated the working method of the DoE report on climate change rather than that of the international assessment at Villach in 1985.

By the end of August 1987, the draft had been circulated among the various agencies of the NCP Policy Board and met with general approval. Döös had anticipated that the State Department would take action to make the draft a reality. Surprised at the days of inaction, Döös decided to take matters into his own hands. On a planned holiday back to Europe, he decided to visit the WMO headquarters in Geneva. He informally discussed the draft proposal with the people familiar with the proposed intergovernmental mechanism and, in particular, Jim Bruce. Bruce had by now been given the job of negotiating the terms of reference with his counterpart at UNEP, Tolba.[64] Döös began by updating Bruce on what was going on in Washington DC: that there was a draft proposal and that this draft had been reviewed by the relevant agencies. According to Döös, upon reviewing the draft, Bruce asked him whether there would be any obstacle to the draft being presented as a WMO proposal. Döös saw none and instead felt that the U.S. agencies could well be happy, which they were, as they might get what they wanted almost without having to ask for it.[65] Crucially, this way of proceeding made any decisions taken over the scope, design and planned activities appear to have been taken multilaterally rather than as a result of U.S. influence.

A couple of months later, on 3 November 1987, Bruce faxed a slightly revised version of the Döös–Hecht draft proposal to Jim Rasmussen, a National Oceanic and Atmospheric Association meteorologist. Rasmussen forwarded this fax to Alan Hecht, Eugene Bierly and Richard Hallgren a few days later. Upon reviewing the proposal, Döös assumed that Bruce wanted to be assured that the revised version was still acceptable to the U.S. agencies.[66] As the largest financial contributor to the WMO, the United States influenced it significantly, but often behind closed doors. U.S. delegations often utilized soft power diplomacy to ensure favourable outcomes. In this instance, U.S. negotiators drew on an extensive network of informal professional connections to translate their own proposal into an apparently less partisan UNEP–WMO proposal resolution.

The back and forth continued, as another draft proposal revised the WMO document just a week later. This draft presented the Global

Weather Experiment Intergovernmental Panel as a useful model for the intergovernmental mechanism. This resulted in the eventual adoption of the assessment mechanism being named the Intergovernmental Panel on Climate Change. These exchanges across the Atlantic finally culminated in the June 1988 decision of the fortieth session of the WMO Executive Council to establish the IPCC with the twin objectives of:

(i) assessing the scientific information that is related to the various components of the climate change issue, such as emissions of major greenhouse gases and modification of the Earth's radiation balance resulting therefrom, and that needed to enable the environmental and socioeconomic consequences of climate change to be evaluated; and

(ii) formulating realistic response strategies for the management of the climate change issue.[67]

This resolution was neither the end of the discussions over what the IPCC would, should and could do, nor was it merely a reflection of the draft proposals described above. Rather, decisions taken in the interim, and even after the IPCC's formation, continued to shape and reshape the new mechanism.

Conclusion

This chapter has detailed the contested and often intertwined scientific and political agendas of climate change between 1985 and 1988. I have argued that the IPCC's specific institutional arrangements are a product of this tussle. The 1985 Villach meeting was a crucial event in setting the agenda during the emergence of climate change as a political issue. UNEP Executive Director Tolba, in particular, used this conference as an opportunity to make the case for a global convention on climate change. Alongside the WMO and ICSU, he facilitated a receptive setting in which scientists articulated overtly political recommendations. These recommendations were influenced by the coincident ozone negotiations and led the Villach delegates towards a particularly technocratic approach to policy making. As a result, the Villach (1985) delegates also advocated a highly 'scientized' consensus-based approach to decision making geared towards narrowing the terrain for political debate – focusing on their objective of a global climate convention.

As the findings of the second Villach Conference were conveyed to the U.S. Secretary of State George Schultz, the spotlight of attention amongst U.S. policy makers in the DoE, the EPA and the State Department was on climate change. While its conclusions did not drive the formation of the IPCC, fears that it would led U.S. officials to act to ensure that the result was favourable to them. The discussions prompted by the 1985 Villach Conference centred on the key issue of a mooted global convention on climate change. In the U.S. NCP Policy Board, these negotiations culminated in resolutions passed by the executive bodies of the WMO and UNEP agreeing to jointly sponsor 'an ad hoc intergovernmental mechanism that could carry out assessments' that would be 'objective, balanced and internationally co-ordinated'.[68]

The NCP Policy Board reached agreement on the proposed intergovernmental assessment mechanism through compromise despite divergent aspirations for the IPCC. On the one hand, the DoE proposed such a mechanism as a means of controlling the trajectory towards a global convention and buying time before seriously engaging with the issue. On the other hand, the EPA saw it as a useful stepping stone towards achieving a global convention by widening the base of countries aware of and interested in the issue. Thus, the assessment mechanism as proposed by the U.S. government was at once a barrier as well as an aid to introducing a global convention on climate change. The idea for a new intergovernmental assessment mechanism was translated to the international stage via the WMO Congress in May 1987, after which the United States retained its influence on the subsequent proposals and counterproposals.

Rather than campaigning overtly to get what they wanted at the WMO meetings, the U.S. delegations utilized soft power diplomacy to ensure favourable outcomes. At the WMO Congress, for instance, U.S. negotiators drew on an extensive network of informal professional connections in order to translate their own proposal into a UNEP–WMO proposal. While Tolba and the Villach delegates did set the political agenda for climate change in the late 1980s, U.S. officials recaptured the issue and successfully moulded the proposal towards their own preferences. Thus, the eventual agreement for an intergovernmental assessment mechanism created a considerably different advisory mechanism from the one envisaged by Tolba in 1985, which was a small technocratic panel reporting directly to UNEP and the WMO. Whilst the U.S.-backed plan did propose greater scientific involvement, it was strictly on the terms outlined by governments. This was intended to be both a means of reducing the influence of UNEP and the scientists and administrators at UNEP, and a way of ensuring greater governmental involvement.

This chapter has shown more broadly some of the limitations of international organizations and scientists when they operate outside of the constraints of governments in trying to shape the political discourse and substantive outcomes. Furthermore, this case study has emphasized just how important the U.S. government was at the time in amplifying or hindering the political momentum generated by environmental international organizations. To a significant degree, the IPCC's establishment was driven by U.S. support, albeit subtle, masked and enacted through intermediaries such as Bo Döös. Accordingly, the U.S. government can be seen as both a powerful laggard and constructive leader on international environmental issues.[69]

David Hirst is a policy analyst at the House of Commons Library, United Kingdom.

Notes

1. Alan Hecht, Interview with David Hirst, 19 July 2012.
2. Shardul Agrawala, 'Context and Early Origins of the IPCC', *Climatic Change* 39(4) (1994): 605–20, 609.
3. Fred Pearce, 'Histories: The Week the Climate Changed', *New Scientist*, 15 October 2005, 52–54, 52.
4. Wendy E. Franz, 'The Development of an International Agenda for Climate Change' Discussion Paper E-97-07, Kennedy School of Government, Harvard University (August 1997): 1–34.
5. Agrawala, 'Context and Early Origins of the IPCC'.
6. World Climate Programme, On the Assessment of the Role of CO_2 on Climate Variations and their Impact. Report of a WMO/UNEP/ICSU Meeting of Experts in Villach, Austria, November 1980, Geneva, WMO, WMO Library and Archive, Geneva.
7. Robert M. White, 'World Climate Conference: Geneva 1979', *WMO Bulletin* 28(3) (1979): 177–78.
8. Bert Bolin, *A History of the Science and Politics of Climate Change* (Cambridge: Cambridge University Press, 2007), 35.
9. Ibid., 35–36.
10. SCOPE, *The Greenhouse Effect, Climatic Change and Ecosystems*, Bert Bolin, Bo Döös, R. A. Warwick and Jill Jäger (eds) (New York: Wiley-Blackwell, 1986), http://www.scopenvironment.org/downloadpubs/scope29/contents.html, accessed 25 May 2016.
11. Ibid.

12. Jerome van der Sluijs, Josee van Eijndhoven, Simon Shackley and Brian Wynne, 'Anchoring Devices in Science for Policy: The Case of Consensus around Climate Sensitivity', *Social Studies of Science* 28(2) (1998): 291–323.
13. Ibid., 315.
14. SCOPE, *The Greenhouse Effect*.
15. James P. Bruce, 'The World Climate Programme: Achievements and Challenges', in *Climate Change: Science Impacts and Policy - Proceedings of the Second World Climate Conference*, J. Jaeger and H.L. Ferguson (eds) (Cambridge: Cambridge University Press, 1991), 151–54, 152.
16. Bolin, *A History of the Science and Politics of Climate Change*, 37.
17. Statement by the WMO/UNEP/ICSU International Conference on the assessment of the role of carbon dioxide and of other greenhouse gases in climate variations and associated impacts in SCOPE, *The Greenhouse Effect*.
18. Franz, 'The Development of an International Agenda for Climate Change', 12.
19. Jim Bruce, Interview with W.E Franz and S. Agrawala. Telephone, April 1997, quoted in Franz, 'The Development of an International Agenda for Climate Change', 13.
20. James Dooge as quoted in WMO, Report of the International Conference on the Assessment of the Role of Carbon Dioxide and of Other Greenhouse Gases in Climate Variations and Associated Impacts, Villach, Austria: 1986, WMO Library and Archive, Geneva.
21. Peter M. Haas, 'Do Regimes Matter? Epistemic Communities and Mediterranean Pollution Control', *International Organization* 43(3) (1989): 377–403.
22. Thomas F. Gieryn, 'Boundary-Work and the Demarcation of Science from Non-science: Strains and Interests in Professional Ideologies of Scientists', *American Sociological Review* 48(6) (1983): 781–95.
23. Ibid.
24. Statement by the WMO/UNEP/ICSU International Conference on the assessment of the role of carbon dioxide and of other greenhouse gases in climate variations and associated impacts in SCOPE, *The Greenhouse Effect*.
25. WMO, Report of the International Conference on the Assessment of the Role of Carbon Dioxide and of Other Greenhouse Gases in Climate Variations and Associated Impacts, 1986.
26. Karen T. Litfin, *Ozone Discourses: Science and Politics in Global Environmental Cooperation* (New York: Columbia University Press, 1994), 57.
27. Peter M. Haas, 'Banning Chlorofluorocarbons: Epistemic Community Efforts to Protect Stratospheric Ozone', *International Organization* 46(1) (1992): 187–224, 201.
28. Robert Watson, Interview with David Hirst, 13 February 2012.
29. Richard E. Benedick, *Ozone Diplomacy: New Directions in Safeguarding the Planet* (Cambridge, MA: Harvard University Press, 1998).
30. Watson, Interview with David Hirst, 13 February 2012.
31. James Losey, Interview with Karen Litfin, 17 September 1990 in Litfin, *Ozone Discourses*, 68.

32. Hecht, Interview with David Hirst.

33. Jill Jaeger, Interview with David Hirst, 20 February 2012.

34. Hecht, Interview with David Hirst.

35. Ibid.

36. Ibid.

37. National Academy of Sciences, *Changing Climate, Report of the Carbon Dioxide Assessment Committee* (Washington DC: National Academy Press, 1983).

38. Ibid., 61.

39. Naomi Oreskes, Erik M. Conway and Matthew Shindell, 'From Chicken Little to Dr. Pangloss: William Nierenberg, Global Warming, and the Social Deconstruction of Scientific Knowledge', *Historical Studies in the Natural Sciences* 38(1) (2008): 109–52, 113.

40. Naomi Oreskes and Erik M. Conway, *Merchants of Doubt: How a Handful of Scientists Obscured the Truth on Issues from Tobacco Smoke to Global Warming* (London: Bloomsbury, 2010).

41. National Academy of Sciences, *Changing Climate*, 4.

42. Alan D. Hecht and Dennis Tirpak, 'Framework Agreement on Climate Change: A Scientific and Policy History', *Climatic Change* 29(4) (1995): 371–402, 381.

43. Stephen Seidel and Dale Keyes, *Can We Delay a Greenhouse Warming? The Effectiveness and Feasibility of Options to Slow a Build-Up of Carbon Dioxide in the Atmosphere* (Washington DC: Environmental Protection Agency, 1983).

44. Ibid.

45. It is unclear exactly why the DoE ignored these conclusions, but it seems likely the divergent priorities of the two organizations – energy vs. environment – go a long way towards explaining this.

46. WMO, 'World Meteorological Congress', www.wmo.int/pages/governance/congress/index_en.html, accessed 25 May 2016.

47. John Zillman, Interview with David Hirst, 26 March 2012.

48. Agrawala, 'Context and Early Origins of the IPCC', 613.

49. John W. Zillman, 'Australian Participation in the Work of the Intergovernmental Panel on Climate Change 1988–2001: Part I', *Energy and Environment* 19(1) (2008): 21–42, 32.

50. Zillman, Interview with David Hirst.

51. WMO, 'Intergovernmental Panel on Climate Change: First Session, Geneva, November 1988', *WMO Bulletin* 38(2) (1989): 113–14.

52. Jim Bruce, Email correspondence with David Hirst, 6 November 2012.

53. Zillman, Interview with David Hirst.

54. UNEP, 'Resolution GC14/20: Global Climate Change', in *UNEP Report of the Governing Council on the Work of its 14th Session* (New York: United Nations, 1987) UN General Assembly 42nd Session, Supplement No. 25 (A/42/25), 72.

55. Hecht, Interview with David Hirst.

56. Eugene Bierly, Interview with David Hirst, 1 August 2012.

57. Bo Döös and Eugene Bierly, 'Bierly – Döös Paper. Items to Be Considered', 10 May 2009, Personal Papers of Eugene Bierly (hereinafter Bierly Papers), 11.

58. Bo Döös and Alan Hecht, 'Intergovernmental Panel on Climate Change' (August 1987), Bierly Papers (Doc A.), 1–6.
59. Döös and Bierly, 'Bierly – Döös Paper', 11.
60. Alan Hecht and Bo Döös, 'Climate Change, Economic Growth and Energy Policy: A Recommended Strategy for the Coming Decades', *Climate Change* 13 (1988): 1–3.
61. Döös and Hecht, 'Intergovernmental Panel on Climate Change', 1–2.
62. Ibid., 1–2.
63. Ibid., 2–4.
64. James P. Bruce, Interview with David Hirst, 1 February 2012.
65. Döös and Bierly, 'Bierly – Döös Paper', 12.
66. James P. Bruce, 'Fax from J. P. Bruce to J. Rasmussen: Intergovernmental Mechanism on Climate Change', 16 November 1987, Bierly Papers.
67. WMO, 'Intergovernmental Panel on Climate Change' Resolution (EC-XL), June 1988, WMO Library and Archive, Geneva.
68. WMO, 'WMO Executive Council: Twenty-Ninth Session, Geneva, June 1987', *WMO Bulletin* 36(4) (1987): 307.
69. Pamela S. Chasek, 'U.S. Policy in the UN Environmental Arena: Powerful Laggard or Constructive Leader?', *International Environmental Agreements: Politics, Law and Economics* 7(4) (2007): 363–387, 364.

Bibliography

Benedick, Richard E., *Ozone Diplomacy: New Directions in Safeguarding the Planet* (Cambridge, MA: Harvard University Press, 1998).
Bierly, Eugene, Interview with David Hirst, 1 August 2012.
Bolin, Bert, *A History of the Science and Politics of Climate Change* (Cambridge: Cambridge University Press, 2007).
Bruce, James P., 'Fax from J. P. Bruce to J. Rasmussen: Intergovernmental Mechanism on Climate Change', 16 November 1987, Bierly Papers.
———. 'The World Climate Programme: Achievements and Challenges', in *Climate Change: Science Impacts and Policy – Proceedings of the Second World Climate Conference*, J. Jaeger and H. L. Ferguson (eds) (Cambridge: Cambridge University Press, 1991): 151–154.
———. Interview with David Hirst, 1 February 2012.
Bruce, James P., Interview with W.E Franz and S. Agrawala. Telephone, April 1997, quoted in Wendy E. Franz, 'The Development of an International Agenda for Climate Change' Discussion Paper E-97-07, Kennedy School of Government, Harvard University (August 1997): 1–34, 13.
———. Email correspondence with David Hirst, 6 November 2012.
Chasek, Pamela S., 'U.S. Policy in the UN Environmental Arena: Powerful Laggard or Constructive Leader?', *International Environmental Agreements: Politics, Law and Economics* 7(4) (2007): 363–387.
Dooge, James, quoted in WMO, Report of the International Conference on the Assessment of the Role of Carbon Dioxide and of Other Greenhouse Gases in

Climate Variations and Associated Impacts, Villach, Austria: 1986, WMO Library and Archive, Geneva.

Döös, Bo and Eugene Bierly, 'Bierly – Döös Paper. Items to Be Considered', 10 May 2009, Personal Papers of Eugene Bierly (Bierly Papers).

Döös, Bo and Alan Hecht, 'Intergovernmental Panel on Climate Change', August 1987, Bierly Papers (Doc A.), 1–6.

Franz, Wendy E., 'The Development of an International Agenda for Climate Change', Discussion Paper E-97-07, Kennedy School of Government, Harvard University (August 1997): 1–34.

Gieryn, Thomas F., 'Boundary-Work and the Demarcation of Science from Non-science: Strains and Interests in Professional Ideologies of Scientists', *American Sociological Review* 48(6) (1983): 781–95.

Haas, Peter M., 'Do Regimes Matter? Epistemic Communities and Mediterranean Pollution Control', *International Organization* 43(3) (1989): 377–403.

———. 'Banning Chlorofluorocarbons: Epistemic Community Efforts to Protect Stratospheric Ozone', *International Organization* 46(1) (1992): 187–224.

Hecht, Alan, Interview with David Hirst, 19 July 2012, Skype.

Hecht, Alan and Bo Döös, 'Climate Change, Economic Growth and Energy Policy: A Recommended Strategy for the Coming Decades', *Climate Change* 13 (1988): 1–3.

Hecht, Alan and Dennis Tirpak, 'Framework Agreement on Climate Change: A Scientific and Policy History', *Climatic Change* 29(4) (1995): 371–402.

Jaeger, Jill, Interview with David Hirst, 20 February 2012, Skype.

Litfin, Karen T., *Ozone Discourses: Science and Politics in Global Environmental Cooperation* (New York: Columbia University Press, 1994).

Losey, James, Interview with Karen Litfin, 17 September 1990, in Karen Litfin, *Ozone Discourses: Science and Politics in Global Environmental Cooperation* (New York: Columbia University Press, 1994), 68.

National Academy of Sciences, *Changing Climate, Report of the Carbon Dioxide Assessment Committee* (Washington DC: National Academy Press, 1983).

Oreskes, Naomi and Erik M. Conway, *Merchants of Doubt: How a Handful of Scientists Obscured the Truth on Issues from Tobacco Smoke to Global Warming* (London: Bloomsbury, 2010).

Oreskes, Naomi, Erik M. Conway and Matthew Shindell, 'From Chicken Little to Dr. Pangloss: William Nierenberg, Global Warming, and the Social Deconstruction of Scientific Knowledge', *Historical Studies in the Natural Sciences* 38(1) (2008): 109–52.

Pearce, Fred, 'Histories: The Week the Climate Changed', *New Scientist*, 15 October 2005, 52–54.

SCOPE, *The Greenhouse Effect, Climatic Change and Ecosystems*, Bert Bolin, Bo Döös, R. A. Warwick and Jill Jäger (eds) (New York: Wiley-Blackwell, 1986), www.scopenvironment.org/downloadpubs/scope29/contents.html, accessed 4 April 2016.

Seidel, Stephen and Dale Keyes, *Can We Delay a Greenhouse Warming? The Effectiveness and Feasibility of Options to Slow a Build-up of Carbon Dioxide in the Atmosphere* (Washington DC: Environmental Protection Agency, 1983).

Shardul Agrawala, 'Context and Early Origins of the IPCC', *Climatic Change* 39(4) (1994): 605–20.

Sluijs, Jerome van der, Josee van Eijndhoven, Simon Shackley and Brian Wynne, 'Anchoring Devices in Science for Policy: The Case of Consensus around Climate Sensitivity', *Social Studies of Science* 28(2) (1998): 291–323.

UNEP, 'Resolution GC14/20: Global Climate Change', in *UNEP Report of the Governing Council on the Work of its 14th Session* (New York: United Nations, 1987), UN General Assembly 42nd Session, Supplement No. 25 (A/42/25).

Watson, Robert, Interview with David Hirst, 13 February 2012.

White, Robert M., 'World Climate Conference: Geneva 1979', *WMO Bulletin* 28(3) (1979): 177–78.

WMO, Report of the International Conference on the Assessment of the Role of Carbon Dioxide and of Other Greenhouse Gases in Climate Variations and Associated Impacts, 1986.

——. 'WMO Executive Council: Twenty-Ninth Session, Geneva, June 1987', *WMO Bulletin* 36(4) (1987): 307.

——. 'Intergovernmental Panel on Climate Change' Resolution (EC-XL), June 1988, WMO Library and Archive, Geneva.

——. 'Intergovernmental Panel on Climate Change: First Session, Geneva, November 1988', *WMO Bulletin* 38(2) (1989): 113–14.

——. 'World Meteorological Congress', www.wmo.int/pages/governance/congress/index_en.html, accessed 25 May 2016.

WMO/UNEP/ICSU International Conference, 'Statement on the Assessment of the Role of Carbon Dioxide and of Other Greenhouse Gases in Climate Variations and Associated Impacts (Villach, Austria, October 1985)', in SCOPE, *The Greenhouse Effect, Climatic Change and Ecosystems*, Bert Bolin, Bo Döös, R.A. Warwick and Jill Jäger (eds) (New York: Wiley-Blackwell, 1986), http://www.scopenvironment.org/downloadpubs/scope29/statement.html, accessed 4 April 2016.

World Climate Programme, On the Assessment of the Role of CO_2 on Climate Variations and their Impact. Report of a WMO/UNEP/ICSU Meeting of Experts in Villach, Austria, November 1980, Geneva, WMO, WMO Library and Archive, Geneva.

Zillman, John W., 'Australian Participation in the Work of the Intergovernmental Panel on Climate Change 1988–2001: Part I', *Energy and Environment* 19(1) (2008): 21–42.

——. Interview with David Hirst, 26 March 2012.

Conclusion

Setting Agendas, Building Institutions and Shaping Binding International Commitments

Wolfram Kaiser and Jan-Henrik Meyer

International organizations (IOs) have been quite successful in addressing environmental degradation in the global twentieth century, especially since the 1972 Stockholm Conference. The 1987 Montreal Protocol constitutes an excellent example of how under certain conditions their work has had a strong impact by establishing binding international commitments. In this case, it took the form of a decision to phase out chlorofluorocarbons (CFCs) and other harmful chemical substances that were depleting the earth's protective ozone layer – a process that notably led to increased incidence of skin cancer and (as most scientists argue) has contributed to global warming.[1] In 2014, however, the latest scientific assessment by the World Meteorological Organization (WMO) and the United Nations Environmental Programme (UNEP) found that this depletion of the ozone layer has apparently stopped and that it is now on track to full recovery.[2]

However, member states remain politically and legally responsible for implementing (even internationally binding) norms and targets propagated by IOs or agreed in international treaties and conventions. This leads to great variation on the ground. To begin with, IO recommendations are subject to sharply diverging interpretations in different cultural and socioeconomic contexts and member states. Thus, forty or more different notions of the United Nations' (UN) concept of 'sustainable development' propagated in the 1987 Brundtland Report already coexisted by the end of the 1980s.[3] Not surprisingly, the implementation of norms and targets at the national level can vary a great deal, too, depending not only on the normative and political commitment of governments, political parties and other participants in policy making, but also their financial resources, legal system and administrative capacities. Even within the European Union (EU) with its 'supranational' environmental laws that have direct effect in the member states, studies have found substantial variation in how effectively they are implemented.[4]

Moreover, the separate institutionalization of environmental policy has not necessarily led to stricter norms and laws or their effective implementation. As Wolfram Kaiser demonstrates in his chapter, IOs with horizontal competences across different policy fields and sectors often segregated environmental matters and policy making from other fields and economic sectors such as steel. As a result, economic and sector-specific policy making sometimes excluded environmental matters entirely from their agenda. In the case of the negotiations about the steel issue in times of economic crisis, this was largely due to the industry interest in avoiding new environmental regulation. Such regulation, steelmakers feared, would have forced them to invest heavily into cleaner technology and put further pressure on profit margins.

Still, IOs have played a crucial role in driving transnational debates about environmental degradation and in steering international and national environmental policy making. The chapters in this book mainly contribute to our understanding of the IOs' working patterns and networking. They bring out several key functions of IOs, such as agenda setting. They also help us to conceptualize temporal change in international environmental protection. With the chapters' findings, as we will see below, the book as a whole mainly contributes to four sets of literature already introduced at the start of the book. Importantly, it also opens up new perspectives for more (collaborative) historical and interdisciplinary research on IOs in environmental protection in the global twentieth century.

To begin with, even more obviously than in the case of their member states, IOs are not unitary actors. They are not cohesive institutions that can always purposefully and strategically work towards clearly defined and consensual objectives, let alone impose them on member states. As a result, it is crucially important not to treat them as 'black boxes'. Hence, the chapters in this book recognize their fragmentation into administrative units and committees, and the role of multiple actors in developing IO policies. The chapters provide ample evidence for how experts, national governments and international non-governmental organizations (INGOs) have sought, from the inside of IOs and from the outside, to influence their activities and transnational discourses and policy making on the environment.

Thus, several chapters highlight the role of experts and expertise in global environmental politics and policy making. They contributed to the increasing interest in the research field of environmental history in experts and their involvement in defining environmental issues, such as the concept of biodiversity.[5] Both technological and scientific expertise already played a prominent role in IO policy making from roughly the middle of the nineteenth century. In those days, however, most experts shared the technocratic internationalist preference for depoliticizing issues to wrestle control over them from governments and diplomats.[6] These experts sought to work behind the scenes to assess and agree on technological options for their crossborder regulation and to address transnational scientific challenges, as in the case of the International Office of Epizootics (now the World Organization for Animal Health) set up in Paris in 1924 to control and prevent the spread of animal diseases through global trade.[7] As becomes clear in Enora Javaudin's chapter, this approach to the environment as primarily a scientific and technical problem still informed some of the scientific debate about environmental issues during the 1960s. Today, environmental policy makers still rely on scientific and technological knowledge and face problems of scientific insecurity and slow scientific consensus building. However, the role of experts transformed during and after the 1960s in line with broader social, political and generational changes in Western societies, including attitudes towards the state, democracy, and political participation and the politicization of the environment.[8]

As part of their attempt to establish the environment as a distinctive field of politics and policy making at the turn of the 1970s, many experts deliberately and strategically began to politicize concrete and visible problems such as air and water pollution. While scientists initially dominated, increasingly lawyers and, notably, economists started to establish themselves and were consulted as environmental experts, as Jan-Henrik Meyer notes in his chapter. Thus, at the outset, Barbara Ward, who worked with Maurice Strong in the UN and also had close contacts with the Vatican (as Luigi Piccioni shows in his chapter), was an economist. Over time, however, she grew into the role of environmental policy expert. These new environmental policy experts, who played a key role in linking different IOs, INGOs and governments in informal network-type relationships, often held strongly ideological worldviews and had wider political commitments associated with them. Influenced by, and embedded in, the new social movements that emerged in the 1960s, they used the media, especially television, in sometimes dramatic ways to draw the transnational public's attention to emerging and initially insufficiently understood environmental issues such as fine particulate matters or acid rain. They highlighted how such environmental degradation directly affected them and their quality of life in order to shore up public support for their advice and political demands. Frequently working with INGOs like Greenpeace and Friends of the Earth,[9] for example, these experts sought to transform societies, economic systems and world politics, or they tried to stop what they considered highly problematic developments, such as the large-scale introduction of nuclear power, which became a paramount environmental concern from the early 1970s onwards.[10] Experts' views and ambitions frequently extended beyond the traditional much narrower environmental conservation agenda, including issues of democratic participation and social and global justice, for example.

Thus, together with INGOs, many environmental experts vehemently promoted new norms both from within IOs and from the outside, such as through events organized on the margins of the Stockholm Conference, for example. However, national governments ultimately still made the decisions in the IOs. Some Western governments took the lead in advocating stricter environmental protection. They drew on their strong political position, particularly in IOs such as the Organisation for Economic Co-operation and Development (OECD), which was limited to Western market economies. It created the first environmental directorate and specialized committee of any IO in 1970.[11] In contrast, power relations in the UN were far more volatile. Here, as Stephen Macekura highlights in his chapter, non-aligned countries like Brazil tried, with some success, to organize the Global South around an agenda prioritizing development over the environment. These countries also argued that the Global North was the primary cause of environmental degradation and, as a result, should shoulder the economic and financial costs of environmental protection. At the same time, under certain conditions, Western governments tried to turn the experts' politicization strategy on its head to regain control. They set up institutions designed to take matters out of the hands of experts, as David G. Hirst shows for the case of the US institutional preference for the Intergovernmental Panel on Climate Change (IPCC) set up in 1988.

This book shows several crucial functions of IOs in environmental protection. The first of these is their capacity to develop and disseminate new political agendas, concepts and policy tools.[12] One especially pertinent example is precisely the concept of 'sustainable development', the origins of which are analysed from different perspectives by Stephen Macekura and Iris Borowy in their chapters. This concept sought to reconcile the environment and development agendas. The cleavage between the economic development needs of the Global South in particular and environmental protection had been at the heart of the negotiations at the 1972 Stockholm Conference. Different IOs subsequently worked hard to make these two core objectives of global and regional IOs conceptually more compatible, which eventually led to the Brundtland Report's definition.

Several chapters in this book provide evidence for the great importance of individual policy entrepreneurs for developing such agendas, concepts and policy tools – a perspective that much of the political science literature on IOs more generally and on environmental governance in particular ignores altogether or downplays. This literature tends to be more interested in structures than individual agency.[13] Identifying and analysing such entrepreneurship marks a particular contribution that historians can make, drawing on multiple sources, including the archives of IOs, national governments and INGOs and private papers. In the most striking case, Strong played a key role as secretary-general of the Stockholm Conference in preparing and completing it successfully despite strong frictions between the Global North and the Global South, broadly speaking. He then directed UNEP from Nairobi before returning to Canada in 1976. He remained a global environmental leader thereafter, drawing on the many contacts and connections he had built, especially in the years from 1970 to 1976.[14]

Similarly, as one of the few female experts in the male-dominated world of IOs, Barbara Ward linked and shaped environmental agendas across different fora (including the Vatican, as Piccioni shows in his chapter) before and after the Stockholm Conference. Not surprisingly, she features in a number of chapters in this book. This finding contributes an interesting aspect to the wider debate about the role of women in the rise of environmentalism – from Rachel Carson to the 'mothers against nuclear power'.[15]

The most successful and influential environmental policy entrepreneurs were able to mediate between different fields, to use Bourdieu's term, and institutions. Strong, for example, worked for private businesses before and after his time at the Stockholm Conference and UNEP. The economist Ward developed into an environmental policy expert. In the case discussed by Michael W. Manulak in his chapter, development economists cooperated at the Founex seminar in June 1971, which provided crucial input for the Stockholm Conference. In this group of experts, straddling the divide between economics and development studies, a small 'clique' of interventionist development experts with shared objectives managed to control the process of deliberation and the policy recommendations.

IOs have had a second crucial function for which the chapters in this book provide ample evidence: to make a major contribution to building new global and regional institutions in the field of the environment. UNEP is the most prominent of these institutions. Its foundation was influenced by institutional competition within the UN system. The older and more established UN organizations, such as the Food and Agriculture Organization, sought to ensure that it would be given the more limited status as a 'programme'. Moreover, while countries from the Global South succeeded in having UNEP's headquarters placed in Nairobi, the capital of Kenya, this was a Pyrrhic victory for the environmental protection agenda within the world organization. UNEP's geographical location actually marginalized the organization somewhat within the wider UN system. Nevertheless, UNEP developed a life of its own and pushed issues, concepts and policy tools that strongly influenced global environmental protection.

At the same time, it would be misleading to focus only on these major IOs and programmes with a global reach. At the time of the Stockholm Conference, many specialized treaties, secretariats and IOs already existed with responsibility for more clearly delineated issues and policy challenges. One example is the 1959 Antarctic Treaty. As discussed by Alessandro Antonello in his chapter, from early 1978 onwards, its member state governments began negotiating an agreement for conserving marine living resources. The main focus was on krill, small zooplankton, which was, and remains, crucial to the food chain in the Southern Ocean ecosystem and its whales in particular, and which countries like the Soviet Union and Japan sought to exploit as a natural resource. As this case shows, IOs, once established, have often shown a high degree of path dependency.[16] Their development and policy options were largely curtailed by the original membership in the IO and earlier agendas. In this case, the Antarctic Treaty's original Western-dominated membership facilitated U.S. policy to limit any access by the Soviet Union to Antarctica. Indeed, the 1980 Convention on the Conservation of Antarctic Marine Living Resources served this strategic Cold War purpose as much as saving krill and other Southern Ocean species.

The chapters in this book also allow tentative conclusions to be made about temporal change in the role of IOs in environmental protection in the global twentieth century. In the first half of the twentieth century, transnational discourses and the work of some specialized IOs focused on nature conservation, which had its origins in the nineteenth century, and issues related to resources and pollution. Throughout the 1960s and culminating in the Stockholm Conference, scientists and other experts, INGOs and proactive governments succeeded in developing the 'human environment' into a guiding concept. In this second phase, frequently apocalyptic scenarios of future environmental degradation like that in the 1972 Club of Rome report appeared to demand immediate and internationally coordinated action[17] – something that greatly facilitated the institutionalization of international environmental policy making, such as in UNEP or the OECD directorate and committee, as well as the European Communities (EC), as Jan-Henrik Meyer discusses in his chapter.

This period of international activism and institutionalization was then followed by a third phase, which lasted well into the 1980s. This was a period characterized by economic difficulties in the Global North, with low growth, rising budget and state deficits, growing unemployment and social frictions and unrest. These difficulties were by no means limited to Western Europe and North America; they also engulfed Eastern Europe in the Council for Mutual Economic Assistance, where the endemic problems of command economies led member states to borrow heavily in order to import crucial goods from the West, not least to appease their populations. This economic crisis led to a renewed focus by governments and IOs on fostering growth as their foremost political concern.[18] Various IOs tried to reconcile the apparent or actual need for growth in the Global North and South with the need for environmental protection. IOs now developed concepts like 'sustainable development' to overcome this cleavage. It became more difficult, however, to get governments to agree to binding targets and more drastic measures that could potentially benefit many, but would also impose costs on some, whether particular countries, sectors, companies or social groups.

The resulting distributive conflicts have parallels in the fourth and contemporary phase of global environmental politics, which began in the late 1980s, when the IPCC was formed, as analysed by David G. Hirst in his chapter. This period has been characterized by the climate change agenda overshadowing all other environmental issues – a challenge that appears to pose an existential threat to the long-term survival of humankind. Prioritizing climate change implies the need to make tradeoffs within environmental policy. Thus, where new dam building, intensive land use for biomass production or the return to nuclear energy are legitimized as necessary measures against climate change,[19] the resulting policy decisions can threaten to undermine achievements in other areas of environmental protection, such as nature conservation. These complex connections also illustrate that the comprehensive and all-encompassing concept of the environment, created in the late 1960s, includes potentially contradictory objectives that policy makers, INGOs and IOs will have to resolve in the future.

With these results, the book makes a major contribution to the four sets of literature to which we set out to contribute collectively. Our findings highlight the need, first of all, to understand the history of global environmental protection as much more than the bargaining of material interests among states or the ideological battle between 'green states' and their unenlightened 'dirty' cousins, as media reporting on climate change negotiations sometimes has it. We show that the politics of global environmental negotiations have been characterized by the involvement of multiple actors.[20] IOs have often been able to exercise soft power through their institutional choices, selection of personnel, networking of people and involvement of scientific experts and INGOs, which strengthened their advocacy role and legitimacy. They have frequently succeeded in embedding their policy priorities in normative discourses about environmental protection that have resonated well with many media and national publics, pressuring sometimes reluctant governments to take action.

Moreover, the activism of many actors, especially INGOs but also experts with strong political agendas, was not just motivated by rational scientific arguments and debates; instead, other factors often played a role. These included emotions such as existentialist fears about the future of humanity, and changing norms and values. In the field of environmental protection, especially in the formative period of its globalization, experts often did not (yet) form 'epistemic communities' characterized by shared scientific beliefs.[21] Their views about global approaches to tackling environmental degradation often depended on their preferences on related issues, such as the development of the Global South.[22]

Moreover, as we saw in the introduction, research on the role of IOs in environmental protection in historical perspective has been limited to date. So far, most studies have concentrated on single IOs. As a result, they tend to have a strong focus on institutional change within these organizations. Alternatively, they look at single environmental issues.

Our book has highlighted the need to go beyond such a limited focus. It is crucial to reconstruct the activities of single IOs within a wider web of IOs active in environmental policy making and protection as well as the multiple roles of the same individual entrepreneurs in different institutional settings and their networks across IOs. Actors from governments to environmental policy entrepreneurs often searched for the best venue to advance their agenda and propagate global solutions. IOs presented themselves as the best sites for making progress with tackling global environmental issues in a kind of competitive bidding process. Cooperation and competition among them created many vectors for transferring policy ideas and solutions. Much still needs to be done to research the exchange relations and transfer processes among IOs, including among regional integration organizations on different continents.

Finally, our book and its findings contribute to developing a more global perspective on environmental policy making and protection. The close interdependence of the environmental and development agendas particularly underlines the need to incorporate actors from the Global South fully into the analysis of IO activism and policy making. These actors often resented what they saw as a neocolonial agenda of the Global North to impose the costs of their own reckless industrialization and exploitation of the environment on developing countries whose first priority they considered to be raising living standards.

While addressing the origins of many debates, key concepts and policy tools, especially from the 1960s through to the late 1980s, this book can only mark a first step in the direction of a more systematic study of the role of IOs in environmental protection in the global twentieth century and beyond. For the moment, many archive-based studies, including many chapters in this book, still deal with one IO or one specific environmental policy issue. Even much more limited studies require broadly global perspectives and ample resources for multi-archival and multilingual research. In the future, however, it would be desirable for more (possibly collaborative) research to compare different IOs and their role in environmental protection and, even more importantly, to study their connections, interactions and transfers. Several chapters in this book provide evidence of different IOs or individual policy entrepreneurs creating networks among people and connections between debates and discourses, as well as fostering the transfer of concepts and policy tools from one IO to another. One pertinent example of such IO activity is the links between the OECD and the EC in the establishment of the polluter pays principle that Jan-Henrik Meyer discusses in his chapter.[23] But much more work needs to be done to research the globalization of environmental protection through such networks and connections.

We also readily admit that our book is limited in another way for practical, not conceptual reasons: perspectives based on the sources of IOs and 'Western' countries still predominantly inform the analysis. It is true that leading Western countries, notably the United States, but also Canada, the United Kingdom, Sweden and others, were able to influence global debates and policy initiatives to a large extent. They did so both through their access to superior scientific resources and their strong role in various IOs due to their economic and military power and funding for international collaboration. More often than not, however, 'Western'-dominated perspectives simply result from a lack of research to date on the priorities and policies of other, non-'Western' actors. This has to do with researchers' language skills and the segmented nature of research communities and networks, but also unsatisfactory archival and access policies, which make the reconstruction of their preferences and priorities much more difficult.

Nonetheless, it would be desirable critically to reconsider our routine assumption of Western dominance and thus decentre the 'West' more radically in analysing the work of IOs in environmental protection. Thus, while Alessandro Antonello makes a convincing case for U.S. and, to a lesser extent, British dominance in the negotiations over the Southern Ocean ecosystem, it would be important to learn more about the motivations of Soviet and Japanese policy – two countries that were, after all, in different Cold War camps from one another – and to find out why in the end they caved in to a conservation-focused policy regime. Future research should focus especially on the impact of the Cold War and East–West relations on the work of IOs in the field of environmental protection,[24] and on the North–South conflict and the nexus between the environment and development, which features prominently in this book and has been so virulent up to the present day.

Moreover, the chapters in this book also focus on the role of IOs at the transnational and global levels. There is a second crucial dimension of the involvement of IOs in environmental politics, however, and this concerns the 'vertical' links in this policy field between IOs, member states, and subnational state and societal actors. Only such a 'vertical' perspective can help us understand better how and why scientists and policy makers in member states sometimes 'uploaded' policy ideas to the IO level, and how they in turn became 'downloaded' to the national and subnational level.[25] In this way we can also understand better the scope and limits of IOs' agendas, concepts and policy tools. As the UN principle of environmental policy integration or environmental 'mainstreaming' shows, the interpretation and implementation of such principles can vary massively across the globe.[26] Furthermore, even in tightly integrated regional integration organizations like the EU, agreeing to legally binding norms does not necessarily mean implementing them effectively on the ground.[27] Future research therefore could also fruitfully focus on this particular 'vertical' dimension of the work of IOs and their impact on legislation and administrative implementation in member states.

Finally, we also advocate the 'mainstreaming' of environmental history in modern and contemporary transnational and global history. Similar to the institutionalization of the environment itself as a policy field, the professionalization and institutionalization of environmental history as a research field seems to have fostered a degree of segregation from other historical subfields that is not beneficial for either. Crucially, this segregation appears to have strengthened shared tacit normative assumptions among some environmental historians that the internationalization of environmental issues and the work by IOs on environmental protection essentially foster human progress.[28] However, some of our book's findings suggest a more nuanced view of the problematic sides of IOs politics and policy making, such as issues of the quality and transparency of governance structures, when small cliques and networks have a disproportionately large influence on IO action, or when some participants in debates and negotiations prevail over others, ostensibly to foster environmental protection, but perhaps sometimes only to secure their own economic, political or institutional interests. Against this background, we hope that our book instigates more critically analytical research in this vein on IOs and environmental protection in the global twentieth century and beyond.

Wolfram Kaiser is Professor of European Studies, University of Portsmouth, United Kingdom, and Visiting Professor at the College of Europe, Bruges, Belgium, and at NTNU Trondheim, Norway.

Jan-Henrik Meyer is a senior researcher at the Max-Planck-Institute for the History of European Law, Frankfurt, Germany and an associate researcher at the Centre for Contemporary History, Potsdam.

Notes

1. Richard Elliot Benedick, *Ozone Diplomacy: New Directions in Safeguarding the Planet* (Cambridge, MA: Harvard University Press, 1991); Karen Litfin, *Ozone Discourses: Science and Politics in Global Environmental Cooperation* (New York: Columbia University Press, 1994).
2. UNEP, 'Ozone Layer on Track to Recovery: Success Story Should Encourage Action on Climate', http://www.unep.org/newscentre/Default.aspx?document ID=2796&ArticleID=10978&l=en, accessed 25 May 2016.
3. David Pearce, Anil Markandya and Edward Barbier, *Blueprint for a Green Economy* (London: Earthscan, 1989), 173–85.
4. Tanja Börzel, Tobias Hofmann, Diana Panke and Carina Sprungk, 'Obstinate and Inefficient: Why Member States Do Not Comply with European Law', *Comparative Political Studies* 43(11) (2010): 1363–90; Tanja Börzel, *Environmental Leaders and Laggards in Europe: Why There is (Not) a Southern Problem* (Aldershot: Ashgate, 2003).
5. Sverker Sörlin, 'Reconfiguring Environmental Expertise', *Environmental Science and Policy* 28(1) (2013): 14–24; Libby Robin, 'The Rise of the Idea of Biodiversity: Crises, Responses and Expertise', *Quaderni. Communication, technologies, pouvoir* 76(3) (2011): 25–37.
6. See Wolfram Kaiser and Johan Schot, *Writing the Rules for Europe: Experts, Cartels, and International Organizations* (Basingstoke: Palgrave Macmillan, 2014).
7. Cornelia Knab, 'Infectious Rats and Dangerous Cows: Transnational Perspectives on Animal Diseases in the First Half of the Twentieth Century', *Contemporary European History* 20(3) (2011): 281–306.
8. See e.g. Adam Rome, *The Genius of Earth Day: How a 1970 Teach-in Unexpectedly Made the First Green Generation* (New York: Hill & Wang, 2013).
9. Frank Zelko, *Make it a Green Peace!: The Rise of Countercultural Environmentalism* (Oxford: Oxford University Press, 2013); Brian Doherty and Timothy Doyle, *Environmentalism, Resistance and Solidarity. The Politics of Friends of the Earth International* (Basingstoke: Palgrave Macmillan, 2013).
10. Jan-Henrik Meyer, '"Where Do We Go from Wyhl?" Transnational Anti-nuclear Protest Targeting European and International Organisations in the 1970s', *Historical Social Research* 39(1) (2014): 212–35, 213, 218.
11. Iris Borowy, *Defining Sustainable Development for Our Common Future: A History of the World Commission on Environment and Development (Brundtland Commission)* (Abingdon: Routledge, 2014).
12. Jan-Henrik Meyer, 'Getting Started: Agenda-Setting in European Environmental Policy in the 1970s', in *The Institutions and Dynamics of the European Community, 1973–83*, Johnny Laursen (ed.) (Baden-Baden: Nomos, 2014), 221–42.
13. Oran R. Young, *On Environmental Governance: Sustainability, Efficiency, and Equity* (Boulder: Paradigm, 2013); Oran R. Young, *The Institutional Dimensions*

of Environmental Change: Fit, Interplay, and Scale (Cambridge, MA: MIT Press, 2002).

14. For Strong's continued involvement in relevant networks advocating sustainable development, see Felix Dodds, Michael Strauss and Maurice F. Strong, *Only One Earth: The Long Road via Rio to Sustainable Development* (Abingdon: Routledge, 2012).

15. Mark Hamilton Lytle, *The Gentle Subversive: Rachel Carson, Silent Spring and the Rise of the Environmental Movement* (New York: Oxford University Press, 2007); Christof Mauch, 'Blick durchs Ökoskop. Rachel Carsons Klassiker und die Anfänge des modernen Umweltbewusstseins', *Studies in Contemporary History* 9(1) (2012): 1–4; Astrid Mignon Kirchhof, 'Frauen in der Antiatomkraftbewegung. Das Beispiel der Mütter gegen Atomkraft', *Ariadne* 64 (2013): 48–57; Jens Ivo Engels, 'Gender Roles and German Anti-nuclear Protest: The Women of Wyhl', in *Le démon moderne. La pollution dans les sociétés urbaines et industrielles d'Europe – The Modern Demon. Pollution in Urban and Industrial European Societies*, Christoph Bernhardt and Geneviève Massard-Guilbaud (eds) (Clermont-Ferrand: Presses Univ. Blaise Pascal, 2002), 407–24.

16. For this concept, see, in relation to regional integration, Paul Pierson, 'Increasing Returns, Path Dependence, and the Study of Politics', *American Political Science Review* 94(2) (2000): 251–267; Paul Pierson, *Politics in Time: History, Institutions, and Social Analysis* (Princeton: Princeton University Press, 2004).

17. Dennis Meadows et al., *The Limits to Growth* (New York: Universe Books, 1972).

18. Matthias Schmelzer, *The Hegemony of Growth: The OECD and the Making of the Economic Growth Paradigm* (Cambridge: Cambridge University Press, 2016).

19. E.g. Ted Nordhaus and Michael Shellenberger, *Break Through: Why We Can't Leave Saving the Planet to the Environmentalists* (Boston, MA: Mariner Books, 2009).

20. Governments nonetheless played an important role not only in decision making in IOs, but also in more informal international policy coordination. For economic policy, see, for example, Emmanuel Mourlon-Druol and Federico Romero (eds), *International Summitry and Global Governance: The Rise of the G-7 and the European Council, 1974–1991* (London: Routledge, 2014).

21. As postulated by Peter M. Haas, 'Introduction: Epistemic Communities and International Policy Coordination', *International Organization* 46(1) (1992): 1–35.

22. Stephen J. Macekura, *Of Limits and Growth: The Rise of Global Sustainable Development in the Twentieth Century* (Cambridge: Cambridge University Press, 2015).

23. See also Jan-Henrik Meyer, 'Appropriating the Environment: How the European Institutions Received the Novel Idea of the Environment and Made

it Their Own', *KFG Working Paper* 31 (2011): 1–33, http://edocs.fu-berlin.de/docs/receive/FUDOCS_document_000000012522, accessed 18 May 2016.
24. Kai F. Hünemörder, 'Environmental Crisis and Soft Politics: Détente and the Global Environment, 1968–1975', in *Environmental Histories of the Cold War*, John R. McNeill and Corinna R. Unger (eds) (Cambridge: Cambridge University Press, 2010), 257–76.
25. On the logic of uploading and downloading issues in international politics, see e.g. Tanja A. Börzel, 'Pace-Setting, Foot-Dragging and Fence-Sitting: Member State Responses to Europeanization', *Journal of Common Market Studies* 40(2) (2002): 193–214.
26. Nicolas de Sadeleer, *Environmental Principles: From Political Slogans to Legal Rules* (Oxford: Oxford University Press, 2002).
27. Tanja Börzel, *Environmental Leaders and Laggards in Europe: Why There is (Not) a Southern Problem* (Aldershot: Ashgate, 2003).
28. E.g. John McCormick, *The Global Environmental Movement* (Chichester: John Wiley, 1995); Anna-Katharina Wöbse, *Weltnaturschutz: Umweltdiplomatie in Völkerbund und Vereinten Nationen 1920–1950* (Frankfurt: Campus, 2011), 327–36. Highlighting, however, the ironies and multiple implications of international action, see Joachim Radkau, *Die Ära der Ökologie. Eine Weltgeschichte* (Munich: Beck, 2011), 580–613.

Bibliography

Benedick, Richard Elliot, *Ozone Diplomacy: New Directions in Safeguarding the Planet* (Cambridge, MA: Harvard University Press, 1991).
Borowy, Iris, *Defining Sustainable Development for Our Common Future: A History of the World Commission on Environment and Development (Brundtland Commission)* (Abingdon: Routledge, 2014).
Börzel, Tanja A., 'Pace-Setting, Foot-Dragging and Fence-Sitting. Member State Responses to Europeanization', *Journal of Common Market Studies* 40(2) (2002): 193–214.
———. *Environmental Leaders and Laggards in Europe. Why There is (Not) a Southern Problem* (Aldershot: Ashgate, 2003).
Börzel, Tanja A., Tobias Hofmann, Diana Panke and Carina Sprungk, 'Obstinate and Inefficient. Why Member States Do Not Comply with European Law', *Comparative Political Studies* 43(11) (2010): 1363–90.
De Sadeleer, Nicolas, *Environmental Principles. From Political Slogans to Legal Rules* (Oxford: Oxford University Press, 2002).
Dodds, Felix, Michael Strauss and Maurice F. Strong, *Only One Earth: The Long Road via Rio to Sustainable Development* (Abingdon: Routledge, 2012).
Doherty, Brian and Timothy Doyle, *Environmentalism, Resistance and Solidarity: The Politics of Friends of the Earth International* (Basingstoke: Palgrave Macmillan, 2014).
Engels, Jens Ivo, 'Gender Roles and German Anti-nuclear Protest: The Women of Wyhl', in *Le démon moderne. La pollution dans les sociétés urbaines et*

industrielles d'Europe – The Modern Demon. Pollution in Urban and Industrial European Societies, Christoph Bernhardt and Geneviève Massard-Guilbaud (eds) (Clermont-Ferrand: Presses Univ. Blaise Pascal, 2002), 407–24.

Haas, Peter M., 'Introduction: Epistemic Communities and International Policy Coordination', *International Organization* 46(1) (1992): 1–35.

Hünemörder, Kai F., 'Environmental Crisis and Soft Politics: Détente and the Global Environment, 1968–1975', in *Environmental Histories of the Cold War*, John R. McNeill and Corinna R. Unger (eds) (Cambridge: Cambridge University Press, 2010), 257–76.

Kaiser, Wolfram and Johan Schot, *Writing the Rules for Europe: Experts, Cartels, International Organizations* (Basingstoke: Palgrave Macmillan, 2014).

Kirchhof, Astrid Mignon, 'Frauen in der Antiatomkraftbewegung. Das Beispiel der Mütter gegen Atomkraft', *Ariadne* 64 (2013): 48–57.

Knab, Cornelia, 'Infectious Rats and Dangerous Cows: Transnational Perspectives on Animal Diseases in the First Half of the Twentieth Century', *Contemporary European History* 20(3) (2011): 281–306.

Litfin, Karen, *Ozone Discourses: Science and Politics in Global Environmental Cooperation* (New York: Columbia University Press, 1994).

Lytle, Mark Hamilton, *The Gentle Subversive: Rachel Carson, Silent Spring and the Rise of the Environmental Movement* (New York: Oxford University Press, 2007).

Macekura, Stephen, *Of Limits and Growth: The Rise of Global Sustainable Development in the Twentieth Century* (Cambridge: Cambridge University Press, 2015).

Mauch, Christof, 'Blick durchs Ökoskop. Rachel Carsons Klassiker und die Anfänge des modernen Umweltbewusstseins', *Studies in Contemporary History* 9(1) (2012): 1–4.

McCormick, John, *The Global Environmental Movement* (Chichester: John Wiley, 1995).

Meadows, Dennis et al., *The Limits to Growth* (New York: Universe Books, 1972).

Meyer, Jan-Henrik, 'Appropriating the Environment: How the European Institutions Received the Novel Idea of the Environment and Made it Their Own', *KFG Working Paper* 31 (2011): 1–33, http://edocs.fu-berlin.de/docs/receive/FUDOCS_document_000000012522", accessed 18 May 2016.

———. 'Getting Started: Agenda-Setting in European Environmental Policy in the 1970s', in *The Institutions and Dynamics of the European Community, 1973–83*, Johnny Laursen (ed.) (Baden-Baden: Nomos, 2014), 221–42.

———. '"Where Do We Go from Wyhl?" Transnational Anti-nuclear Protest Targeting European and International Organisations in the 1970s', *Historical Social Research* 39(1) (2014): 212–35.

Mourlon-Druol, Emmanuel and Federico Romero (eds) *International Summitry and Global Governance: The Rise of the G-7 and the European Council, 1974–1991* (London: Routledge, 2014).

Nordhaus, Ted and Michael Shellenberger, *Break Through: Why We Can't Leave Saving the Planet to the Environmentalists* (Boston, MA: Mariner Books 2009).

Pearce, David, Anil Markandaya and Edward Barbier, *Blueprint for a Green Economy* (London: Earthscan, 1989).
Pierson, Paul, 'Increasing Returns, Path Dependence, and the Study of Politics', *American Political Science Review* 94(2) (2000): 251–67.
———. *Politics in Time: History, Institutions, and Social Analysis* (Princeton: Princeton University Press, 2004).
Radkau, Joachim, *Die Ära der Ökologie. Eine Weltgeschichte* (Munich: Beck, 2011).
Robin, Libby, 'The Rise of the Idea of Biodiversity: Crises, Responses and Expertise', *Quaderni. Communication, technologies, pouvoir* 76(3) (2011): 25–37.
Rome, Adam, *The Genius of Earth Day: How a 1970 Teach-in Unexpectedly Made the First Green Generation* (New York: Hill & Wang, 2013).
Schmelzer, Matthias, *The Hegemony of Growth: The OECD and the Making of the Economic Growth Paradigm* (Cambridge: Cambridge University Press, 2016).
Sörlin, Sverker, 'Reconfiguring Environmental Expertise', *Environmental Science and Policy* 28(1) (2013): 14–24.
UNEP, 'Ozone Layer on Track to Recovery: Success Story Should Encourage Action on Climate', 2014, http://www.unep.org/newscentre/Default.aspx?DocumentID=2796&ArticleID=10978&l=en, accessed 25 May 2016.
Wöbse, Anna-Katharina, *Weltnaturschutz: Umweltdiplomatie in Völkerbund und Vereinten Nationen 1920–1950* (Frankfurt: Campus, 2011).
Young, Oran R., *The Institutional Dimensions of Environmental Change: Fit, Interplay, and Scale* (Cambridge, MA: MIT Press, 2002).
———. *On Environmental Governance: Sustainability, Efficiency, and Equity* (Boulder: Paradigm, 2013).
Zelko, Frank, *Make it a Green Peace!: The Rise of Countercultural Environmentalism* (Oxford: Oxford University Press, 2013).

Index

www.ingramcontent.com/pod-product-compliance
Lightning Source LLC
Chambersburg PA
CBHW070902030426
42336CB00014BA/2292